APPLYING PIC18 MICROCONTROLLERS

Architecture, Programming, and Interfacing
Using C and Assembly

APPLYING PIC18 MICROCONTROLLERS

Architecture, Programming, and Interfacing Using C and Assembly

BARRY B. BREY
DeVry University

Upper Saddle River, New Jersey
Columbus, Ohio

Library of Congress Cataloging-in-Publication Data

Brey, Barry B.
  Applying PIC18 microcontrollers architecture, programming, and interfacing using
C and Assembly / Barry B. Brey.
      p. cm.
  Includes bibliographical references and index.
  ISBN 0-13-088546-0
  1. Programmable controllers.   2. C (Computer program language)   3. Assembler language
(Computer program language)   I. Title.
  TJ223.P76B72 2008
  629.8'95—dc22

                                                              2006038207

Editor in Chief: Vernon Anthony
Acquisitions Editor: Jeff Riley
Editorial Assistant: Lara Dimmick
Production Coordination: Shelley Creager, Techbooks
Production Editor: Holly Shufeldt
Design Coordinator: Diane Ernsberger
Cover Designer: Ali Mohrman
Cover photo: Corbis
Production Manager: Matt Ottenweller
Director of Marketing: David Gesell
Senior Marketing Manager: Ben Leonard
Marketing Assistant: Les Roberts

This book was set in Times by Techbooks. It was printed and bound by R. R. Donnelley & Sons
Company. The cover was printed by Phoenix Color Corp.

Pearson Prentice Hall™ is a trademark of Pearson Education, Inc.
Pearson® is a registered trademark of Pearson plc
Prentice Hall® is a registered trademark of Pearson Education, Inc.

Pearson Education Ltd.
Pearson Education Singapore Pte. Ltd.
Pearson Education Canada, Ltd.
Pearson Education—Japan

Pearson Education Australia Pty. Limited
Pearson Education North Asia Ltd.
Pearson Educación de Mexico, S.A. de C.V.
Pearson Education Malaysia Pte. Ltd.

PEARSON
Prentice
Hall

10 9 8 7 6 5 4 3 2 1
ISBN-13: 978-0-13-088546-3
ISBN-10: 0-13-088546-0

This textbook is dedicated to my children
who have brought me much joy:
Brenda, Gary, Renee, Leslie, and David

PREFACE

Microcontrollers are used in a wide variety of applications in automobiles, appliances, industrial controls, medical equipment, and so forth. This textbook provides a comprehensive glimpse into the architecture, programming, and interfacing of this modern marvel. The Microchip PIC18 family of microcontrollers is used throughout this text to explain the architecture, programming, and interfacing of the microcontroller. The PIC18 is chosen because it is the newest 8-bit microcontroller available from Microchip, and what is learned about this family is applicable to the earlier families of microcontrollers from PIC as well as other microcontrollers from other manufacturers. The level of this textbook is appropriate for educational programs in almost any field of technology or science, and also valuable to the experienced practitioner as a reference and to the hobbyist interested in learning about microcontrollers.

The author incorporates many of the techniques learned in the past 40 years of working with and teaching electrical/mechanical computers, electronic computers, microprocessors, and microcontrollers to provide as complete an experience as possible in this field of study. To his credit are 31 textbooks, written in the past 22 years, in the fields of digital electronics, microprocessors, and embedded systems including editions in many languages used throughout the world at major universities.

ORGANIZATION

The microcontroller is a complete computer system on an integrated circuit containing memory for programs, as well as data storage and integrated input/output, devices. As with any computer system, the microcontroller is programmed to perform a dedicated function.

This text provides a fairly complete look at microcontroller programming in both assembly and C language. Although C language is often used, assembly language programming still appears within C language programs as blocks of assembly code to perform tasks. In order to understand programming, this book first introduces general computer architecture and then concentrates on the PIC18 family architecture. Chapter 1 introduces much of the jargon used in the field of microprocessors and microcontrollers.

Once the architecture is explained, the next few chapters deal with assembly language programming for controlling the microcontrollers. Many common programming algorithms are presented along with programming examples that show how various tasks are performed in a microcontroller-based system. Assembly language programming is described in Chapters 3 and 4, with Chapter 4 geared to readers who want an in-depth view of assembly language programming. For the programmer mainly interested in C language, Chapter 4 is optional. To gain the most from this text, it is recommended that the learner have some knowledge of introductory computer programming and basic electronics.

After programming is described in fair detail, the hardware interaction of the microcontroller is investigated, along with timing and details of interfacing to common electronic components. Many practical sample interfaces are presented along with the software drivers required to use them, including interrupt processing for many peripheral components. Common devices include keyboards, keypads, switches, relays, barcode readers, infrared remote controls, solenoids, sensors, motors, LCD and LED displays, ADCs, DACs, PWMs, as well as a variety of other interfacing components. For connectivity, the USB (universal serial bus) and the CAN (controller area network) are discussed and illustrated with sample applications. Throughout this book an attempt was made to introduce and completely explain as many devices as possible and show many complete system applications. By doing so, applications can be constructed by cutting and pasting code from the many examples with minimal changes.

APPROACH

Each chapter contains an introduction that describes the basic concepts explained in the chapter. The body of the chapter illustrates these concepts using many practical sample systems that are complete, including schematic diagrams, discussions about the operation of the interface, and a complete software listing for each interfacing example. Through these examples, learners will gain a good understanding of the microcontroller and its operations in a wide variety of systems.

Each chapter ends with a numeric summary that reviews the material discussed in the chapter. All programs from the text are contained on a companion CD-ROM for use in constructing new system software. This greatly reduces the amount of time required to construct a software system. Finally, to reinforce the concepts presented in each chapter, questions and problems are included to provide experience with problem solving and the development of microcontroller-based systems. Upon completion of this text the learner will be proficient in microcontroller interfacing, programming, and system design.

This text can be used in a two-course sequence where the first course mainly provides practice with the assembler and C language compiler and the second course concentrates on developing systems using the PIC microcontroller. As a capstone, a senior project course could highlight what a student has learned through the design and construction of a significant project based on the microcontroller.

SUPPLEMENTS

To access supplementary materials online, instructors need to request an instructor access code. Go to **www.prenhall.com,** click the **Instructor Resource Center** link, and then click **Register Today** for an instructor access code. Within 48 hours after registering you will receive a confirming e-mail including an instructor access code. Once you have received your code, go to the site and log on for full instructions on downloading the materials you wish to use.

KEEP IN TOUCH

I have been teaching computer, microcontroller, and microprocessor technology for over 35 years and continue to do so on a daily basis, even during the summer. To maintain contact and provide additional information I have had a presence in the electronic media since 1985, first on CompuServe (which is now owned by America Online), and now on the Internet, where I have had a Web page since 1995. Visit my Web page and view information about many topics and also technical short reports that may aid in learning some aspects of computers, microcontrollers, and microprocessors.

My Web page address: http://members.ee.net/brey

My e-mail address: bbrey@ee.net

CONTENTS

CHAPTER 1

Introduction to Computer Architecture

The PIC (*programmable interface controller*) microcontrollers are a series of RISC (*reduced instruction set computer*) integrated circuits produced by Microchip Technology Incorporated (http://www.microchip.com). Variations are currently available in package sizes of as small as 18 pins and as large as 128 pins. A series of products are produced for the PIC microcontroller called the *BASIC Stamp®* from Parallax, Incorporated (http://www.parallax.com), which is programmed in the BASIC language or the Java language instead of assembly language. This chapter introduces the microcontroller and system architecture as the basis for controlling machinery using the microcontroller. Also defined are many of the terms used with microcontroller technology as an introduction to this fascinating field of study.

Upon completion of this chapter you will be able to:

1. Define the terms used in the field of microcontrollers.
2. Describe the purpose of each component part of a computer system.
3. Relate the operation of each section of a computer system and its interrelation to other system components.
4. Define data types used with microcontrollers.
5. Convert between common number systems.

1-1 BASIC COMPUTER ARCHITECTURE

The most common computer architecture is represented by the block diagram illustrated in Figure 1-1. The first evidence of this architecture was proposed and used by Charles Babbage in his *Analytical Engine* in 1856 (http://www.cbi.umn.edu) and continues as the basic system architecture used in most modern digital computers. The Analytical Engine was a mechanical computer system powered by a hand crank. In more recent years, before the rediscovery of the work of Charles Babbage, the architecture of a computer was attributed to John von Neumann, as he described it in the spring of 1945, and today is often referred to as the *von Neumann architecture*. Chapter 2 explains a different architecture called the *Harvard architecture*, which is not as common as the von Neumann architecture, but used inside the PIC family of microcontrollers.

FIGURE 1-1   Block diagram of a computer system.

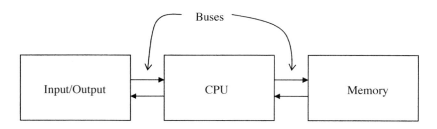

CPU

The block diagram of a computer system, although simple, contains three main blocks that represent the architecture of most modern digital computer systems. At its core is the CPU (*central processing unit*)—the controlling element of the system. Early CPUs were mechanical systems, such as the analytical engine, calculators, and accounting machines, commonly used into the 1970s. Early electronics digital computers were originally constructed with vacuum tubes and later with transistors. Modern computer systems are constructed with complementary MOSFET (*CMOS*) technology integrated circuits.

    Central processors can be of any size and have been anything from 4 bits in width to 64 bits. (A *bit* is a binary digit with the value of zero or one). Most electronic digital computer systems are based on the binary number system because the circuitry, which is currently in use, accurately represents only binary data. Examples of the 4-bit processors are early microprocessors such as the 4004 or 4040 from Intel Corporation. Currently, the latest microprocessors are 64 bits in width from AMD (*Advanced Micro Devices*) and Intel. The size of the CPU is determined by its memory address width and its arithmetic and logic unit, which is responsible for performing arithmetic and logic operations on integer numbers in the system. A 4-bit microprocessor performs these operations on 4-bit integers (called *nibbles*), whereas a 64-bit microprocessor performs them on integers of a maximum size of 64-bits. The 64-bit machines also perform arithmetic and logic on 32-bit, 16-bit, and 8-bit numbers. Today the most common smallest directly addressable number is usually an 8-bit number called a *byte*. A microprocessor or microcontroller can be of any width from 1 bit to any number of bits, although most microprocessors are a multiple of 8 bits and many microcontrollers are either 8 bits or 16 bits.

    Another variable in the CPU is its clock speed. The clock speed varies considerably and is currently a few megahertz for microcontrollers to several gigahertz for more powerful microprocessors used in desktop computer systems. A *megahertz* system (MHz) is one based on a clocking frequency of a million pulses per second called Hertz (Hz), whereas a *gigahertz* system (GHz) is one based on a clocking frequency of a billion pulses a second. *Heinrich Rudolf Hertz* (http://www.ideafinder.com/history/inventors/hertz.htm) is honored by name to define the number of pulses per second of an alternating current (originally called cycles per second or cps). Although the clock frequency of a microprocessor or microcontroller may not indicate the number of instructions it executes per second, it directly determines the speed of the system. The exact number of instructions executed per second is based on the internal design of the CPU, the complexity of the instructions that it executes, and the program being executed. One thing is certain: If a 10-MHz machine is compared to a 2-GHz machine, the 2-GHz machine will be the faster of the two machines.

    The type of microprocessor can also determine its speed. A RISC (*reduced instruction set computer*) is a machine that executes one instruction per clock and contains only basic instructions. A CISC (*complex instruction set computer*) executes many more different instructions than a RISC machine, but some of the instructions require more than one clock to complete. Today these terms blur, because most newer microprocessors use a combination of CISC and RISC technologies to perform most instructions in a single clock and some, such as the Pentium® family from Intel, perform many instructions in as little as one-third of a clock. Realize that these newer machines often contain multiple integer units that function simultaneously.

CPU Task. The CPU is responsible for executing a series of sequential instructions organized into a grouping of instructions called a *program*. In fact, the only thing that the CPU is capable of doing is executing instructions from the time it is powered until the time it is switched off. It relentlessly fetches instructions from the memory system and executes them. This *fetch* and *execute* sequence continues for as long as the CPU is powered. The idea of the CPU and the way that it operates, first proposed by Charles Babbage, has not changed since, and will likely remain the same for quite some time. Newer machines execute more than one instruction simultaneously because many have multiple arithmetic and logic processors located within the CPU. Some of the latest machines also have multiple cores. A multiple core machine is a machine that has more than one microprocessor in the integrated circuit.

The program in a computer system is a collection of instructions aligned in a sequential fashion, which is executed by the CPU in a likewise manner. For example, suppose a program is needed to add a 6 to a 2. The first step (or instruction) obtains the number 6. The second obtains the number 2. The third step adds the 6 and 2 together. The final operation stores the sum somewhere in the memory system. As illustrated in this example, the simple task of adding 6 and 2 together is broken down into steps called *instructions*. These instructions form a program that is executed by the microprocessor or microcontroller.

1. Get a 6.
2. Get a 2.
3. Add 6 and 2.
4. Store the sum in memory.

Each step of this simple program is converted to the language that the CPU understands, called *machine language*. Machine language is a series of numbers that represent instructions. In a hypothetical CPU, the "get instruction" might be the number 00, the add instruction might be the number 01, and the store instruction might be the number 02. The program, in numeric machine language, appears as 00 06 00 02 01 02 00, or as shown in Example 1-1.

EXAMPLE 1-1

```
00 06
00 02
01
02 00
```

As can be seen in this example, writing numeric machine code is cumbersome and cryptic. Luckily, no one writes programs in numeric machine code, although they did in the early days of computing. Instead, a tool called an *assembler* is used to write the program in its symbolic form using *mnemonic opcodes* such as GET and ADD. The assembler, which is also a program, converts the symbolic assembly language program into numeric machine code. An assembly language program that generates the program in Example 1-1 might appear as shown in Example 1-2.

EXAMPLE 1-2

```
GET  6
GET  2
ADD
STORE 0
```

Even this is cumbersome, but at least it is more readable than a series of numbers. Because an assembly language program can be quite long and challenging to write, modern programming typically uses a *high-level language* to generate the machine code. A high-level language is a program that accepts as its input a pseudo-English-like language and then converts it into numeric machine code. Some of these languages are BASIC, C, C++, or Java. The same program might appear in C language as depicted in Example 1-3.

EXAMPLE 1-3

```
char answer;          //set aside a place for the answer
answer = 6 + 2;       //perform the addition and store sum in answer
```

As seen from this C program, there is less typing, it requires no knowledge of machine code, or numeric memory locations, and is fairly easy to read. The machine code generated by the C language program may not be as efficient as the code generated by an assembler, but it is much easier to write, understand, maintain, and requires less time and skill to write. It is for these reasons that software is often developed using a high-level language. The *software* of a computer system is its programs and the *hardware* is the computer system circuitry. At times software is referred to as *variable hardware* because it changes or modifies the way the hardware functions. This is especially true in a microcontroller-based system.

It is interesting to note that Augusta Ada Byron (later named the Countess of Lovelace after a marriage to Lord Byron) earned the reputation for being the first computer programmer. She earned this distinction because of the programs that she wrote in numeric machine language for Charles Babbage and his Analytical Engine. A computer language called Ada, which is used by the U.S. Department of Defense, is named in her honor.

CPU Function.   A CPU performs three main tasks in a computer system: (1) data transfer, (2) arithmetic and logic, and (3) program flow control. Data transfers represent the most common CPU task. Most of a CPU's time is spent transferring data. Data transfers include fetching an instruction from memory, transferring data between registers or memory locations, and transferring the result from some arithmetic or logic operation between memory and the CPU or between an I/O (*input/output*) device and the CPU. The instruction fetch portion of a data transfer operation is the most common and important operation performed by the CPU. The instruction fetch allows the CPU to execute the program from the memory system at a high rate of speed. This *stored program* concept makes the computer very powerful and was first envisioned by Charles Babbage. Most programs transfer data for at least 50% of the time and often a much higher percentage. Table 1-1 lists some of the most common data transfer operations performed by most CPUs.

A CPU spends much less time performing arithmetic and logic operations on numbers than data transfers. Often a program requires that the CPU devotes a small percentage of its time to

TABLE 1-1   Most common data transfer operations.

Operation	Comment
Memory → Opcode Register	Fetch an operation to opcode register
Memory → Data Register	Fetch memory data to a data register
Data Register → Memory	Store register data in a memory location
Data Register → Data Register	Transfer contents of a data register to another data register
Data Register → I/O	Transfer contents of a data register to an I/O device
I/O → Data Register	Transfer contents of an I/O device to a data register

TABLE 1-2 Common arithmetic and logic operations.

Operation	Comment
Addition	Between registers, a register and memory, or with immediate data
Subtraction	Between registers, a register and memory, or with immediate data
Multiplication	Between registers, a register and memory, or with immediate data; result can be double the size
Division	Between registers, a register and memory, or with immediate data; dividend can be double the size (some microcontrollers do not have division)
Negation	The sign of a number is changed
AND	Logical bit-wise AND between registers, a register and memory, or with immediate data
OR	Logical bit-wise OR between registers, a register and memory, or with immediate data
Exclusive-OR	Logical bit-wise Exclusive-OR between registers, a register and memory, or with immediate data
NOT	Logical bit-wise NOT between registers, a register and memory, or with immediate data

arithmetic and logic operations. Table 1-2 lists typical arithmetic and logic operations performed by the CPU. These operations are normally performed by an integer arithmetic unit. Real numbers cannot be operated on by the CPU without a program or special processing unit called an *arithmetic unit* or *numeric processor*. Microprocessors often contain a numeric processor as well as the integer processor, whereas the microcontroller seldom contains a numeric processor.

The most powerful function performed by the CPU is its ability to modify the flow of a program through the use of simple numeric decisions. *Program flow control* instructions allow a section or sections of code to be used multiple times in a program. Program flow control instructions also allow the flow of a program to be transferred to a function or procedure. There are unconditional and conditional program flow control instructions. The unconditional program flow control instructions are the GOTO and function CALL instructions. The conditional program flow control instructions allow a number to be tested for a condition to determine if the flow of the program is modified.

Conditional program flow control instructions test a number to determine if it's zero, if a carry occurred after addition, or if the result of an operation is negative or positive. Table 1-3 lists many of the commonly testable conditions in most CPUs. The conditional program flow instructions represent a small percentage of the instructions in a typical program, yet these instructions are what make a CPU a powerful component in a computer system with the perceived ability to think. For example, how can a computer determine if the number typed on a keyboard is a 3? It does this by subtracting a 3 from the number typed and then it tests to see if the result is zero. If the result it is zero, the flow of the program is modified to perform some task in response to the 3 that was typed. The illusion here is that the computer is able to think and reason that a

TABLE 1-3 Conditions tested by many CPUs.

Condition	Comment
Zero	Is a number zero or not zero?
Carry	Did a carry occur or no carry?
Sign	Is a number positive or negative?
Overflow	Some CPUs test for an arithmetic overflow.

3 was typed on the keyboard, when in reality all that was done was to subtract a 3 from the number that was typed and test the result for a zero. What makes the computer system a powerful tool? The software, which is written by programmers, is the key because it makes the computer appear to think and reason.

Memory

The *memory* is an extremely important component of a computer system because it provides a place to store programs and the data used by a program. Without the memory, a computer system would be no more powerful or faster than a simple four-function calculator. With memory, the computer accesses the instructions in a program from the memory at a very high rate of speed. Incidentally, Charles Babbage's Analytical Engine contained memory that stored 1000 numbers that were each 20 decimal digits wide. It too executed a program stored in its memory.

Two types of memory are used with microprocessors and microcontrollers: *read-only memory* (ROM) and read/write memory, typically called *random access memory* (RAM). The ROM in a microcontroller-based system stores static data as a program and as constant data used by the program. Most embedded microcontroller-based or microcontroller-based systems store only one program in the memory that functions as an operating system. The RAM in a microcontroller-based system stores dynamic data for the system. It is important to note that a disk drive and similar devices are not considered memory, but are considered I/O devices even though they store programs and data. The distinction is made because the disk drive and similar devices are treated differently when connected to the system. The *driver* (a program that controls an I/O device) for a disk drive must send quite a bit of information to the disk drive to cause it to operate. The difference is that the ROM or RAM memory is not accessed by a program, but by direct connection through a series of wires to the CPU and is directly controlled by the signals that emanate from the CPU.

ROM. The ROM in a system is an *erasable/programmable read-only memory* (EPROM), an *electrically erasable/programmable read-only memory* (EEPROM or E^{2}PROM), or occasionally, a factory-programmed read-only memory. Both the EPROM and EEPROM store data for approximately 20 years or until erased. The factory-programmed ROM is permanent. The EPROM can be erased up to approximately 100 times, whereas the EEPROM can be erased from 10,000 to 1,000,000 times, depending on its manufacturer and when it was manufactured.

The EPROM has been in use for many years and must be removed from the system to erase. The EPROM is erased by placing it under a high-intensity ultraviolet light for anywhere from 10 to 20 minutes. EPROM memory is available with speeds of up to about 100 ns for a read access. The EEPROM is erased electrically without removing it from the system. EEPROM memory is often called *Flash memory* and also has an access time of about 100 ns. The main difference is that the erase time for an EEPROM is much shorter than for the EPROM. Erase times are typically in the milliseconds for the EEPROM and minutes for the EPROM. Recent common applications for the EEPROM include the system BIOS in a personal computer; USB flash drives (called *JumpDrives*™, *PocketDrives*™, *PenDrives*™, or *ThumbDrives*™), that have replaced the floppy disk drive in most computer systems; and MP3 audio players.

RAM. The RAM in a system is either SRAM (*static RAM*) or DRAM (*dynamic RAM*). Both memory types are *volatile memory,* which fail to retain data after removed from power. The SRAM stores data in memory cells that are continuously powered; this makes them fast, but consumes a fair degree of power. The SRAM devices have access times of 1 ns or less, which makes them excellent candidates for computer system memory except for the power consumption. The DRAM stores data in a memory cell that is a capacitor. Because capacitors retain a charge for an indefinite time, a DRAM must be periodically rewritten to assure the integrity of the data. The process of periodically rewriting the data to a DRAM is called a *refresh*. A DRAM usually retains data for 2 to 4 ms before it must be refreshed. Refreshing a DRAM requires time from the system. Also, data cannot be stored and retrieved at nearly the same speed as an SRAM. Access times for the DRAM are around 40 to 50 ns, whereas access times for the SRAM are 1 ns or less. Today we use DRAM for large memory systems and SRAM for small systems. This means that most microcontrollers do not use DRAM memory because the memory size is small. The technology is changing, and it is hoped that soon SRAM power consumption will be reduced enough to use SRAM for large memory systems, which should yield a 50-time boost in the speed of system memory.

I/O

The *input/output* (I/O) in a system is a CPU's connection to other machines and humans. Without I/O, the CPU is able to solve problems, but it cannot provide the result to humans or other machines. Because of the diversity of I/O devices connected to a CPU, much of the work with microprocessors or microcontrollers involves interfacing and controlling, through software drivers, the I/O devices in a computer system. Today, I/O devices perform just about any task or function imaginable. Much of this text is devoted to interfacing the microcontroller to many of these I/O devices and providing software drivers to control them. Some common I/O devices are keyboards for inputting information and LCD or video displays for outputting data. In general, anything that produces or accepts an electrical signal can be and often is interfaced to the computer as an I/O device.

Buses

The connections between the blocks in the block diagram of a computer system are called buses. A *bus* is a common set of wires that carry a specific type of information. A computer system has three buses: (1) address, (2) data, and (3) control. The address bus carries address information for selecting a memory location or an I/O location, the data bus carries the data that is transferred between the memory or I/O and the microcontroller, and the control bus controls whether the memory and I/O is read or written. It is through these three buses that the microcontroller is able to execute programs from the memory system and control the I/O devices attached to the system.

Address Bus. All computers have an *address bus* with signal lines that are almost always labeled A0, A1, A2, and so forth. The A0 signal is the least significant address bus bit position and the A1 signal is the next to the least significant address bus bit, and so forth. The number of address bus pins or signals varies between computers. For example, if a computer has a 16-bit address bus (numbered from A0 to A15), it addresses 64K of memory or 64 × 1,024 (65,536) locations in the memory system. A *computer K* (pronounced *kay*) is 1,024, or 2^{10}. We know this, because the number of locations is 2 raised to the number of address pins in a system, or 2^{16} in a system with a 16-bit address bus. This is the number of binary combinations (64K) for a 16-bit address. Likewise, a system that contains a 20-bit address bus addresses 1M (pronounced *1 Meg*) memory locations (2^{20}). A 1M memory contains 1K times 1K locations or 1,048,576 locations. Memory addresses range in size from 12 bits (4K) on some microcontrollers to as many as 40 bits (1T, or *1 Tera*) on the latest Pentium 64-bit microprocessors (EMT-64) from Intel Corporation. Table 1-4 lists the address and memory sizes for the microprocessors and microcontrollers.

Memory is numbered in hexadecimal (*number base* or *radix 16*) from location zero to the maximum number for the address width. For example, a 4K memory, which has a 12-bit address, is numbered from memory location 000 to FFF in hexadecimal. It contains 1000 hexadecimal locations. Likewise a 64K memory, which has a 16-bit address, is addressed at locations 0000 to FFFF hexadecimal and contains 10,000 hexadecimal locations. A 4-bit binary number is required to represent each hexadecimal digit of the address. Hexadecimal numbers are often denoted by using a 0x in front of the number or the letter H following the number. Examples are 0x3A and 3AH.

TABLE 1-4   Address and memory sizes.

Address Size	Memory Capacity	Capacity in Hexadecimal	Note
10 bits	1K (1,024) or 2^{10}	400	Kilo
			Kibi
12 bits	4K (4,096) or 2^{12}	1000	
16 bits	64K (65,536) or 2^{16}	1 0000	
20 bits	1M (1,048,576) or 2^{20}	10 0000	Meg
			Mebi
30 bits	1G (1,073,741,824) or 2^{30}	4000 0000	Gig
			Gibi
40 bits	1T (1,099,511,627,766) or 2^{40}	100 0000 0000	Tera
			Tebi
50 bits	1P (1K * 1T) or 2^{50}	4 0000 0000 0000	Peta
			Pebi
60 bits	IE (1K * 1P) or 2^{60}	1000 0000 0000 0000	Exa
			Exbi

Note: Kibi, Mebi, Gibi, Tebi, Pebi, and Exbi are proposed names.

Data Bus.   The *data bus* in a computer system is usually a 2-way bus or a bidirectional bus that carries data between the CPU and the memory and I/O system. The width of the data bus, which is numbered from D0 to Dn, varies. As with the memory address, the zero bit (D0) is the least significant data bus bit. An 8-bit computer usually has an 8-bit data bus, a 16-bit computer usually has a 16-bit data bus, and so forth. In some machines, the data bus is twice the width of the CPU to accomplish data transfers at higher speeds. The Pentium microprocessor from Intel is a 32-bit machine that uses a 64-bit data bus.

Control Bus.   Memory and I/O connected to the microprocessor need to be controlled. The *control bus* performs this task through a *read signal* often called $\overline{RD}$ and a *write signal* often called $\overline{WR}$. Note that the overbar indicates a signal is an active low signal. For example, the RD (read) signal causes a read when it becomes a logic zero. At times the # symbol is used to indicate that a signal is active low, as in #RD. If the microprocessor fetches an instruction from memory, it places the memory address on the address bus, the read signal is forced low to inform the memory to do a read operation, and the data are then transferred from the memory to the microprocessor through the data bus. Likewise, the same steps are performed for a write, except that the microprocessor issues a write signal in place of the read signal.

Any other control bus signal is specific to a particular microprocessor. Some of the additional control bus signals provide interrupt inputs and various other control functions besides memory and direct I/O control to the computer system.

Microprocessor and Microcontroller

Charles Babbage built his computer from the technology of his time, which was mechanical. His machine was constructed with gears and levers using odometer-style mechanisms to store 1000, 20-digit decimal numbers in its memory. I/O was accomplished through holes punched on a paper tape. These early ideas of mechanical computing were carried forward for quite some time until more modern computers first appeared in the twentieth century.

The IBM Corporation built a vast industry and empire based on punched cards that survived in computing applications well into the 1980s. These punched cards (called *Hollerith cards* named for Herman Hollerith, the founder of IBM) were mechanical devices that functioned as input/output media to mechanical computers in the 1950s and then electronic computers in the 1960s. The early mechanical computers were often called *accounting machines* and were programmed through a series of jumper wires. These early machines gave way to electronic computers, which were often called *mainframes*. The mainframe computers were expensive and large, and with the advent of modern integrated circuits, became obsolete when the microprocessor overtook the mainframe beginning in the 1980s.

The microprocessor (4004) was invented by Intel in 1971 as a 4-bit CPU that operated with a 20-KHz clock and addressed a 4-bit-wide memory system with a 12-bit memory address (4K). The technology progressed quickly and, by the end of the 1970s, the size of the microprocessor had become 16 bits at Intel with a 1M-byte memory and a 6-MHz clock (the 8086 microprocessor). Motorola also was producing microprocessors and it too had advanced the technology with a 32-bit microprocessor that addressed a 16M-byte memory using an 8-MHz clock (the 68000 microprocessor). Clearly it appears that Motorola had the advantage, yet it did not become dominant. This probably occurred because IBM decided to use the Intel microprocessor in its personal computer. The architecture, and Intel, boomed because of the IBM moniker, and Motorola processors were relegated to the Apple Macintosh computer, which did not enjoy the success of the personal computer from IBM. Recently Apple computer announced that it is using the Pentium microprocessor in its new line of Apple computers.

The *microcontroller,* which is a self-contained computer system, also appeared at Intel in 1977 as an 8-bit machine (8048) that contained a microprocessor, memory, and I/O connections. The microcontroller was designed to control machinery and the microprocessor was designed as a replacement for the CPU in a mainframe. The first major applications of the microcontroller were in the Epson FX-80 dot matrix printer and in the keyboard on an IBM personal computer. The printer was built around the 8048 microcontroller and brought low-cost printers to the marketplace for the first time. The keyboard contained a universal peripheral interface (UPI), the 8042, which appeared in 1977 as a microcontroller used to read keystrokes from the computer user. This gave rise to the term *embedded controller,* or *embedded microprocessor,* because the computer system was hidden or embedded in another product, such as a printer. From outward appearances the FX-80 was a printer, even though internally it was a computer system. The same is true for the keyboard; from all appearances the keyboard was a keyboard, but internally it was a computer system.

When Intel abandoned the microcontroller and universal peripheral interface in favor of the microprocessor in the late 1980s, Microchip Incorporated was formed to fill a void in the microcontroller arena in 1989. Since its inception, Microchip has become the leading provider of 8-bit microcontroller technology. Even the name of their device, a PIC (peripheral interface controller) was initiated by Intel years earlier as the closely named universal peripheral interface (UPI) in the 8042.

1-2   ## NUMBER SYSTEMS

The use of the microprocessor or microcontroller requires a working knowledge of binary, decimal, and hexadecimal numbering systems. This section provides a background for those who are unfamiliar with these numbering systems. Conversions between decimal and binary, decimal and hexadecimal, and binary and hexadecimal are described.

Digits

Before numbers are converted from one number base to another, the digits of a number system must be understood. Early in our education, it was learned that a decimal (*base 10*) number is constructed with 10 digits: 0 through 9. The first digit in any numbering system is always zero.

For example, a base 8 (octal) number contains 8 digits: 0 through 7; a base 2 (binary) number contains 2 digits: 0 and 1. If the base of a number exceeds 10, the additional digits use the letters of the alphabet, beginning with an A. For example, a base 12 number contains digits 0 through 9, followed by A for 10 and B for 11. Note that a base 10 number does contain a 10 digit, just as a base 8 number does not contain an 8 digit. The most common numbering systems used with modern computers are decimal, binary, and hexadecimal (base 16). (Many years ago octal numbers were popular.) Each of these number systems are described and used in this section.

Positional Notation

Once the digits of a number system are understood, larger numbers are constructed by using positional notation. In elementary school, it was learned that the position to the left of the units position is the tens position, the position to the left of the tens position is the hundreds position, and forth. An example is the decimal number 132: This number has 1 hundred, 3 tens, and 2 units. What probably was not learned in elementary school was the exponential value of each position: The unit's position has a weight of 10^0, or 1; the tens position has a weight of 10^1, or 10; and the hundreds position has a weight of 10^2, or 100. The exponential powers of the positions are critical for understanding numbers in other numbering systems. The position to the left of the radix (*number base*) point, called a decimal point only in the decimal system, is always the units position in any number system. For example, the position to the left of the binary point is always 2^0, or 1; the position to the left of the octal point is 8^0, or 1. In any case, any number raised to its zero power is always 1, or the unit's position.

The position to the left of the units position is always the number base raised to the first power; in a decimal system, this is 10^1, or 10. In a binary system, it is 2^1, or 2; and in an octal system, it is 8^1, or 8. Therefore, an 11 decimal has a different value or number of units than an 11 binary. The decimal number is composed of 1 ten plus 1 unit, and has a value of 11 units; whereas the binary number 11 is composed of 1 two plus 1 unit, for a value of 3 units. The 11 octal has a value of 9 units.

In the decimal system, positions to the right of the decimal point have negative powers. The first digit to the right of the decimal point has a value of 10^{-1}, or 0.1. In the binary system, the first digit to the right of the binary point has a value of 2^{-1}, or 0.5. In general, the principles that apply to decimal numbers also apply to numbers in any other number system.

Example 1-4 shows 110.101 in binary (often written as 110.101_2). It also shows the power and weight or value of each digit position. To convert a binary number to decimal, add weights of each digit to form its decimal equivalent. The 110.101_2 is equivalent to a 6.625 decimal ($4 + 2 + 0.5 + 0.125$). Notice that this is the sum of 2^2 (or 4) plus 2^1 (or 2), but 2^0 (or 1) is not added because there are no digits under this position. The fractional part is composed of 2^{-1} (.5) plus 2^{-3} (or .125), but there is no digit under the 2^{-2} (or .25) so .25 is not added.

EXAMPLE 1-4

```
Power            2²    2¹    2⁰    2⁻¹    2⁻²    2⁻³
Weight           4     2     1     .5     .25    .125
Number           1     1     0  .  1      0      1
Numeric Value    4  +  2  +  0  +  .5  +  0   +  .125 = 6.625
```

Suppose that the conversion technique is applied to a base 6 number, such as 25.2_6. Example 1-5 shows this number placed under the powers and weights of each position. In the example, there is a 2 under 6^1, which has a value of 12 (2×6), and a 5 under 6^0, which has a value of 5 (5×1). The whole number portion has a decimal value of $12 + 5$, or 17. The number to the right of the hex point is a 2 under 6^{-1}, which has a value of .333 ($2 \times .167$). Therefore the number 25.2_6 has a value of 17.333 in decimal.

EXAMPLE 1-5

```
Power               6¹        6⁰       6⁻¹
Weight              6         1       .167
Number              2         5    .   2
Numeric Value      12    +    5   +  .333  =  17.333
```

Conversion to Decimal

The prior examples have shown that to convert from any number base to decimal, determine the weights or values of each position of the number, and then sum the weights to form the decimal equivalent. Suppose that a 125.7_8 octal is converted to decimal. To accomplish this conversion, first write down the weights of each position of the number. This appears in Example 1-6. The value of 125.7_8 is 85.875 decimal, or 1×64 plus 2×8 plus 5×1 plus $7 \times .125$.

EXAMPLE 1-6

```
Power               8²        8¹       8⁰        8⁻¹
Weight             64         8        1        .125
Number              1         2        5    .    7
Numeric Value      64    +   16    +   5   +   .875  =  85.875
```

Notice that the weight of the position to the left of the units position is 8. This is 8 times 1. Then notice that the weight of the next position is 64, or 8 times 8. If another position existed, it would be 64 times 8, or 512. To find the weight of the next higher-order position, multiply the weight of the current position by the number base (or 8, in this example). To calculate the weights of position to the right of the radix point, divide by the number base. In the octal system, the position immediately to the right of the octal point is $\frac{1}{8}$, or .125. The next position is $\frac{.125}{8}$, or .015625, which can also be written as $\frac{1}{64}$. Also note that the number in Example 1-6 can be written as the decimal number $85\frac{7}{8}$.

Example 1-7 shows the binary number 11011.0111 written with the weights and powers of each position. If these weights are summed, the value of the binary number converted to decimal is 27.4375.

EXAMPLE 1-7

```
Power          2⁴      2³     2²     2¹     2⁰     2⁻¹     2⁻²      2⁻³        2⁻⁴
Weight         16       8      4      2      1     .5      .25     .125      .0625
Number          1       1      0      1      1  .   0       1        1          1
Numeric Value  16   +   8  +   0  +   2  +   1  +   0   +  .25   +  .125   +  .0625  =  27.4375
```

It is interesting to note that 2^{-1} is also $\frac{1}{2}$, 2^{-2} is $\frac{1}{4}$, and so forth. It is also interesting to note that 2^{-4} is $\frac{1}{16}$, or .0625. The fractional part of this number is $\frac{7}{16}$, or .4375 decimal. Notice that 0111 is a 7 in binary code for the numerator and the rightmost 1 is in the $\frac{1}{16}$ position for the denominator. Other examples: the binary fraction of .101 is $\frac{5}{8}$ and the binary fraction of .001101 is $\frac{13}{64}$.

Hexadecimal numbers are often used with computers. The 6A.C is illustrated with its weights in Example 1-8. The sum of its digits is 106.75, or $106\frac{3}{4}$. The whole number part is represented with 6×16 plus 10 (A) $\times 1$. The fraction part is 12 (C) as a numerator and 16 (16^{-1}) as the denominator, or $\frac{12}{16}$, which is reduced to $\frac{3}{4}$.

EXAMPLE 1-8

```
Power              16¹       16⁰      16⁻¹
Weight             16        1       .0625
Number              6        A    .   C
Numeric Value      96    +  10   +  .75  =  106.75
```

Conversion from Decimal

Conversions from decimal to other number systems are more difficult to accomplish than conversion to decimal. To convert the whole number portion of a number to decimal, divide by 1 radix. To convert the fractional portion, multiply by the radix.

Whole Number Conversion from Decimal.   To convert a decimal whole number to another number system, divide by the radix and save the remainders as significant digits of the result. An algorithm for this conversion is as follows:

1. Divide the decimal number by the radix (number base).
2. Save the remainder (first remainder is the least significant digit).
3. Repeat steps 1 and 2 until the quotient is zero.

For example, to convert a 10 decimal to binary, divide it by 2. The result is 5 with a remainder of 0. The first remainder is the unit's position of the result (in this example, a 0). Next divide the 5 by 2. The result is 2 with a remainder of 1. The 1 is the value of the two's (2^1) position. Continue the division until the quotient is a zero. Example 1-9 shows this conversion process. The result is written as 1010_2 from the bottom to the top.

EXAMPLE 1-9

```
2) 10          remainder = 0
  2) 5         remainder = 1
   2) 2        remainder = 0
    2) 1       remainder = 1              result = 1010
       0
```

To convert a 10 decimal into base 8, divide by 8, as shown in Example 1-10. A 10 decimal is a 12 octal.

EXAMPLE 1-10

```
8) 10          remainder = 2
  8) 1         remainder = 1             result = 12
     0
```

Conversion from decimal to hexadecimal is accomplished by dividing by 16. The remainders will range in value from 0 through 15. Any remainder of 10 through 15 is then converted to the letters A through F for the hexadecimal number. Example 1-11 shows the decimal number 109 converted to a 6D hexadecimal.

EXAMPLE 1-11

```
16) 109        remainder = 13  (D)
   16) 6       remainder = 6             result = 6D
      0
```

Converting from a Decimal Fraction.   Conversion from a decimal fraction to another number base is accomplished with multiplication by the radix. For example, to convert a decimal fraction into binary, multiply by 2. After the multiplication, the whole number portion of the result is saved as a significant digit of the result, and the fractional remainder is again multiplied by the radix. When the fraction remainder is zero, multiplication ends. Note that some numbers are never-ending (repetend). That is, a zero is never a remainder. An algorithm for conversion from a decimal fraction is as follows:

1. Multiply the decimal fraction by the radix (number base).
2. Save the whole number portion of the result (even if zero) as a digit. Note that the first result is written immediately to the right of the radix point.
3. Repeat steps 1 and 2, using the fractional part of step 2 until the fractional part is zero.

Suppose that a .125 decimal is converted to binary. This is accomplished with multiplications by 2, as illustrated in Example 1-12. Notice that the multiplication continues until the fractional remainder is zero. The whole number portions are written as the binary fraction (0.001) in this example.

EXAMPLE 1-12

```
     .125
x       2
    0.25          digit is 0

     .25
x      2
    0.5           digit is 0

      .5
x      2
    1.0           digit is 1          result = 0.001₂
```

This same technique is used to convert a decimal fraction into any number base. Example 1-13 shows the same decimal fraction of .125 from Example 1-12 converted to octal by multiplying by 8.

EXAMPLE 1-13

```
     .125
x       8
    1.0           digit is 1          result = 0.1₈
```

Conversion to a hexadecimal fraction appears in Example 1-14. Here, a decimal .046875 is converted to hexadecimal by multiplying by 16. Note that a .046875 is a 0.0C in hexadecimal.

EXAMPLE 1-14

```
    .046875
x        16
    0.75          digit is 0

     .75
x     16
   12.0           digit is 12(C)      result = 0.0C₁₆
```

Binary-Coded Hexadecimal

Binary-coded hexadecimal (BCH) is used to represent hexadecimal data in binary code. A binary-coded hexadecimal number is a hexadecimal number written so that each digit is represented by a 4-bit binary number. The values for the BCH digits appear in Table 1-5. Note that we often represent a hexadecimal number as 0x8A where the 0x is a signal to the computer that the number that follows is hexadecimal.

Hexadecimal numbers are represented in BCH code by converting each digit to BCH code with a space between each coded digit. Example 1-15 shows 2AC converted to BCH code. Note that each BCH digit is separated by a space.

TABLE 1-5 Binary-coded hexadecimal (BCH) code.

Hexadecimal Digit	BCH Code
0	0000
1	0001
2	0010
3	0011
4	0100
5	0101
6	0110
7	0111
8	1000
9	1001
A	1010
B	1011
C	1100
D	1101
E	1110
F	1111

EXAMPLE 1-15

```
2AC = 0010 1010 1100
```

The purpose of BCH code is to allow a binary version of a hexadecimal number to be written in a form that can easily be converted between BCH and hexadecimal. Example 1-16 shows a BCH coded number converted back to hexadecimal code.

EXAMPLE 1-16

```
1000 0011 1101 . 1110 = 83D.E
```

Complements

At times, data are stored in complement form to represent negative numbers. Two systems are used to represent negative data: *radix* and radix − 1 complements. The earliest system was the radix − 1 complement, in which each digit of the number is subtracted from the radix − 1 to generate the radix − 1 complement to represent a negative number.

Example 1-17 shows how the 8-bit binary number 01001100 is one's (radix − 1) complemented to represent it as a negative value. Notice that each digit of the number is subtracted from one to generate the radix − 1 (one's) complement. In this example, the negative of 01001100 is 10110011. The same technique can be applied to any number system, as illustrated in Example 1-18, in which the fifteen's (radix − 1) complement of a 5CD hexadecimal is computed by subtracting each digit from a fifteen.

EXAMPLE 1-17

```
  1111 1111
− 0100 1100
  1011 0011
```

EXAMPLE 1-18

```
  15 15 15
−  5  C  D
   A  3  2
```

Today, the radix − 1 complement is not used by itself; it is used as a step for finding the radix complement. The radix complement is used to represent negative numbers in modem computer systems. (The radix − 1 complement was used in the early days of computer technology.) The main problem with the radix − 1 complement is that a negative or a positive zero exists; in the radix complement system, only a positive zero can exist.

To form the radix complement, first find the radix − 1 complement, and then add a one to the result. Example 1-19 shows how the number 0100 1000 is converted to a negative value by two's (radix) complementing it.

EXAMPLE 1-19

```
  1111 1111
− 0100 1000
  1011 0111    (one's complement)
+         1
  1011 1000    (two's complement)
```

To prove that 0100 1000 is the inverse (negative) of 1011 1000, add them together to form an 8-digit result. The ninth digit is dropped and the result is zero because 0100 1000 is a positive 72, whereas 1011 1000 is a negative 72. The same technique applies to any number system. Example 1-20 shows how the inverse of a 345 hexadecimal is found first by fifteen's complementing the number, and then by adding one to the result to form the sixteen's complement. As before, if the original 3-digit number 345 is added to the inverse of CBB, the result is a 3-digit 000. As before, the fourth bit (carry) is dropped. This proves that 345 is the inverse of CBB. Additional information about complements, of one's and two's is presented with signed numbers in the next section.

EXAMPLE 1-20

```
  15 15 15
−  3  4  5
   C  B  A    (fifteen's complement)
+        1
   C  B  B    (sixteen's complement)
```

1-3 COMPUTER DATA FORMATS

Successful programming requires a precise understanding of data formats. This section describes many common computer data formats as they are used with the PIC family of microcontrollers. Commonly, data appear as ASCII, BCD, signed and unsigned integers, and occasionally as floating-point numbers (real numbers). Other forms are available but are not presented here because they are not commonly found.

ASCII Data

ASCII (*American Standard Code for Information Interchange*) data represent alphanumeric characters in the memory of a computer system (see Table 1-6). The standard ASCII code is a 7-bit code, with the eighth and most significant bit used to hold parity in some antiquated systems. If ASCII data are used with a printer, the most significant bits are 0 for alphanumeric printing and 1 for graphics printing. In the personal computer, an extended ASCII character set is selected by placing a 1 in the left-most bit. Table 1-7 shows the extended ASCII character set, using code 0x80–0xFF. The extended ASCII characters store some foreign letters and punctuation, Greek

TABLE 1-6   ASCII code.

First	X0	X1	X2	X3	X4	X5	X6	X7	X8	X9	XA	XB	XC	XD	XE	XF
								Second								
0X	NUL	SOH	STX	ETX	EOT	ENQ	ACK	BEL	BS	HT	LF	VT	FF	CR	SO	SI
1X	DLE	DC1	DC2	DC3	DC4	NAK	SYN	ETB	CAN	EMS	SUB	ESC	FS	GS	RS	US
2X	SP	!	"	#	$	%	&	'	(	)	*	+	-	,	.	/
3X	0	1	2	3	4	5	6	7	8	9	:	;	<	=	>	?
4X	@	A	B	C	D	E	F	G	H	I	J	K	L	M	N	O
5X	P	Q	R	S	T	U	V	W	X	Y	Z	[	\	]	^	_
6X	`	a	b	c	d	e	f	g	h	i	j	k	l	m	n	o
7X	p	q	r	s	t	u	v	w	x	y	z	{	|	}	~	⣿

characters, mathematical characters, box-drawing characters, and other special characters. Note that extended characters can vary from one printer to another. The list provided is designed to be used with the IBM ProPrinter, which also matches the special character set found with most word processors.

The ASCII control characters, also listed in Table 1-6, perform control functions in a computer system, including clear screen, backspace, line-feed, and so forth. To enter the control codes through the computer keyboard, hold down the Control key while typing a letter. To obtain the control code 0x01, type a Control-A; a 0x02 is obtained by a Control-B, and so forth. Note that the control codes often appear on the screen as ^A for Control-A, ^B for Control-B, and so forth. Also note that the carriage return code (CR) is the Enter key on most modem keyboards. The purpose of CR is to return the cursor or print-head to the left margin. Another code that appears in many programs is the line feed code (LF), which moves the cursor down one line.

To use Table 1-6 or 1-7 for converting alphanumeric or control characters into ASCII characters, first locate the alphanumeric code for conversion. Next, find the first digit of the hexadecimal ASCII code. Then find the second digit. For example, the capital letter A is ASCII code 0x41, and the lowercase letter *a* is ASCII code 0x61.

ASCII data are most often stored in memory by using a special directive to the assembler program called *declare byte,* or DB, or by the DATA directive for strings. (The assembler is a program that is used to program a computer in its native binary machine language.) The DB

TABLE 1-7   Extended ASCII code, as printed by the IBM ProPrinter.

First	X0	X1	X2	X3	X4	X5	X6	X7	X8	X9	XA	XB	XC	XD	XE	XF
							Second									
0X		☺	☻	♥	♦	♣	♠	●	◘	○	◙	♂	♀	♪	♫	☼
1X	►	◄	↕	‼	¶	§	▬	↨	↑	↓	→	←	∟	↔	▲	▼
8X	Ç	ü	é	â	ä	à	å	ç	ê	ë	è	ï	î	ì	Ä	Å
9X	É	æ	Æ	ô	ö	ò	û	ù	ÿ	Ö	Ü	¢	£	¥	P$_t$	ƒ
AX	á	í	ó	ú	ñ	Ñ	ª	º	¿	⌐	¬	½	¼	¡	«	»
BX	░	▒	▓	│	┤	╡	╢	╖	╕	╣	║	╗	╝	╜	╛	┐
CX	└	┴	┬	├	─	┼	╞	╟	╚	╔	╩	╦	╠	═	╬	╧
DX	╨	╤	╥	╙	╘	╒	╓	╫	╪	┘	┌	█	▄	▌	▐	▀
EX	α	β	Γ	π	Σ	σ	µ	γ	Φ	Θ	Ω	δ	∞	φ	∈	∩
FX	≡	±	≥	≤	⌠	⌡	÷	≈	°	∙	·	√	ⁿ	²	■	

directive and the DATA directive, and several examples of their usage with ASCII-coded characters, are listed in Example 1-21. Notice how each character using DB is surrounded by apostro-
phes (')—never use the quote (") unless you are using DATA.

EXAMPLE 1-21

```
DB      'B'
DB      'r'
DB      'e'
DB      'y'
DB      0x00
DB      'Brey', 0

DATA    "Brey", 0
DATA    "Barry", 0
```

BCD (Binary-Coded Decimal) Data

Binary-coded decimal (BCD) information is stored in either packed or unpacked forms.
Packed BCD data are stored as two digits per byte and unpacked BCD data are stored as one
digit per byte. The range of a BCD digit extends from 0000_2 to 1001_2, or 0–9 decimal.
Unpacked BCD data are returned from a keypad or keyboard. Packed BCD data are used for
some of the instructions included for BCD addition and subtraction in the instruction set of
the microprocessor.

Table 1-8 shows some decimal numbers converted to both the packed and unpacked BCD
forms. Applications that require BCD data are point-of-sales terminals and almost any other device that performs a minimal amount of simple arithmetic. If a system requires complex arithmetic, BCD data are seldom used because there is no simple and efficient method of performing
complex BCD arithmetic.

Example 1-22 shows how to use the assembler to define both packed and unpacked BCD
data. Example 1-23 shows how to do this using C/C++ and char or bytes. In all cases, the convention of storing the least-significant data first is followed. This means that to store 83 into
memory, the 3 is stored first, followed by the 8. Also note that with packed BCD data, the 0x
(hexadecimal) precedes the number to ensure that the assembler stores the BCD value rather than
a decimal value for packed BCD data. Notice how the numbers are stored in memory as unpacked, one digit per byte; or packed, two digits per byte.

EXAMPLE 1-22

```
;Unpacked BCD data (least-significant data first)
;
NUMB1   DB      3,4,5           ;defines number 543
NUMB2   DB      7,8             ;defines number 87
                                ;
;Packed BCD data (least-significant data first)
                                ;
NUMB3   DB      0x37,0x34       ;defines number 3437
NUMB4   DB      3,0x45          ;defines number 4503
```

TABLE 1-8 Packed and unpacked BCD data.

Decimal	Packed		Unpacked		
12	0001 0010		0000 0001	0000 0010	
623	0000 0110	0010 0011	0000 0110	0000 0010	0000 0011
910	0000 1001	0001 0000	0000 1001	0000 0001	0000 0000

EXAMPLE 1-23

```
//Unpacked BCD data (least-significant data first)
//
char Numb1 = 3,4,5;          //defines number 543
char Numb2 = 7,8;            //defines number 87
//
//Packed BCD data (least-significant data first)
//
char Numb3 = 0x37,0x34;      //defines number 3437
char Numb4 = 3,0x45;         //defines number 4503
```

Byte-Sized Data

Byte-sized data are stored as *unsigned integers* and *signed integers*. Figure 1-2 illustrates both the unsigned and signed forms of the byte-sized integer. The difference in these forms is the weight of the leftmost bit position. Its value is 128 for the unsigned integer and minus 128 for the signed integer. In the signed integer format, the leftmost bit represents the sign bit of the number, as well as a weight of minus 128. For example, 0x80 represents a value of 128 as an unsigned number; as a signed number, it represents a value of −128. Unsigned integers range in value from 0x00–0xFF (0–255). Signed integers range in value from −128 to 0 to +127.

Although negative signed numbers are represented in this way, they are stored in the two's complement form. The method of evaluating a signed number by using the weights of each bit position is much easier than the act of two's complementing a number to find its value. This is especially true in the world of calculators designed for programmers.

Whenever a number is two's complemented, its sign changes from negative to positive or positive to negative. For example, the number 00001000 is a +8. Its negative value (–8) is found by two's complementing the +8. To form a two's complement, first one's complement the number. To one's complement a number, invert each bit of a number from zero to one or from one to zero. Once the one's complement is formed, the two's complement is found by adding a one to the one's complement. Example 1-24 shows how numbers are two's complemented using this technique.

EXAMPLE 1-24

```
+8 = 00001000
     11110111 (one's complement)
+           1
-8 = 11111000 (two's complement)
```

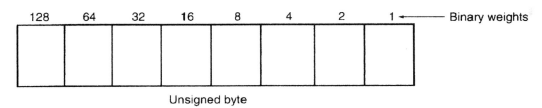

Unsigned byte

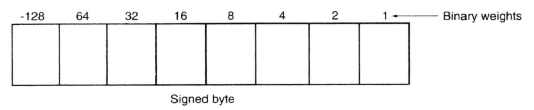

Signed byte

FIGURE 1-2   Byte-sized (8-bit) data.

Another, and probably simpler, technique for two's complementing a number starts with the rightmost digit. Start by writing down the number from right to left. Write the number exactly as it appears until the first one. Write down the first one, and then invert all bits to its left. Example 1-25 shows this technique with the same number as in Example 1-24.

EXAMPLE 1-25

```
+8 = 00001000
         1000   (write number to first 1)
     1111       (invert the remaining bits)
-8 = 11111000
```

To store 8-bit data in memory using the assembler program, use the DB or DATA directive as in prior examples or char as in C/C++ examples. Example 1-26 lists many forms of 8-bit numbers stored in memory using the assembler program. Byte data may also be stored in memory using the DATA directive or the DE (*declare EEPROM memory data*) directives. Notice in the example that a hexadecimal number is defined with the letter 0x preceding the number, and that a decimal number is written with a period preceding the number. Example 1-27 shows the same byte data defined for use with a C/C++ program.

EXAMPLE 1-26

```
;Unsigned byte-sized data
;

DATA1    DB    .254    ;define 254 decimal
DATA2    DB    0x87    ;define 87 hexadecimal
DATA3    DB    .71     ;define 71 decimal

;
;Signed byte-sized data
;

DATA5    DB    .100    ;define 100 decimal
DATA6    DB    0xFF    ;define -1 decimal
DATA7    DB    .56     ;define 56 decimal
```

EXAMPLE 1-27

```
//Unsigned byte-sized data
//

unsigned char Data1 = 254;      //define 254 decimal
unsigned char Data2 = 0x87;     //define 87 hexadecimal
unsigned char Data3 = 71;       //define 71 decimal

//
//Signed byte-sized data
//

char Data4 = -100;              //define -100 decimal
char Data5 = +100;              //define +100 decimal
char Data6 = -1;                //define -1 decimal
char Data7 = 56;                //define 56 decimal
```

Word-Sized Data

A word (16 bits) is formed with two bytes of data. The least significant byte is always stored in the lowest-numbered memory location in the PIC18 family, and the most significant byte is stored in the highest. This method of storing a number is called the *little endian* format. An alternate method, used with all other PIC family members, is called the *big endian* format. In the big endian format, numbers are stored with the lowest location containing the most significant data.

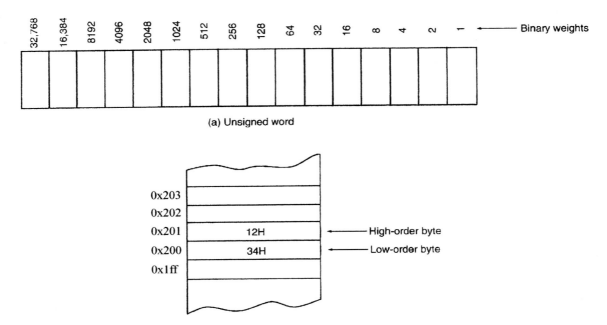

(a) Unsigned word

(b) **The contents of data memory location 0x200 and 0x201 are the word 0x1234.**

FIGURE 1-3   Word-sized (16-bit)data.

Figure 1-3a shows the weights of each bit position in a word of data, and Figure 1-3b how the number 1234H appears when stored in the memory location 3000H and 3001H. The only difference between a signed and an unsigned word is the leftmost bit position. In the unsigned form, the leftmost bit is unsigned and has a weight of 32,768; in the signed form, its weight is a −32,768. As with byte-sized signed data, the signed word is in two's complement form when representing a negative number. Also, notice that the low-order byte is stored in the lowest-numbered memory location (3000H) and the high-order byte is stored in the highest-numbered location (3001H) as it would be in the PIC18 family.

Example 1-28 shows several signed and unsigned word-sized data stored in memory using the assembler program. Example 1-29 shows how to store the same numbers in a C/C++ program, which uses the short directive to store a 16-bit integer. Notice that the *declare word* directive, or DW, causes the assembler to store words in the memory instead of bytes, as in prior examples. The DATA directive can also be used to define a word. When using the PIC C18 C language complier, int is also used to declare a 16-bit integer.

EXAMPLE 1-28

```
;Unsigned word-sized data
;

DATA1   DW    .2544        ;define 2544 decimal
DATA2   DW    0x87AC       ;define 87AC hexadecimal
DATA3   DW    .710         ;define 710 decimal


;
;Signed word-sized data
;

DATA4   DW    .13400       ;define 13400 decimal
DATA5   DW    +198         ;define +198 decimal
DATA6   DW    0xFFFF       ;define -1 decimal
```

EXAMPLE 1-29

```
//Unsigned word-sized data
//

unsigned short Data1 = 2544;        //define 2544 decimal
unsigned short Data2 = 0x87AC;      //define 87AC hexadecimal
unsigned int Data3 = 710;           //define 710 decimal

//
//Signed word-sized data
//

short Data4 = 13400;                //define 13400 decimal
short Data5 = +198;                 //define +198 decimal
int Data6 = -1;                     //define -1 decimal
int Data7 = 3;                      //define +3
```

Real Numbers

Real numbers are occasionally encountered in applications and are presented here because occasionally embedded controllers must deal with real numbers. A real number, or a floating-point number, as it is often called, contains two parts: a mantissa, significand, or fraction; and an exponent. Figure 1-4 depicts both the 4- and 8-byte forms of real numbers as they are stored in any Intel system. Note that the 4-byte number is called single-precision and the 8-byte form is called double-precision. The form presented here is the same form specified by the IEEE standard, IEEE-754, version 10.0. This standard has been adopted as the standard form of real numbers with virtually all programming languages and many applications packages. This standard also applies to the data manipulated by the numeric coprocessor in the personal computer. Figure 1-4a shows the single-precision form that contains a sign-bit, an 8-bit exponent, and a 24-bit fraction (mantissa). Note that because applications often require double-precision floating-point numbers they are also presented in Figure 1-4b.

Simple arithmetic indicates that it should take 33 bits to store all three pieces of data. Not true—the 24-bit mantissa contains an *implied* (hidden) one-bit that allows the mantissa to represent 24 bits while being stored in only 23 bits. The hidden bit is the first bit of the normalized real number. When normalizing a number, it is adjusted so that its value is at least 1, but less than 2. For example, if a 12 is converted to binary (1100_2), it is normalized and the result is 1.1×2^3. The whole number 1 is not stored in the 23-bit mantissa portion of the number; the 1 is the hidden one-bit. Table 1-9 shows the single-precision form of this number and others.

The exponent is stored as a biased exponent. With the single-precision form of the real number, the bias is 127 (0x7F) and with the double-precision form, it is 1023 (0x3FF). The bias and exponent are added before being stored in the exponent portion of the floating-point number.

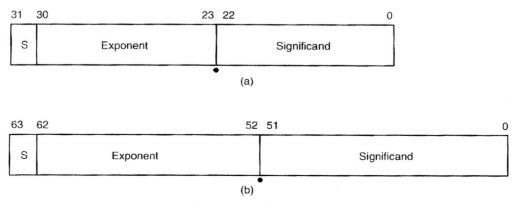

FIGURE 1-4 Floating-point data: (a) single-precision and (b) double-precision.

TABLE 1-9   Single-precision real numbers.

Decimal	Binary	Normalized	Sign	Biased Exponent	Mantissa
+12	1100	1.1×2^3	0	10000010	10000000 00000000 00000000
−12	1100	1.1×2^3	1	10000010	10000000 00000000 00000000
+100	1100100	1.1001×2^6	0	10000101	10010000 00000000 00000000
−1.75	1.11	1.11×2^0	1	01111111	11000000 00000000 00000000
+0.25	0.01	1.0×2^{-2}	0	01111101	00000000 00000000 00000000
+0.0	0	0	0	00000000	00000000 00000000 00000000

The previous example, has an exponent of 2^3, represented as a biased exponent of $127 + 3$ or 130 (0x82) in the single-precision form, or as 1026 (0x402) in the double-precision form.

There are two exceptions to the rules for floating-point numbers. The number 0.0 is stored as all zeros. The number infinity is stored as all ones in the exponent and all zeros in the mantissa. The sign-bit indicates either a positive or a negative infinity.

In C/C++ a single-precision number is defined as *float* and a double-precision number is defined as *double,* as shown in Example 1-30. There is no way to define these numbers with the PIC assembler unless they are first converted to byte or word format by hand. The PIC C18 compiler uses float and double, but both types are actually stored and used as 32-bit single-precision numbers.

EXAMPLE 1-30

```
//Single-precision real numbers
//

float Numb1 = 1.234;
float Numb2 = -23.4;
float Numb3 = 4.3e2;

//
//Double-precision real numbers

double Numb4 = 123.4;
double Numb5 = -23.4;
```

FIGURE 1-5   Microchip and IEEE formats for floating-point numbers.

IEEE-754

s eee eeee	e fff ffff	ffff ffff	ffff ffff

Microchip

eeee eeee	s fff ffff	ffff ffff	ffff ffff

s = sign bit

e = exponent

f = fraction

Microchip uses a different format for a floating-point number in the C18 C language compiler. Figure 1-5 illustrates the difference between the microchip floating-point format and the IEEE floating-point format. The difference is the placement of the sign-bit in the number. This difference does not affect the operation of the C compiler.

1-4   SUMMARY

1. Charles Babbage envisioned and constructed the first computing system, the Analytical Engine. Later, John von Neumann refined the architecture of the computer.
2. The first programmer was Augusta Ada Byron (later the Countess of Lovelace) who earned the reputation as the first programmer because of the programs that she wrote for Charles Babbage's Analytical Engine.
3. The world's first microprocessor, the Intel 4004, was a 4-bit microprocessor—a programmable controller on a chip. It addressed a mere 4,096 four-bit memory locations.
4. The CPU is the controlling element in a computer system. The CPU performs data transfers, does simple arithmetic and logic operations, and makes simple decisions. The CPU executes programs stored in the memory system to perform complex operations in short periods of time.
5. All computer systems contain three buses to control memory and I/O. The address bus is used to request a memory location or I/O device. The data bus transfers data between the microprocessor and its memory and I/O spaces. The control bus controls the memory and I/O, and requests the reading or writing of data.
6. Numbers are converted from any number base to decimal by noting the weights of each position. The weight of the position to the left of the radix point is always the units position in any number system. The position to the left of the units position is always the radix times one. Succeeding positions are determined by multiplying by the radix. The weight of the position to the right of the radix point is always determined by dividing by the radix.
7. Conversion from a whole decimal number to any other base is accomplished by dividing by the radix. Conversion from a fractional decimal number is accomplished by multiplying by the radix.
8. Hexadecimal data are represented in hexadecimal form or in a code called binary-coded hexadecimal (BCH). A binary-coded hexadecimal number is one that is written with a 4-bit binary number that represents each hexadecimal digit.
9. The ASCII code is used to store alphabetic or numeric data. The ASCII code is a 7-bit code; it can have an eighth bit that is used to extend the character set from 128 codes to 256 codes. The carriage return (Enter) code returns the print head or cursor to the left margin. The line feed code moves the cursor or print head down one line.
10. Binary-coded decimal (BCD) data are sometimes used in a computer system to store decimal data. These data are stored either in packed (two digits per byte) or unpacked (one digit per byte) form.
11. Binary data are stored as a byte (8 bits) and word (16 bits) in a computer system. These data may be unsigned or signed. Signed negative data are always stored in the two's complement form. Data that are wider than 8 bits are always stored using the little endian format.
12. Floating-point data are used in computer systems to store whole, mixed, and fractional numbers. A floating-point number is composed of a sign, a mantissa, and an exponent.
13. The assembler directives DB (declare byte) and DW (declare word) are used to store bytes and words of data in the memory system.

1-5   QUESTIONS AND PROBLEMS

1. Who developed the Analytical Engine?
2. The world's first microprocessor was developed in 1971 by _____.

3. Who was the Countess of Lovelace?
4. What is a von Neumann machine?
5. What company developed the first microcontroller?
6. What is the acronym CISC?
7. What is the acronym RISC?
8. A binary bit stores a(n) _____ or a(n) _____.
9. A computer K is equal to _____ bytes.
10. A computer M is equal to _____ K bytes.
11. A computer G is equal to _____ M bytes.
12. What is a nibble?
13. Draw the block diagram of a computer system.
14. List the three buses found in all computer systems.
15. Which bus transfers the memory address to the I/O device or to the memory?
16. Which control signal causes the memory to perform a read operation?
17. What is the stored program concept?
18. What is the difference between an EPROM and an EEPROM?
19. If a memory has a 14-bit address, how many locations does it contain?
20. How many memory locations are found in a 4K × 8 memory?
21. What is an SRAM?
22. What type of memory device would be used to store dynamic data?
23. Define the purpose of the following assembler directives:
    a. DB
    b. DATA
    c. DW
24. Define the purpose of the following C/C++ directives:
    a. char
    b. short
    c. int
    d. float
    e. double
25. If a memory address bus contains the following signal lines, determine the number of memory locations that can be addressed.
    a. 12
    b. 14
    c. 16
    d. 18
    e. 32
26. Convert the following binary numbers into decimal:
    a. 1101.01
    b. 111001.0011
    c. 101011.0101
    d. 111.0001
27. Convert the following octal numbers into decimal:
    a. 234.5
    b. 12.3
    c. 7767.07
    d. 123.45
    e. 72.72
28. Convert the following hexadecimal numbers into decimal:
    a. A3.3
    b. 129.C

    c. AC.DC

    d. FAB.3

    e. BB8.0D

29. Convert the following decimal integers into binary, octal, and hexadecimal:

    a. 23

    b. 107

    c. 1238

    d. 92

    e. 173

30. Convert the following decimal numbers into binary, octal, and hexadecimal:

    a. 0.625

    b. 0.00390625

    c. 0.62890625

    d. 0.75

    e. 0.9375

31. Convert the following hexadecimal numbers into binary-coded hexadecimal code (BCH):

    a. 23

    b. AD4

    c. 34.AD

    d. BD32

    e. 234.3

32. Convert the following binary-coded hexadecimal numbers into hexadecimal:

    a. 1100 0010

    b. 0001 0000 1111 1101

    c. 1011 1100

    d. 0001 0000

    e. 1000 1011 1010

33. Convert the following binary numbers to the one's complement form:

    a. 1000 1000

    b. 0101 1010

    c. 0111 0111

    d. 1000 0000

34. Convert the following binary numbers to the two's complement form:

    a. 1000 0001

    b. 1010 1100

    c. 1010 1111

    d. 1000 0000

35. How is a byte declared using the assembler?

36. Convert the following words into ASCII-coded character strings using the assembler:

    a. FROG

    b. Arc

    c. Water

    d. Well

37. What is the ASCII code for the Enter key and what is its purpose?

38. Use an assembler directive to store the ASCII-character string 'What time is it?' in the memory.

39. Convert the following decimal numbers into 8-bit signed binary numbers:

    a. +32

    b. −12

    c. +100

    d. −92

40. Convert the following decimal numbers into signed binary words:
    a. +1000
    b. −120
    c. +800
    d. −3212
41. Use an assembler directive to store byte-sized −34 into the memory.
42. Create a byte-sized variable called Fred1 and store a −34 in it in C/C++.
43. Show how the following 16-bit hexadecimal numbers are stored in the memory system (show both the big and little endian forms):
    a. 0x1234
    b. 0xA122
    c. 0xB100
44. What is the difference between the big endian and little endian formats for storing numbers that are larger than eight bits in width?
45. Use an assembler directive to store a 123A hexadecimal into the memory.
46. Convert the following decimal numbers into both packed and unpacked BCD forms:
    a. 102
    b. 44
    c. 301
    d. 1000
47. Convert the following binary numbers into signed decimal numbers:
    a. 10000000
    b. 00110011
    c. 10010010
    d. 10001001
48. Convert the following BCD numbers (assume that these are packed numbers) to decimal numbers:
    a. 1000 1001
    b. 0000 1001
    c. 0011 0010
    d. 0000 0001
49. Convert the following decimal numbers into single-precision floating-point numbers:
    a. +1.5
    b. −10.625
    c. +100.25
    d. −1200
50. Convert the following IEEE-754, version 10.0, single-precision floating-point numbers into decimal numbers:
    a. 0 10000000 11000000000000000000000
    b. 1 01111111 00000000000000000000000
    c. 0 10000010 10010000000000000000000
51. Use the Internet to write a short report about any one of the following computer pioneers:
    a. Charles Babbage
    b. Augusta Ada Byron
    c. John von Neumann
52. Use the Internet to write a short report about any one of the following computer languages:
    a. COBOL
    b. ALGOL
    c. FORTRAN
    d. PASCAL
53. Use the Internet to write a short report detailing the features of the PIC family of microcontrollers.

CHAPTER 2

PIC18 Family Architecture and Program Development

Before the microcontroller is programmed or interfaced, its architecture must be understood. This chapter presents the architecture of the PIC family of microcontrollers, concentrating on the PIC18 family members from Microchip Technology Incorporated. The PIC18 family is the newest PIC family available. Also introduced are the integrated development environment (IDE) for the PIC microcontroller and the assembly/link process.

Upon completion of this chapter you will be able to:

1. Describe the internal architecture of the PIC microcontroller and the interaction of its components.
2. Detail the system architecture of a PIC-based microprocessor system.
3. Explain the purpose of each register in the general programming model of the microcontroller.
4. Program a simple I/O port for input or output operation.
5. Use the integrated development environment (IDE) for programming the PIC18 family.
6. Simulate the execution of a program for the PIC microcontroller.
7. Describe the function of assembly language directives.
8. Explain the assembly and link process.

2-1   PIC18 ARCHITECTURE

The architecture of the PIC-based microcontroller system is slightly different from the architecture or general computer system presented in Chapter 1. Because the PIC contains the memory and I/O for most applications, the PIC microcontroller is just a single block in the PIC-based system. Additional memory or I/O capabilities can be added to the system, but in many cases the memory and I/O provided on the PIC is sufficient for most applications.

Figure 2-1 illustrates a sample system that uses the PIC as its controller. This system contains an LCD display for displaying information, a keyboard for entering data, an optical card reader for reading information from a card, and a serial interface to a host computer for uploading data and downloading information to the controller. This sample system contains no external memory because the memory is inside of the microcontroller. The entire system contains only the I/O devices and the microcontroller, which makes it appear different from the block diagram of a computer system presented in Chapter 1. If the microcontroller were drawn in the style of Figure 1-1, there would be two blocks, one for the PIC and one for I/O.

FIGURE 2-1 Simple micro-controller-based system.

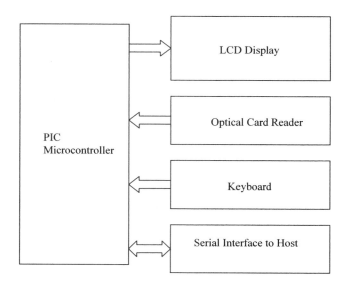

What is the purpose of this simple system? It could be used to obtain information from workers at a factory and function as a time clock. The worker shows up for work, slides his ID card through the optical card reader, and the system logs his ID number into a remote computer through the serial interface to the host computer system. The host then verifies acceptance of the ID card and displays "accepted" or some error or other message on the LCD display. This system contains two input devices (keyboard and optical card reader), an output device (LCD display), and a bidirectional I/O device (serial interface to the host).

The PIC microcontroller contains: memory for program and data storage, I/O connections to interface with the card reader, LCD display, keyboard, serial interface (internal to the controller), and even a timer that can be used as a time-of-day clock and calendar for the system. This level of integration makes the microcontroller a powerful and inexpensive way to design a system. The PIC microcontroller costs just a few dollars and is programmed in assembly language, C, BASIC, or Java.

Overview of the Internal PIC Architecture

Now that we have some idea of the power of the PIC as the controller in a system, we need to understand the internal structure of the device. As mentioned, the PIC contains memory and a CPU. It also contains the I/O interfacing components needed to control devices in the system. Before the microcontroller can be programmed, this internal structure must be understood.

Figure 2-2 depicts the internal architecture of the PIC18 family of controllers. Notice that the internal architecture is similar to the block diagram of a computer system as presented in Figure 1-1. The main difference is that the memory in the PIC is divided into two distinct sections: one for storing the program, and one for storing data. This architecture is called the **Harvard architecture** devised in 1944 for an electromechanical computer (Mark I) developed by Howard Hathaway Aiken who in 1947 said, "Only six electronic digital computers would be required to satisfy the computing needs of the entire United States." The Harvard architecture splits the system memory into two sections to improve system performance. In the traditional von Neumann architecture system, both the data and instructions are fetched through the same bus, which often causes delays because of contentions whenever an instruction manipulates memory data. Data and I/O are normally 8 bits in width, whereas instructions are often wider than 8 bits. In the Harvard architecture implemented in the PIC microcontroller, the data storage and I/O buses are 8 bits wide and the program storage is 16 bits wide. This organization allows instructions to be fetched from the program storage in one read cycle or operation to improve efficiency. Note that some family PIC microcontrollers use a 12-bit wide program memory, but here a 16-bit program memory is detailed for the PIC18 family.

FIGURE 2-2 Internal structure of the PIC 18 family of microcontrollers.

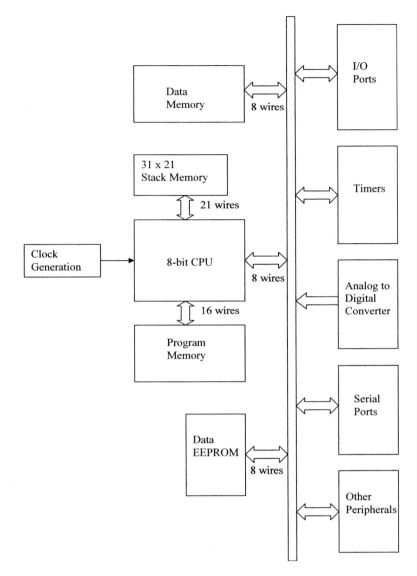

Because the program memory and data memory are separate, opcode fetches overlap executions, improving the efficiency of the microcontroller. This effect is called *pipelining* and is illustrated in Figure 2-3. Notice that during the first instruction cycle clock, instruction 1 is fetched from the memory, then during the second instruction cycle clock, instruction 2 is fetched while instruction 1 executes. This overlapping or pipelining allows most instructions to execute in a single clock. Overlapped fetch and execution occurs only in systems that have a separate program and data memory using the Harvard Architecture. Microprocessors such as the Pentium from Intel actually use the same architecture internally through a data cache and program cache that are used to implement pipelining and the Harvard architecture internally from these caches.

Program Memory. The **program memory** is 16 bits wide and each location holds most single word instructions. Note that some instructions require 32 bits of memory or two consecutive memory locations, but most instructions are 16 bits wide. The program memory is divided into two parts: One section is the on-chip program memory and the other is the external program memory, if needed. The amount of on-chip program memory varies between family members

FIGURE 2-3 Pipelining instructions and executions in the microcontroller.

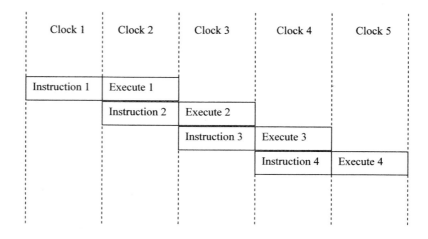

from as little as 4K bytes to as much as 128K bytes. The on-chip program memory is either flash memory that can be reprogrammed many times in the 18F family or **one-time-programmable** (OTP) memory in the 18C family. The flash program memory is usually programmed by an external programmer, but can be self-programming in many versions of the microcontroller. If it is self-programmable, the device usually has a boot block that contains a program called a bootstrap loader that is not erased and reprogrammed when the device is flashed. More coverage of self-programming and the bootstrap loader is explained in Chapter 10. A one-time-programmable memory is like an EPROM except there is no quartz window for erasure so it cannot be reprogrammed, it can be programmed only once. Table 2-1 lists all of the current PIC18 family members and the amount of internal program memory found in each.

The program memory is addressed through a 21-bit program address held in a register called a **program counter**. In Chapter 1 we learned that a memory address of 21 bits (2^{21}) addresses 2M of

TABLE 2-1   Amount of program memory within the PIC18 family members.

Part Number(s)	Memory Size in Bytes
1220, 1230, 2220, 4220	4K
1320, 1330, 2320, 2331, 4320, 4331, 6310, 6390, 8310, 8390	8K
2439, 4439	12K
242, 248, 442, 2410, 2420, 2431, 2480, 4410, 4420, 4431, 4480, 6410, 6490, 8410, 8490	16K
2455, 2539, 4455, 4539	24K
252, 258, 452, 458, 658, 858, 2510, 2520, 2550, 2580, 4510, 4520, 4550, 6520, 65J10, 8520, 85J10	32K
2515, 2525, 2585, 4515, 4525, 4585, 6525, 6527, 6585, 65J15, 8525, 8527, 8585, 85J15	48K
2610, 2620, 2680, 4610, 4620, 4680, 66J10, 6620, 6621, 6622, 6680, 8620, 8621, 8622, 8680, 86J10	64K
6627, 66J15, 8627, 86J15	96K
6720, 6722, 67J10, 8720, 8722, 87J10	128K

Note: All numbers are preceded with PIC18F or PIC18C; the PIC18F devices have flash memory and the PIC18C devices are one-time programmable (OTP) devices. The PIC18C601 and PIC18C801 contain no internal program memory.

FIGURE 2-4 PIC18 family program memory map.

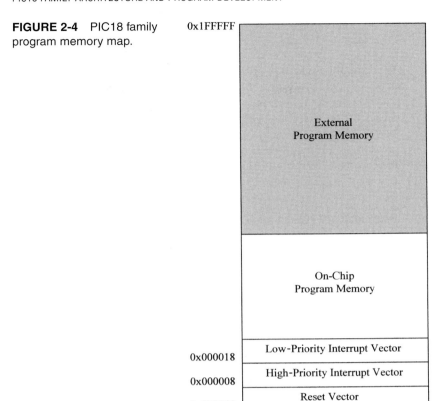

0x1FFFFF

External
Program Memory

On-Chip
Program Memory

Low-Priority Interrupt Vector

0x000018

High-Priority Interrupt Vector

0x000008

Reset Vector

0x000000

memory. In the PIC18 family, the program memory space begins at address 0x000000 and extends to 0x1FFFFF. Suppose a PIC18F4520 is selected for use in a system. This microcontroller has 32K of program memory so the first 32K of the program memory, which is numbered in bytes, begins at address 0x000000 and ends at 0x007FFF. Recall from Chapter 1 that 4K of memory requires 0x1000 locations, therefore a 32K span of memory requires 32K ÷ 4K = 8 or 0x8000 locations.

Figure 2-4 illustrates the structure of the program memory. Some of the locations have dedicated tasks that are important. One such dedicated location is the **reset vector address**, at location 0x000000. The reset vector address is where a program begins execution after a reset. Because there are only 8 bytes of memory between this location and the high-priority interrupt vector address at location 0x000008, it is customary to branch or GOTO another location to continue the program from the reset vector address. The branch is normally accomplished by the unconditional GOTO instruction in a program. The interrupt vector addresses, discussed in Chapter 6 along with interrupts, are where the low- and high-priority interrupts access their interrupt service procedures in the program memory. The remainder of the program memory is available for program storage and also static data storage. Static data storage is where constants and tables of nonchanging data are stored and accessed as data by a program.

Data Memory. The **data memory** is either SRAM *or* SRAM and EEPROM. The data memory (sometimes called a **register file**) provides a place to store transitory data as an application executes and is lost (except for the EEPROM data) when power is removed from the system. The data memory SRAM is accessed through a 12-bit address so the largest data memory available is 3968 bytes at address 0x000 through 0xF7F. The data memory SRAM also contains addresses that are used to program the special functions of the microcontroller. These **special function registers** (SFR) are at data memory addresses 0xF80 through 0xFFF or the upper 128 bytes of the data memory

space. Some versions of the microcontroller use additional data memory for additional special function registers. The other data memory locations are called **general function registers** (GFR) located
at addresses 0x000 through 0xF7F. The GFRs are also sometimes called **register file locations**.

If a PIC microcontroller contains EEPROM, the EEPROM data memory is accessed through
the special function registers in a separate address space devoted to the EEPROM. The size of the
EEPROM is determined by the PIC part number and ranges from 0 bytes to a maximum of 1024
bytes. According to Microchip, a location in the EEPROM can be written up to one million times.
The EEPROM locations should only store information that does not change frequently. It seems that
rewriting a location a million times would allow the EEPROM to be used for anything, but remember the microcontroller might function at 40 MHz in a system and in a few seconds it is possible
to write to an EEPROM memory location many millions of times. Table 2-2 lists the various PIC18
family members and the amount of RAM and EEPROM data storage available to each device.

Stack Memory. In addition to the program, data SRAM, and EEPROM memory, there is also a
small, 31- by 21-bit wide SRAM memory location called a **stack memory**. The stack memory,
as discussed in later sections, holds only return addresses from functions. The stack is 21 bits
wide because it is designed to store program memory addresses. Because the maximum size for
the program memory is 2M bytes, which requires a 21-bit address for access, the stack is 21 bits
in width. Why is it 31 locations deep? Microchip must have decided that a 31-deep stack was
large enough to function in most applications. The only way to use a location on the stack is to
call a function. If a function calls another function, two locations on the stack are used to store two
return addresses. To use all the stack locations requires that functions are nested to a depth of 31,

TABLE 2-2   Data RAM and EPROM in various 18F family members.

Part Number(s)	RAM Size in Bytes	EEPROM Size in Bytes
1230, 1330	256	128
1320	256	256
2220, 2320, 4220, 4320	512	256
2439, 4439	640	256
2410, 4410, 6310, 6390, 6410, 6490, 8310, 8390, 8410, 8490	768	0
242, 248, 442, 448, 2331, 2420, 2431, 2480, 4331, 4420, 4431, 4480	768	256
2539, 4539	1408	256
2510, 4510	1536	0
252, 258, 452, 458, 601, 658, 801, 858, 2520, 2580, 4520, 4580	1536	256
65J10, 65J15, 66J10, 85J10, 85J15, 86J10	2048	0
2455, 2550, 4455, 4550	2048	256
6520, 8520	2048	1024
2585, 2680, 4585, 4680, 6585, 6680, 8585, 8680	3328	1024
6525, 6620, 6621, 6720, 8525, 8620, 8621, 8720	3840	1024
2515, 2610, 4515, 4610, 66J15, 67J10, 86J15, 87J10	3968	0
2525, 2620, 4525, 4620, 6527, 6622, 6627, 6722, 8527, 8622, 8627, 8722	3968	1024

Note: None of the 18C versions contain EEPROM. The number of locations does not include the SFR space.

which occurs very rarely, if ever. Essentially, the only way that the stack can overflow is if a program contains an error that constantly calls functions, which do not return. Recursion is one possible cause of this type of error. If developing recursive software, care must be exercised.

I/O Ports.   The **I/O ports** are used to interface the microprocessor to the outside world. Each I/O port is normally 8 bits in width and can be programmed for inputting or outputting information. The number of bits and I/O ports provided varies in different PIC18 family members and is generally determined by the number of pins on the integrated circuit. Table 2-3 illustrates the number of I/O pins found on various PIC18 family members. The table indicates the number of port bits, but when programming the microcontroller, the ports are organized as Port A for the first 8 bits, Port B for the second 8 bits, and so forth. For example, the 18F1320 has only Ports A and B. The I/O port programming is accomplished through the special function registers located at the top of the data memory at addresses 0xF80 through 0xFFF.

The I/O structure of various family members varies widely, but all devices have at least Port A and Port B. In the first few chapters of this book some sample programs use Ports A and B. An actual development board is suggested for experimenting with the basic I/O functions using these two ports. Which development system chosen for experimentation is up to the learner and could be any that are available from Microchip or any other company. In these first few chapters it is assumed that an I/O port is available with at least four light emitting diodes (LEDs) and a few switches. Later chapters investigate additional I/O modules within the PIC and devices.

To use an I/O port for simple I/O is a fairly easy task. The TRISA register determines the direction (input or output) of each Port A pin. (TRISA is for Port A, TRISB is for Port B, and so forth). The TRIS register for a port is the **data direction register** for the port. When the microcontroller is

TABLE 2-3   I/O port connections on various PIC18 family members.

Part Number(s)	I/O Port Bits
1220, 1230, 1320, 1330	16
2439, 2539	21
258, 2331, 2431	22
242, 248, 252, 258, 1320, 2220	23
2220, 2320, 2410, 2420, 2455, 2480, 2510, 2515, 2520, 2525, 2550, 2580, 2585, 2610, 2620, 2680	25
601	31
4439, 4539	32
442, 448, 452, 458, 4220, 4320, 4331, 4420, 4431, 4455, 4550	34
4220, 4320, 4410, 4480, 4510, 4515, 4520, 4525, 4580, 4585, 4610, 4620, 4680	36
801	42
6390, 6490, 65J10, 65J15, 66J10, 66J15, 67J10	50
658, 6520, 6620, 6720	52
6525, 6585, 6621, 6680	53
6310, 6410, 6527, 6622, 6627, 6722	54
8390, 8490, 85J10, 86J10, 86J15	66
858, 8520, 8620, 8720	68
8525, 8585, 8621, 8680	69
8310, 8410, 8527, 85J15, 8622, 8627, 87J10	70
8722	72

TABLE 2-4   Timers in the PIC18 family members.

Part Number(s)	8-Bit Timers	16-Bit Timers
C242, C252, C442, C452	3	1
1230, 1330	0	2
2439, 2539, 4439, 4539	0	3
242F, 248, 252F, 258, 448, 452F, 458, 601, 658, 801, 858, 1220, 1320, 2220, 2320, 2331, 2410, 2420, 2431, 2455, 2480, 2510, 2515, 2520, 2525, 2550, 2580, 2610, 2620, 2680, 4220, 4320, 4331, 4410, 4420, 4431, 4455, 4480, 4510, 4515, 4520, 4525, 4550, 4580, 4585, 4610, 4620, 4680, 6310, 6390, 6410, 6490, 8310, 8390, 8410, 8490	1	3
6520, 6525, 6527, 6585, 65J10, 65J15, 6620, 6621, 6622, 6627, 6680, 66J10, 66J15, 6720, 6722, 67J10, 8520, 8525, 8527, 8585, 85J10, 85J15, 8620, 8621, 8622, 8627, 8680, 86J10, 86J15, 8720, 8722, 87J10	2	3

reset, the TRIS registers are all programmed for input operation. This protects any circuitry attached to the I/O port pin. A logic zero in a bit of the TRISA register sets the corresponding Port A bit as an output bit, and a logic one in a bit sets the corresponding Port A bit as an input bit. For example, to program Port A bits 0 through 3 (pins labeled RA0 through RA3) as output bits and Port A bits 4 through 7 (pins labeled RA4 through RA7) as input bits, the TRISA register is loaded with 0xF0.

Once the direction of a port's pins is programmed, the port is accessed by its name. The PORTA register is used to communicate with Port A. For example, if a 0x28 is written to the PORTA register, the PORTA pins (if programmed as output pins) will receive the data. Likewise, to read data from the Port A pins, read the PORTA register. It is also possible to access a single Port A pin, as described in later sections, with the PORTAbits directive in C language.

Timers.   **Timers** are programmable modulus counters. A timer can count events and clock pulses, and perform a variety of services for a program. The timers are often programmed to fire (time out) after a certain number of clock pulses have occurred. Various PIC family members have between two and five timers. Table 2-4 lists the various PIC18 family members with the number of timers found in each.

Timers are used for a variety of events in a microprocessor, from causing interrupts at a periodic rate for implementing real-time clocks, to measuring unknown frequencies or counting events. Timers are also used to generate signals for other devices in a system. Timers are explained in complete detail later in the text when the internal I/O is discussed and used in many diverse applications.

Other Internal I/O Devices.   Various models of the PIC18 family also contain other I/O devices that are programmed through the special function registers (SFRs). These other devices include analog-to-digital converters, pulse-width modulators, and serial communications ports of several different types. Because we have yet to look at programming the PIC18, these internal I/O devices are not explained here, but are described in complete detail with many examples later in this book.

Basically, I/O ports are programmed through a TRIS register. A TRIS register determines the direction of the I/O pins (IN or OUT) for a given port. If a bit in the TRIS register is programmed with a zero, the corresponding port pin is programmed as an output pin. If a bit in the TRIS register is programmed with a one, the corresponding port pin is programmed as an input pin. Each port has TRIS registers labeled TRISA for Port A, TRISB for Port B, and so forth. The data are read or written to a PORT register. PORTA is used for Port A, PORTB is used for Port B, and so forth. To select the analog operation of a pin for the analog-to-digital converter or digital operation, the ADCON1 register is used to specify which Port A and Port B are digital or analog. The **default** setting is analog.

2-2   PROGRAMMING MODEL

Before programming the microcontroller, the internal register file set available to the programmer must be known. This register set controls the microcontroller. The microcontroller is an 8-bit device so many of the registers are 8 bits in width or a multiple of 8 bits. Figure 2-5 shows the general programming model of the PIC18 family. These are the registers most commonly used with most instructions for programming and do not include all of the special function registers that control I/O devices and the CPU function. These remaining special function registers are discussed with I/O devices in later chapters.

Register File

The **register file area** or **data memory**, located in the onboard SRAM, is a set of 8-bit-wide general purpose registers (GPR) used to store dynamic data. As described earlier, various models of the PIC18 family contain different amounts of data memory, with all models containing at least

FIGURE 2-5 General programming model of the PIC18 family of microcontrollers.

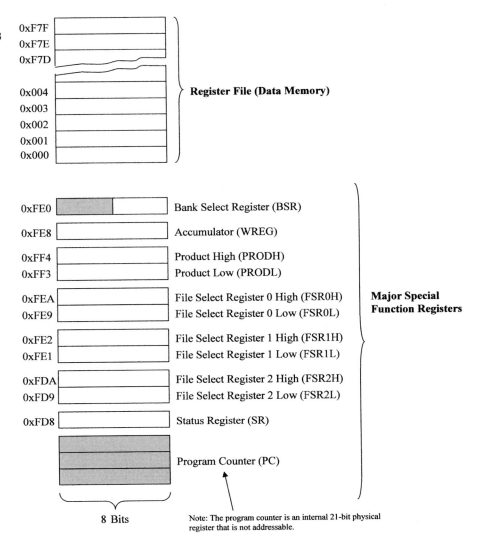

general register file locations 0x000 through 0x07F, plus the special function register at 0xF80 through 0xFFF or 256 bytes of SRAM.

The register file is accessed through a 12-bit address with the most significant 4 bits containing a bank location. **Data memory banks** each contain 256 bytes of data memory. Data bank 0 (0000_2) contains registers 0x000 through 0x0FF, data bank 1 (0001_2) contains register locations 0x100 through 0x1FF, and so forth. In many cases, programming uses locations 0x000 through 0x07F from data bank 0 and locations 0xF80 through 0xFFF from data bank 15. These areas of memory are together called the **access bank**. The access bank is addressed without using a data bank register, so it is easier to access and more efficient to access in a program. (This split of the access bank can vary in different PIC18 family members. The PIC 18F2480, for example, uses 0x00 through 0x5F from data bank 0 and locations 0xF60 through 0xFFF for the special function registers.) These access bank addresses are accessed using a single 8-bit address. To access locations in the data RAM outside of the access bank, use a combination of an 8-bit address and the 4-bit **bank select register** (BSR). For example, to address location 0x432, the bank address is 4 and the 8-bit memory address is 0x32. A bit in an instruction, called the **a-bit**, selects the access bank (when a = 0) or the bank indicated by the bank select register (when a = 1). How is data memory location 0x092 accessed? It is outside of the access bank so the only way to access location 0x092 is by addressing it with the bank select register set to 0 and an 8-bit address of 0x92 with the **a-bit** also set to 1. Figure 2-6 depicts access bank and data bank addressing.

Special Function Registers

The **special function registers (SFR)** are used to perform a variety of special tasks in a microcontroller. The bank select register is one of these special function registers, but other special function registers access memory indirectly, hold a product after a multiplication, indicate the status of the outcome of an instruction, accumulate results from arithmetic and logic operations, and address a location in a program. All of the special function registers are located at the top of the data memory in the access bank. All special function registers have a name and an address, where either can be used for accessing them.

Accumulator (WREG).   The **accumulator** or, as it is often called, the **working register** (W register or WREG) is an 8-bit register that is accessed by many instructions. This register is most likely called an accumulator because it is where results accumulate for many instructions. Most CPUs contain an accumulator as the main working register. In the PIC18 family, the working register (WREG) is located in the SFR area at address 0xFE8. Although this register is assigned an address, many instructions access it without using its address by its name or implicitly as part of an instruction. This implicit form of addressing is why WREG is called the working register or accumulator. The literal or immediate instructions address the WREG as part of the instruction without the need to specify its address. Many other instructions also use the W register.

Bank Select Register (BSR).   The **bank select register** (which is 4 bits wide) plus an 8-bit address combine to form the 12-bit data memory address when the a-bit in the instruction is a logic one as illustrated in Figure 2-6. This allows access to any register in any bank in the data memory. If the a-bit in the instruction is a logic zero, access is restricted to the access bank which is usually comprised of locations 0x000 through 0x07F and locations 0xF80 through 0xFFF. The 8-bit addresses 0x00 through 0x7F select data memory locations 0x000 through 0x07F, and the 8-bit addresses 0x80 through 0xFF select data memory locations 0xF80 through 0xFFF, for the access bank. This description applies to the standard instruction set. (If the extended instruction set is selected, the banks and a-bit do not function in the same fashion. A section in Chapter 10 details the extended instruction set and until then, the text uses the

(a) if a-bit = 0 (access bank)

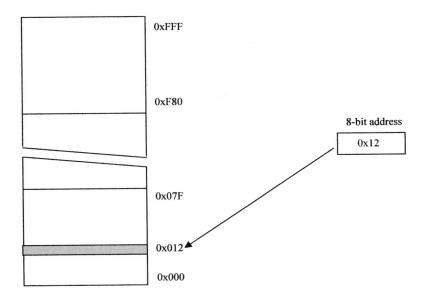

(b) if a-bit = 1 (bank selection)

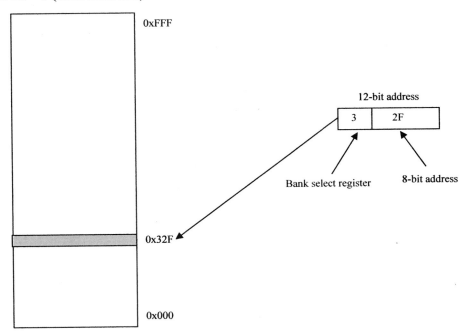

FIGURE 2-6   Register file bank selection.

FIGURE 2-7 Binary bit pattern of a byte operation instruction illustrating the placement of the a-bit.

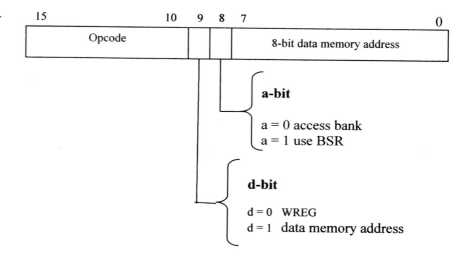

standard instruction set.) Refer to Figure 2-7 for the bit pattern of a byte-oriented instruction for the placement of the a-bit. Also notice that another bit called the d-bit or destination bit selects whether the destination of the instruction is the 8-bit address (when d = 1) or the WREG (when d = 0).

Example 2-1 illustrates the first short sequence of PIC assembly language instructions that add a 6 and 2 together and store the sum into access bank data register address 0x000. Recall that this same task was illustrated in Chapter 1. How is this accomplished? First, a literal instruction (MOVLW) places a 0x06 into the WREG. A MOV does not actually move anything. This has confused new programmers for many years. **A MOV is a COPY**. A MOV copies a 6 into the W register. The letter L in the instruction indicates literal and the letter W indicates the WREG. Therefore, a MOVLW copies a literal value into the W register. **Literal instructions** are used when a **constant** of a known value if needed in a program. The second instruction is another literal instruction. The ADDLW instruction adds a 0x02 (in the instruction) to the 0x06 in WREG to form the sum of 0x08 in the W register. (Notice how the sum **accumulates** in this accumulator or W register). Finally, a byte-oriented move (MOVWF) transfers the answer from the W register to data memory register 0x000. The letter F in a MOVWF instruction designates a register file location. This means a MOVWF copies the contents of WREG into a register file location, in this case, location 0x00. Notice that the MOVWF instruction has two numbers following it. The first is the register file location and the second is the a-bit (in this example a 0, so the access bank is addressed). As an alternative, the word ACCESS (all capitals letters) can be used in place of the 0 for the second parameter as in MOVWF 0x00, ACCESS. This is definitely clearer, but requires more typing so many programmers skip typing the word ACCESS into a program and just type the zero.

EXAMPLE 2-1

```
        MOVLW   0x06            ;place a 0x06 into W
        ADDLW   0x02            ;add a 0x02 to W
        MOVWF   0x00, 0         ;copy W to access bank register 0x00

;       OR                      another version using the ACCESS keyword

        MOVLW   0x06            ;place a 0x06 into W
        ADDLW   0x02            ;add a 0x02 to W
        MOVWF   0x00, ACCESS    ;copy W to access bank register 0x00
```

An **opcode** is an operation code that instructs the microcontroller to perform some operation. In Example 2-1, the literal data of 0x06 is moved into WREG with the opcode MOVLW.

The literal data is called an **operand**. The operand is acted upon by the opcode. An instruction has only a single opcode, but may have anywhere from no operands to many operands as dictated by the instruction.

Suppose the same operation as in Example 2-1 is needed, but instead of placing the result into data register 0x000, the result must be placed into data register 0x200. Address 0x200 is in data memory bank 2. The program sequence listed in Example 2-2 performs this operation by using the bank selection register to address the bank with the MOVLB (move literal to BSR) instruction. Notice how the second operand (a-bit) in the MOVWF instruction designates the bank selection register for the instruction instead of the access bank. Because the instruction normally (by default) places data into the data bank location, the 1 can be dropped in favor of the MOVWF 0x00 instruction. As an alternative to the 1 for data bank access, the word BANKED can be used. All three versions of the program perform exactly the same task.

EXAMPLE 2-2

```
        MOVLW   0x06            ;place a 0x06 into W
        ADDLW   0x02            ;add a 0x02 to W
        MOVLB   2               ;load BSR with bank 2
        MOVWF   0x00, 1         ;copy W to data register 0x00
                                ;of bank 2 or address 0x200

;       OR                      using the BANKED keyword

        MOVLW   0x06            ;place a 0x06 into W
        ADDLW   0x02            ;add a 0x02 to W
        MOVLB   2               ;load BSR with bank 2
        MOVLF   0x00, BANKED    ;copy W to data register 0x00
                                ;of bank 2 or address 0x200

;       OR                      without any bank indication

        MOVLW   0x06            ;place a 0x06 into W
        ADDLW   0x02            ;add a 0x02 to W
        MOVLB   2               ;load BSR with bank 2
        MOVWF   0x00            ;copy W to data register 0x00
                                ;of bank 2 or address 0x200
```

The previous section introduced simple I/O. Suppose that a development system is available that has two LED diodes connected to Port A bits 0 and 1. (These are the RA0 and RA1 pins). The task is to place a 0 on bit 0 and a 1 on bit 1. This is accomplished by programming the ADCON1 register for digital pins, TRISA register with 0s in at least bits 0 and 1, and then sending 0x02 to the PORTA register. Example 2-3 shows this sequence of instructions. For the MOVWF instructions, the access bank bit is not needed when using a named special function register such as TRISA or PORTA. The named register must be in uppercase.

EXAMPLE 2-3

```
MOVLW   0x7F
MOVWF   ADCON1          ;select all digital pins for ports

MOVLW   0x00            ;place 0x00 in Port A direction register
MOVWF   TRISA           ;to select output operation

MOVLW   0x02
MOVWF   PORTA           ;place 0x02 in Port A
```

Product Registers.   The **product registers** hold the result or product after the multiply instruction executes. A special register is needed for multiplication because a product is always twice the width of the multiplier. This is even true in the decimal number system when the number 5 is multiplied by 5. The result of 25 is twice the width of a 5. The MULLW instruction multiplies the W register by the literal value. Because this is an 8-bit microcontroller and the multiplication is

8 bits, the product is 16 bits in width and stored in the PRODL and PRODH registers. The PRODL register holds the low-order 8 bits of the result (L is low), and the PRODH register holds the high-order 8 bits of the result (H is high). The numbers are always unsigned numbers because the instruction set has no signed multiplication instruction available.

Example 2-4 shows how a 3 is multiplied by a 100 decimal and how the product is stored in 0x010 and 0x011. A decimal number is denoted, in an assembly language program, by preceding it with a period. Note that the result is stored using the little endian format where the least significant part of the product is stored in data register 0x010 and the most significant part in 0x011. The MOVFF instructions copy the product into locations 0x010 and 0x011. The MOVFF instruction (move register file to register file) allows any data register to be copied into any data register. The MOVFF instruction does not use banking. After this program executes, data register 0x010 contains a 0x2C and data register 0x011 contains a 0x01. The 0x12C is a 300_{10}. Where is the a-bit? The MOVFF instruction is a 32-bit instruction that does not have the a-bit. Instead, it has enough bit space to store both the source and destination addresses in the 32 bits. An address is 12 bits so only 24 bits of the 32 bits are used to store the address; the remaining bits are used to store the instruction's numeric opcode.

EXAMPLE 2-4

```
MOVLW   3                    ;place 3 into W
MULLW   .100                 ;multiply by 100 decimal
MOVFF   PRODL,   0x010       ;save low product in 0x010
MOVFF   PRODH,   0x011       ;save high product in 0x011
```

File Select Registers (FSR).   The **file select registers** are used to indirectly address or index the data memory. The PIC18 family contains three file select registers (FSR0, FSR1, and FSR2) that each hold a 12-bit data memory location. To illustrate how the FSR registers function, see Example 2-5, which performs the same operation as Example 2-4. The difference is that in Example 2-5, the FSR0 register indexes (addresses) are data registers 0x10 and 0x11. The LFSR 0, 0x10 instruction loads FSR0 with a 0x10. Whenever the file select registers address memory, the operands INDF0, INDF1, or INDF2 are used in the program to specify **indirect address** 0, 1, or 2. (INDF0 uses FSR0, INDF1 uses FSR1, and INDF2 used FSR2). The POSTINC0 (post increment FSR0), in the MOVFF instruction, adds one to FSR0 after using it to address a data memory location. A POSTINC1 adds a 1 to FSR1 and a POSTINC2 adds a 1 to FSR2.

EXAMPLE 2-5

```
MOVLW   3                        ;place 3 into W
MULLW   0x64                     ;multiply by 100
LFSR    0,0x010                  ;address location 0x010 with FSR0
MOVFF   PRODL, POSTINC0          ;save low product, increment FSR0
MOVFF   PRODH, INDF0             ;save high product at 0x011
```

Status Register (SR).   The **status register** (SR) indicates the outcome of an operation. The status bits are often tested in programs after an operation. Figure 2-8 illustrates the contents of the status register. Only five of the eights bits indicate status and are labeled N (negative), OV (overflow), Z (zero), DC (digit carry), and C (carry). The status bits are tested by the conditional branch instructions that are most often used to form the if-then-else construct in programming as well as other programming constructs. The status register bits will normally change only for an arithmetic or logic operation in the PIC.

- N (*negative*)—The negative bit is a logic 1 if the result of an arithmetic or logic operation is negative and a 0 if the result is positive.
- OV (*overflow*)—The overflow bit is a logic 1 if the result of an arithmetic operation overflows the signed contents of an 8-bit answer.

FIGURE 2-8 Status register.

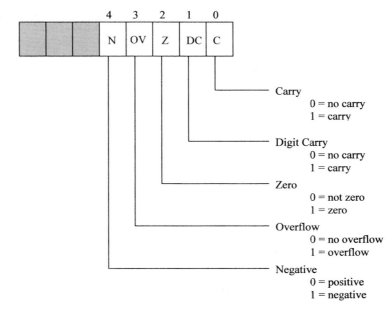

- Z (*zero*)—The zero bit is a logic 1 if the result of an arithmetic or logic operation is a *zero*. If the result is not zero, then the zero bit is a zero.
- DC (*digit carry*)—The digit carry bit is a half-carry. It holds any carry between the lower half (4 bits) of the result and the upper half (4 bits) of a byte and is used only by the DAW. instruction.
- C (*carry*)—The carry status register bit holds a carry from the most significant bit of the result.

Program Counter (PC). The **program counter**, although not directly addressable as PC, is a very important register. The program counter, a 21-bit register in the PIC18 family, addresses the next location in the program memory and allows instructions to be accessed sequentially from the program memory. The program counter is a counter, but does not count programs. It counts up through the memory to access the next instruction in a program, which is probably why it is named the program counter. In some machines, this register is more appropriately called the instruction address register (IAR).

The flow of a program is modified if the contents of the program counter are changed. To change the program counter contents, a GOTO or conditional branch instruction moves a new number into the program counter, causing the program flow to change. A function CALL and RETURN also changes the contents of the program counter. The CALL places the return address (the address of the instruction following the CALL) onto the stack and then performs the GOTO function. The RETURN retrieves the return address from the stack and places it into the program counter. This causes the instructions immediately following the CALL to be executed next.

2-3   INTEGRATED DEVELOPMENT SYSTEM (IDE)

Before much else is accomplished, the **integrated development system** (IDE) of the PIC microcontroller must be understood. The IDE allows programs to be written in assembly language or in C language. It also allows the program to be tested or simulated using the debugging tools

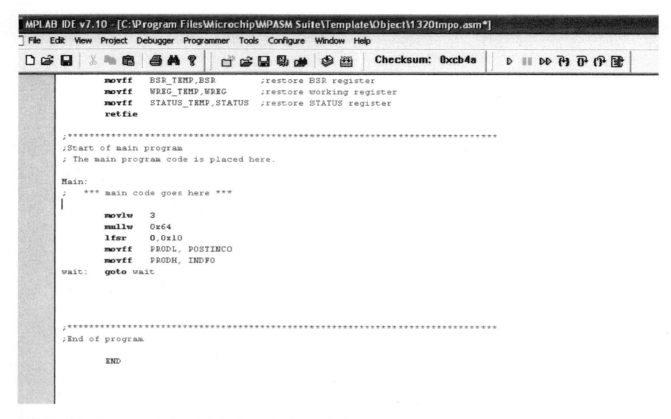

```
MPLAB IDE v7.10 - [C:\Program Files\Microchip\MPASM Suite\Template\Object\1320tmpo.asm*]

 File  Edit  View  Project  Debugger  Programmer  Tools  Configure  Window  Help

                                                Checksum: 0xcb4a

        movff    BSR_TEMP,BSR          ;restore BSR register
        movff    WREG_TEMP,WREG        ;restore working register
        movff    STATUS_TEMP,STATUS    ;restore STATUS register
        retfie


;**************************************************************************
;Start of main program
; The main program code is placed here.

Main:
;    *** main code goes here ***


        movlw    3
        mullw    0x64
        lfsr     0,0x10
        movff    PRODL, POSTINCO
        movff    PRODH, INDFO
wait:   goto wait




;**************************************************************************
;End of program

        END
```

FIGURE 2-9   Screen shot of the IDE for the PIC microcontroller.

incorporated into the IDE. The microcontroller can be flashed or programmed if a programmer is attached to the IDE, and emulated in the circuit if an emulator is attached to the IDE. The IDE is a complete development system for the PIC microcontroller and serves as an excellent learning tool because of its debugging capability.

IDE Overview

Figure 2-9 illustrates a screen shot of the IDE, which is available at http://www.microchip.com as a free download. This Windows operating system based interface is easy to use to create and debug programs. It is also possible to simulate the program in the PC with a simulator that is part of the IDE. In addition, hardware emulators are available that are controlled from the IDE for a more accurate simulation in real-time.

To use the IDE to create a project, the first step is to select the microcontroller used in the project by its type. This is done by clicking "Configure" on the menu bar located at the top of the IDE screen. Next, click on "Select Device . . ." in the drop-down menu that appears. The pop-up dialog box, as shown in Figure 2-10, now appears on the screen. Select the microcontroller that is being used for the current project from the devices listed. This configures the IDE for the microcontroller type used for the project and must be done first for proper operation of the IDE.

A few more steps are required to completely initialize the IDE. Next, locate and click "Project" on the menu bar and select "Project Wizard" from the drop-down menu. The pop-up dialog box that appears is illustrated in Figure 2-11.

FIGURE 2-10 Selecting the microcontroller for the IDE.

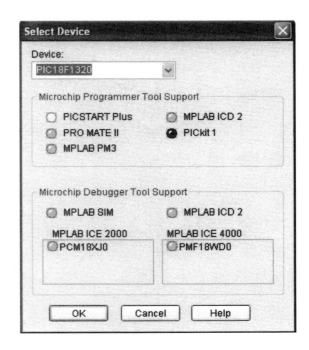

In the project wizard, click on the Next button. The pop-up dialog box illustrated in Figure 2-12 appears. Again, select the microcontroller for the project. This is redundant, but the microcontroller must again be selected at this point from the Device Select menu.

The next prompt asks for the tool set being used to develop the software for the microcontroller. Here the "Microchip MPASM Toolsuite" is selected for developing an assembly language program for the project (see Figure 2-13). The MPASM is the Microchip PIC assembler program. The path names should be correct and not need to be modified if the IDE was properly installed.

FIGURE 2-11 IDE project wizard.

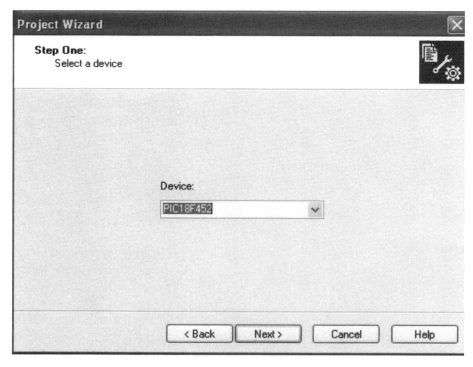

FIGURE 2-12   Selecting the microcontroller.

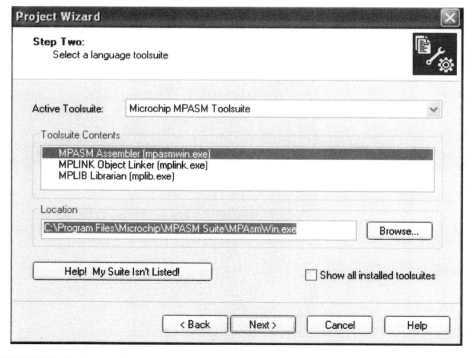

FIGURE 2-13   Tool selection menu.

FIGURE 2-14 Project name menu.

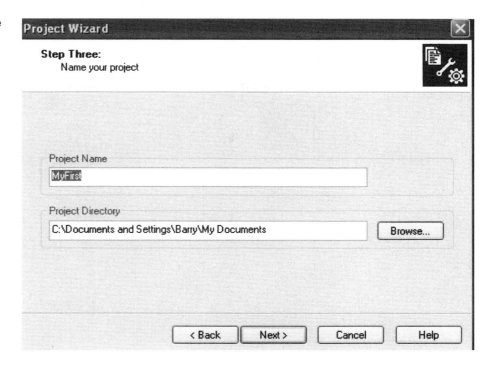

The next menu asks for the project name and path, as illustrated in Figure 2-14. Choose a project name that describes the project currently under development.

The final, and most complicated, step is selecting the linker files for the microcontroller and also the program template file so the program can be written. Figure 2-15 shows a screen shot of this final step of setting up the IDE to develop the program for the project. To accomplish

FIGURE 2-15 Selecting the linker files and project template file.

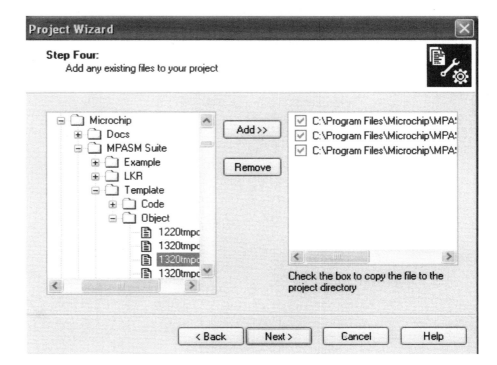

FIGURE 2-16   Project files menu.

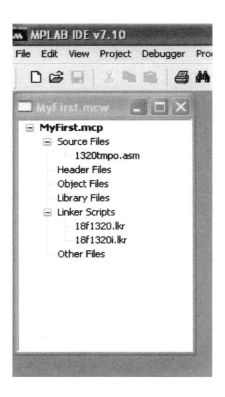

this, locate the Microchip directory; this is most likely under "Program Files" on the C drive. Open the folder under Microchip called MPASM Suite. Under this folder, locate the template file for the microcontroller used in the project in the "Object folder." The file will have a name like "1320tmpo.asm," for example, if the 18F1320 microcontroller is used for the project because of its low cost (less than $3). Click on the file name and add it to the list by clicking "Add." Now, open the "LKR" folder and find the linker files for the selected microcontroller. Two of these are added to the file list for the linker: "18f1320.lkr" and "18f1320i.lkr" for the 18F1320 microcontroller. Once all three files are listed in the file list box, check on the "add to project check box" next to each file name and then click Next. The last dialog box has a button labeled "Finish" to create the project. Click on it and the new project is successfully created.

To take a look at the stock code (template file) inserted by the project wizard, click on View in the menu bar and select "Project." Figure 2-16 depicts the screen that will appear when you click on project view. This small menu displays the files that are included in the project. The file most interesting at this point is the source file that is the template for the system. To view this file, double-click on the file name, in this example, "1320tmpo.asm." This seems like a lot of work to obtain the project and develop software, but the steps are simple to follow and the amount of information that must be provided is fairly short. Once a few programs are developed, these steps become innate.

Example 2-6 shows a listing of what is in the project template file. Much of what is in the file will be a mystery at this point, but will be explained later. This code contains lengthy comments that describe the purpose of each section, as well as some instructions that are needed to set up a basic program for the PIC. The semicolon indicates a comment, and anything following a semicolon to the end of the line does not generate any code. Most of this example is nothing but comments.

EXAMPLE 2-6

```
;****************************************************************************
;   This file is a basic template for creating relocatable assembly code for  *
;   a PIC18F1320. Copy this file into your project directory and modify or     *
;   add to it as needed. Create a project with MPLINK as the language tool     *
;   for the hex file. Add this file and the 18F1320.LKR file to the project.   *
;                                                                              *
;   The PIC18FXXXX architecture allows two interrupt configurations. This      *
;   template code is written for priority interrupt levels and the IPEN bit    *
;   in the RCON register must be set to enable priority levels. If IPEN is     *
;   left in its default zero state, only the interrupt vector at 0x008 will     *
;   be used and the WREG_TEMP, BSR_TEMP, and STATUS_TEMP variables will not     *
;   be needed.                                                                 *
;                                                                              *
;   Refer to the MPASM User's Guide for additional information on the          *
;   features of the assembler and linker.                                      *
;                                                                              *
;   Refer to the PIC18F1220/1320 Data Sheet for additional information on the  *
;   architecture and instruction set.                                         *
;****************************************************************************
;                                                                              *
;   Filename:                                                                  *
;   Date:                                                                      *
;   File Version:                                                              *
;                                                                              *
;   Author:                                                                    *
;   Company:                                                                   *
;                                                                              *
;****************************************************************************
;                                                                              *
;   Files required: P18F1320.INC                                              *
;                   18F1320.LKR                                               *
;                                                                              *
;****************************************************************************

        LIST P=18F1320, F=INHX32    ;directive to define processor and file format
        #include <P18F1320.INC>     ;processor specific variable definitions


;****************************************************************************
;Configuration bits
; The __CONFIG directive defines configuration data within the .ASM file.
; The labels following the directive are defined in the P18F1320.INC file.
; The PIC18F1220/1320 Data Sheet explains the functions of the configuration
; bits. Change the following lines to suit your application.

        __CONFIG _CONFIG1H, _IESO_OFF_1H & _FSCM_OFF_1H & _HS_OSC_1H
        __CONFIG _CONFIG2L, _BOR_OFF_2L & _PWRT_OFF_2L
        __CONFIG _CONFIG2H, _WDT_OFF_2H
        __CONFIG _CONFIG3H, _MCLRE_OFF_3H
        __CONFIG _CONFIG4L, _DEBUG_OFF_4L & _LVP_OFF_4L & _STVR_OFF_4L
        __CONFIG _CONFIG5L, _CP0_OFF_5L & _CP1_OFF_5L
        __CONFIG _CONFIG5H, _CPB_OFF_5H & _CPD_OFF_5H
        __CONFIG _CONFIG6L, _WRT0_OFF_6L & _WRT1_OFF_6L
        __CONFIG _CONFIG6H, _WRTC_OFF_6H & _WRTB_OFF_6H & _WRTD_OFF_6H
        __CONFIG _CONFIG7L, _EBTR0_OFF_7L & _EBTR1_OFF_7L
        __CONFIG _CONFIG7H, _EBTRB_OFF_7H


;****************************************************************************

;Variable definitions
; These variables are needed only if low-priority interrupts are used.
; More variables may be needed to store other special function registers used
; in the interrupt routines.
```

```
                UDATA
WREG_TEMP       RES    1    ;variable in RAM for context saving
STATUS_TEMP     RES    1    ;variable in RAM for context saving
BSR_TEMP        RES    1    ;variable in RAM for context saving
                UDATA_ACS
EXAMPLE                RES   1    ;example of a variable in access RAM
```

;***

;EEPROM data
; Data to be programmed into the Data EEPROM is defined here.

```
DATA_EEPROM     CODE   0xf00000
                DE     "Test Data",0,1,2,3,4,5
```

;***

;Reset vector
; This code will start executing when a reset occurs.

```
RESET_VECTOR    CODE   0x0000
                goto   Main     ;go to start of main code
```

;***

;High-priority interrupt vector
; This code will start executing when a high-priority interrupt occurs or
; when any interrupt occurs if interrupt priorities are not enabled.

```
HI_INT_VECTOR   CODE   0x0008
                bra    HighInt  ;go to high-priority interrupt routine
```

;***

;Low priority interrupt vector
; This code will start executing when a low-priority interrupt occurs.
; This code can be removed if low-priority interrupts are not used.

```
LOW_INT_VECTOR  CODE   0x0018
                bra    LowInt   ;go to low-priority interrupt routine
```

;***

;High-priority interrupt routine
; The high-priority interrupt code is placed here.

```
                CODE

HighInt:

;   *** high-priority interrupt code goes here ***

                retfie  FAST
```

;***

;Low-priority interrupt routine
; The low-priority interrupt code is placed here.
; This code can be removed if low-priority interrupts are not used.

```
LowInt:
                movff   STATUS,STATUS_TEMP   ;save STATUS register
                movff   WREG,WREG_TEMP            ;save working register
                movff   BSR,BSR_TEMP         ;save BSR register

;   *** low priority interrupt code goes here ***

                movff   BSR_TEMP,BSR         ;restore BSR register
                movff   WREG_TEMP,WREG           ;restore working register
                movff   STATUS_TEMP,STATUS   ;restore STATUS register
                retfie
```

;***

```
;Start of main program
; The main program code is placed here.
Main:
;    *** main code goes here ***
;**************************************************************************
;End of program
                END
```

The first two actual lines of code are the LIST directive and the #include statement. The LIST directive informs the assembler that the microcontroller (P) is the 18F1320 and the file format is the Intel Hex 32-bit format, which is used to store the program produced when the assembly language program is compiled and linked. The hex file is used to flash or program the program memory inside the PIC microcontroller.

The second section of code configures the microcontroller. These bits are explained later, and the default configuration selected by the IDE is okay at this point and for many applications.

The next few sections of code define any user variable for data memory locations and also any data to be stored on the EEPROM. Here the EEPROM data is the character string "Test Data" followed by the numbers 0, 1, 2, 3, 4, and 5. Recall that the EPROM is used for storing persistent data that is changed only occasionally. This persistent data is stored without the application of power.

The most important code is at the reset vector location of 0x0000. Recall that the microcontroller starts executing software from location 0x0000. Also recall that there are only 8 bytes of memory between the reset vector address and the high-priority interrupt vector address. For this reason a GOTO instruction is usually placed at the reset vector address. The only main program continues at Main in this example. The GOTO instruction changes the program counter to address Main so the program continues at Main. Notice that there is no code at Main, just a comment that states: ***main code goes here***.

A Sample Program. Now that the project template is displayed on the screen of the IDE, some code or program needs to be placed into the Main section of the template to learn how to program the microcontroller and use the IDE to test programs.

Replace the line in Main (;   *** main code goes here ***) with the assembly language code in Example 2-7. This program does not do much, but provides the basis for learning how to use the IDE to test a program.

EXAMPLE 2-7

```
        MOVLW  0x01      ;load W with 0x01
        ADDLW  0x02      ;add a 0x02 to W
Wait:   GOTO   Wait      ;get stuck here
```

The program in Example 2-7 loads W with a literal 0x01 and then adds a literal 0x02 to it. To see this program in action, enter the code in Example 2-7 into the project template in Main, replacing the comment ***main code goes here***. Once these lines of code are entered in the project template, click "Project" on the menu bar. Next, click on "Build All" in the drop-down list that appears. "Build all" converts the assembly language program into a program called an object program in numeric machine code that the microcontroller can execute. Next click on "View" and then click "Special Function Registers." Find the W register (WREG) at location 0xFE8. On the menu bar, click on "Debugger" and select the MPLAB SIM entry. (See Figure 2-17 for a screen shot.) *MPLAB SIM* is a program that simulates the PIC microcontroller inside of the PC. To run the program, a step at a time, press the F7 key on the keyboard or click on the icon just to the right of the green double arrows called "Step into." For each step, a green arrow scrolls down through

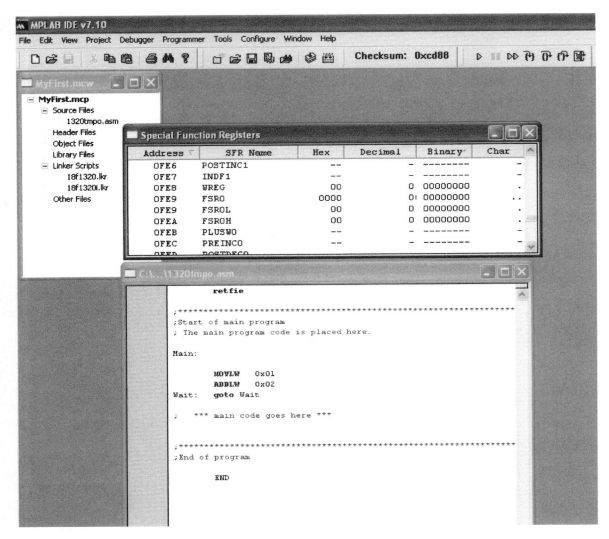

FIGURE 2-17   Screen shot of Example 2-6 and the SFR window.

the program showing exactly where program execution is located. The contents of WREG also change as this is done. To see the task again, click on the yellow icon to the right of the green double arrow icon and the program will restart from the beginning so it can be executed again. Notice that the program is executed in the PC and no microcontroller is needed to learn the assembler or how to program the microcontroller or even debug an application. The MPLAB SIM (simulator) simulates the microcontroller on the personal computer. Other functions of the simulator are to generate an instruction trace (under the View menu), and when I/O ports are used to look at the I/O data with the "Logic Analyzer" entry under the View menu.

Placing the Program on the Microcontroller

In order to place the program on the microcontroller, a programmer is required. Some PIC18-based microcontroller demonstration boards contain a program for downloading software and programming the microcontroller. Programmers are available from Microchip that will program many of the PIC18 family members. The IDE supports the following programmers: PICSTART

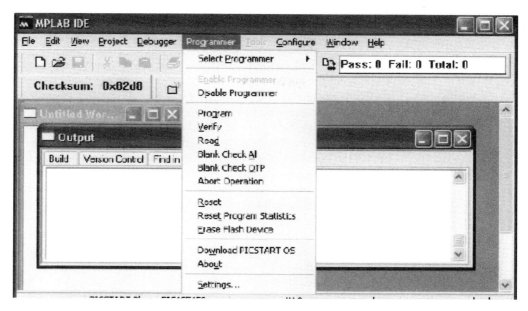

FIGURE 2-18   Programming drop-down menu from IDE.

plus, ICD 2, PM 3, and PROMATE II. Additionally, many other programmers that also program EPROMs, EEPROMS, and other microcontrollers are commonly available. If one of the IDE-supported programmers is available, the IDE can control them and program the microcontroller directly from the IDE.

Once the program is built and debugged in the IDE, the programmer is selected to transfer the hex file to the microcontroller. Most programmers verify that the program is programmed correctly on the microcontroller. This textbook does not attempt to explain the operation of any particular programmer. The documentation and software for programming the microcontroller is provided with the programmer and the task of programming the microcontroller (sometimes called flashing) is very easy to accomplish.

If using one of the IDE-supported programmers, once configured, the programming, verifying, and so forth, functions are provided in a drop-down menu on the IDE menu bar. Figure 2-18 illustrates this drop-down menu when the programmer is installed and attached to the host PC. If an IDE-supported programmer is not available, the hex object file must be transported to the programmer and used to flash the microcontroller.

2-4   ASSEMBLY LANGUAGE, THE ASSEMBLER, AND THE LINKER

Before the assembler is learned and used effectively for programming, some basic information about the assembler and linker programs is required. This section explains the role of these tools in system development and describes the basic assembler syntax.

Assembler and Linker

The **assembler** is a program that converts assembly language (**symbolic machine code**) from a **source file** into numeric machine code stored in an **object file**. As with any CPU, the microcontroller understands only **numeric machine code**. Most assembler programs accomplish the conversion from source file to object file in two passes through the source file. The first pass generates a **symbol table** that stores each label and its address in the program. The first pass also looks for typing errors called **syntax errors** and reports them as error messages. The second pass occurs if there were no errors during the first pass. During this second pass, the addresses retrieved from the symbol table and the codes for each instruction create the object program in numeric machine language.

The source file for the assembler is usually a file that uses the file extension of .asm, and the object file generated by the assembler usually has the extension of .o. File names must follow the convention of the operating system. In most cases, the operating system is Windows so just about anything is allowed except for a few special characters. The object file is usually not in a form that is executable. The executable program file is a file for flashing the microcontroller, usually in the Intel hex format, which is generated by another program called a linker.

The linker program combines multiple object files from the assembler and C language into a single executable file. In the case of the PIC microcontroller, this executable file is usually in the Intel hexadecimal format used with the PIC programmer and often has the extension .hex. It also allows library files (.lib) and linker script files (.lkr) to be used in the linking process. The linker also generates a .coff file (common object file format) and a .cod file (symbolic and debug file) for the development system and debugging process. In addition to the .coff and .cod files, a .lst file (complete program listing file) is generated that displays both the symbolic and numeric programs side-by-side and is accessed in the IDE as a disassembly file in the View menu. These files are accessed from within the IDE and are not normally used individually. Even the programmer, which uses the .hex file, is usually controlled from the IDE.

Assembly Language Statements

Lines of assembly code are called **statements**. A statement contains four distinctive fields of information: label, opcode, operand, and comment. The **label** holds a symbolic memory address so that a statement can be referenced from somewhere else in the program. The GOTO Main statement at the reset vector is an instruction that references a label located elsewhere in a program. The **opcode** contains the assembly language instruction or a directive that directs the operation of the microcontroller, assembler, or linker program. The word GOTO in GOTO Main is an example of an opcode. The **operand** is a literal, a register file address, or some other information used by the opcode to perform a task. In the GOTO Main statement, the word Main is the operand used by the opcode GOTO. The **comment** field is optional, but usually indicates some information about the program or is used to identify a section of code. Comments must begin with a semicolon. Figure 2-19 illustrates a line of code with information in each field of an assembly language statement.

Label. The **label** is normally present only in lines of code that are referred to at some other point in a program. A label can be anything except a reserved token such as an opcode or a register name. A label must begin with a letter (A–Z) or with the underbar as in _Start. A label can contain any letter, a number, an underbar, or a question mark. Labels usually end with a colon, but this is optional. Table 2-5 shows several forms of allowable labels for the assembler.

FIGURE 2-19 Assembly language statement.

Label	Opcode	Operand	Comment
Start:	MOVLW	0x00	;load WREG with 0x00

TABLE 2-5 Sample labels.

Sample Label	Comment
START:	Valid label
Start:	Valid label
Start&Stop:	Invalid because of the &
4Me:	Invalid, labels must start with a letter or underbar
Start_Stop?:	Valid label
Addlw:	Invalid because addlw is an opocde

Opcode.   The **opcode** field must contain a valid opcode for the microcontroller. The opcode field can also contain a **directive**, which is a special command to the assembler or linker. The opcode field may also contain a macro name. A **macro** is a group of instructions predefined by the user and identified by a unique token or name. An opcode can be upper- or lowercase, but most of this text uses uppercase to prevent errors in reading the letter l and the number 1. For example, the movlw command might be misread as a mov1w command in lowercase form where the MOVLW instruction is more easily recognized compared with a MOV1W. The IDE uses the courier font to display a program and in the courier font, a number 1 and a lowercase letter l are identical. Because of this, transcription errors are common when lowercase letters are used with numbers.

Operand.   The **operand**, the information operated upon by the opcode, is placed to the right of the opcode and may contain none, or any number of fields separated by commas. Table 2-6 shows several instructions that use different numbers of operands.

Comments.   **Comments** are anything that begin with a semicolon. Anything that follows a semicolon is ignored until the end of the line. Each comment line must begin with a semicolon because there is no way to continue a comment through several lines as there is in other languages. A good programming practice is to comment on sections of code and not on individual instructions. It is also a good programming practice to separate sections of code by placing at least one blank line before and after each section. Good commenting and programming practice can be seen back in Example 2-6. Notice how each section of the code is separated from the rest by blank lines and comment statements using asterisks. This is important because assembly language is cryptic, and comments help to explain sections of code.

Directives

Directives are special commands to the assembler that may or may not generate machine code. Directives are classed by their function and are explained in functional groupings in this textbook. Not all the directives are explained at this time, some are explained later. The directives most often used for programming are explained in this section.

TABLE 2-6 Sample instruction operands.

Opcode with Operand	Comment
RETURN	No operand
MOVLW 0x00	One operand (0x00)
MOVFF 0x10, 0x11	Two operands (0x10 and 0x11)
DE 1, 2, 4, 6, 9	Multiple operands
CALL Home_ET	One operand

TABLE 2-7   Object file directives.

Directive	Comment
ACCESS_OVR	Object file overlay access RAM
CODE	Begins a block of program code
CODE_PACK	Begins a packed code block
EXTERN	Declares that a label is defined in some other program module
GLOBAL	Declares that a label is available to other program modules
IDATA	Beings a section of initialized data
IDATA_ACS	Begins a section of initialized data in access RAM
UDATA	Begins a section of uninitialized data
UDATA_ACS	Begins a section of uninitialized access data
UDATA_OVR	Begins a section of uninitialized overlay data

Object File Directives.   **Object file directives** control how the assembler generates code in the object file. Refer back to Example 2-6, and notice that some directives control the way the object file is created. Table 2-7 lists many of the object file directives.

The most common object file directives are CODE, UDATA, and UDATA_ACS. The **CODE** directive defines different sections of code. The label is the name of the code section and the operand is the absolute address of the code section. For example, to place code at the reset vector address, use the statement RESET_VECTOR CODE 0x0000 as illustrated in Example 2-6. This informs the assembler that the reset vector begins at program address 0x0000. If CODE is used without a label or operand, then the section of code is usually the main body of the program and the assembler decides the address. Example 2-8 shows how the CODE directive might be used in a system for program development.

EXAMPLE 2-8

```
RESETS_HERE    CODE    0x0000       ;reset to here
       GOTO    MyStuff
HIGH_INT       CODE    0x0008       ;interrupt high
       GOTO    HighInterrupt
LOW_INT        CODE    0x0018       ;interrupt low
       GOTO    LowInterrupt
               CODE                 ;code without a predefined address
HighInterrpt:
       ; **** high-interrupt code here ****
LowInterrupt:
       ; **** low-interrupt code here ****
MyStuff:
       ; **** the application goes here ****
END     ;end of file
```

The **UDATA** and **UDATA_ACS** directives reserve space in the data memory for variables. Programs are much easier to read if labels, rather than numeric addresses, are used to identify locations in the data memory. Refer to Example 2-6 and notice that memory is reserved for three variables in UDATA memory, which begins at data register address 0x080, and one variable is defined in UDATA_ACS memory, which begins at data register address 0x000. Recall that the access bank memory (ACS in a directive) is at register addresses 0x000 through 0x7F. The remainder of the access bank contains the special function registers located at addresses 0xF80 through 0xFFF as designated by the UDATA directive.

The IDATA and IDATA_ACS directives are similar to UDATA and UDATA_ACS, except they are used to initialize memory with data instead of reserving space in the memory. Microchip admits that the IDATA and IDATA_ACS directives may not function correctly if using an emulator. For this reason they should not be used to initialize data. If you need to place data in the memory, write code that will initialize data.

The CODE directive is also used to initialize data EEPROM memory as shown in Example 2-6. When storing data in the EEPROM, use an address of 0xF00000 as illustrated in Example 2-9. Make sure to use the IDE directive to store data in the EEPROM as mentioned in Chapter 1. Note that address 0xF00000 is not an address in the program memory, nor is it the actual address of the EEPROM. It is an address that signals the IDE that this is EEPROM data and nothing more.

EXAMPLE 2-9

```
EEPROM_MEM   CODE   0xF00000
       DE    "This is for my EEPROM!", 0    ;C-style null string
```

Control Directives.   **Control directives** are the next most common group of assembler directives that control the assembly and at times the link process. The most common control directives are #include, end, processor, equ, and set. Table 2-8 lists the control directives for the PIC microcontrollers.

Not all the control directives are explained here, only the ones most often used in programming. For a more complete description of the control directives not defined in this textbook, refer to the MPASM documentation located on the Microchip website.

The **#include directive** includes code files often called header files in the C language. Example 2-6 shows an example of its usage where it is used to add the contents of the P18F1320.INC file to the code listing. The P18F1320.INC file is an include file that contains information (mostly equ directives) that define the configuration of the PIC18F1320, its registers, and other information about the microcontroller.

The **end directive**, although it might seem intuitive, must be placed at the end of the program file. If not, the file will not assemble. Do not place an end directive in an include file or the assembly language process called a build will stop at the end of the include file. The end directive is a signal to the assembler to end the assembly process and nothing that follows it is assembled.

The **processor directive**, which identifies the processor used with the project file, may or may not be used in a file. If the processor type is selected in the configure section of the IDE, then it does not need to be included in the program file. In many cases, the processor directive is not included in the program file.

The **equ** or equate *directive* equates values or labels to a label. If you view one of the header files provided in the IDE, you will notice that there are many equates used to equate the

TABLE 2-8 Control directives.

Directive	Comment
#define	Defines a text substitution label
#include	Includes a source file
#undefine	Removes a substitution label
Constant	Declares a symbol constant
End	Ends the program file (required)
Equ	Equates a constant
Org	Sets the origin of a block
Processor	Selects the processor type
Radix	Specifies the default radix
Set	Defines an assembler variable
Variable	Declares a symbol variable

TABLE 2-9 List directives.

Directive	Comment
error	Issues an error message
errorlevel	Sets the error message level
list	Sets the processor and output file type and other defaults of a program listing
messg	Sets a user message
nolist	Disables listing
page	Inserts a new page in the listing
space	Inserts blank lines in a listing
subtitle	Specifies a program subtitle
title	Specifies a program title

SFR addresses to labels. For example, WREG is equated to 0xFE8 so WREG can be used in a program by name instead of by number. The include file and the equates make the assembly language task easier and more readable.

The **set directive** is used to set a label equal to a value that can be changed later by another set to the same label. The set directive is used less than the equ directive in programming and is similar in function to an equ.

List Directives. The **list directives** are used to control the listing process. Table 2-9 depicts the available list directives. Out of all of these, the most commonly used directive is list. The list directive controls the general program listing for a selected microprocessor. This statement usually appears as in Example 2-6 where the P=18F1320 selects the processor type for the listing and the F=INHX32 selects the type of output file. In this example, the output file is in Intel hexadecimal 32-bit format.

Data Directives. The **data directives**, which describe data, appear in a program in many places and are probably the most used type of assembler directive. Many of these are defined in the first chapter with data definitions, but a few are described here. The data directives generally define the type of data. Refer to Table 2-10 for a complete listing of the data directives.

TABLE 2-10 Data directives.

Directive	Comment
__badram	Identifies unimplemented RAM
__badrom	Identifies unimplemented ROM
__config	Sets processor configuration bits
config	Sets processor configuration bits for the PIC18 family
__idloca	Sets processor ID locations
__maxram	Sets maximum RAM
__maxrom	Sets maximum ROM
cblock	Defines a block of constant data
da	Stores strings in the program memory (PIC12/16)
data	Creates numeric and text data
db	Declares bytes
de	Declares EEPROM data
dt	Defines tables (PIC16/12)
dw	Declares words
endc	Ends automatic block constants
fill	Fills memory with a contant
res	Reserves memory

The most commonly used data directives of the PIC18 family are db, de, dw, and res as described in Chapter 1. The only other data directive commonly used in a PIC18 program is the config directive. Do not mix __config with config. The __config directive can be used with the PIC18 family as illustrated in Example 2-6. This directive sets the microprocessor configuration for the project. These bits and their functions are explained in later sections of this textbook.

2-5   SUMMARY

1. The PIC architecture uses Harvard architecture to split the memory into program memory and data memory. This architecture provides for more efficient program execution.
2. The PIC is a self-contained computer system that contains memory and I/O. This reduces the cost of system implementation.
3. Program memory is available within the PIC in sizes of 4K bytes to 128K bytes, depending on the family members selected for an application. Because most instructions are 2 bytes in width, this allows a program to have between 2K and 64K instructions. All program memory is accessed via a 21-bit address.
4. Data memory is organized in a register file. All data memory is 8 bits in width and varies in size from 256 bytes to 3840 bytes plus an additional 128 bytes for the special function registers. All data memory is accessed via a 12-bit address.
5. The special function registers (SFR) are used as a working register (W); to hold the product after a multiplication (product registers); and indirectly address a data register through the file select registers (FSR), the status bits in the status register, or the program counter to keep track of the location in a program. There are also many other special function registers for controlling the I/O devices of the microcontroller, as discussed in later chapters.
6. The direction of an I/O port is programmed through the TRIS register for the ports, where a 1 programs a corresponding bit as an input bit and a 0 programs a corresponding bit as an output bit. Port A uses TRISA, Port B uses TRISB, and so forth.
7. Data are read or written to I/O port pins by using the PORTA register for Port A or the PORTB register for Port B.
8. The software development system used to generate programs for the PIC microcontroller is called the integrated development environment or IDE. The IDE provides an editor to enter program code, access to the programmer to program the code onto the PIC microcontroller, access to the in-circuit emulator, and access to a simulator to allow a program to be emulated in the personal computer before it is programmed.
9. The IDE is initialized to develop a new project by selecting the processor type from the Configure/Select Device menu and by running the project wizard located under the Project menu.
10. An assembly language statement is constructed with four fields: label, opcode, operand, and comment.
11. The label field is a symbolic memory address to identify a location in the memory. The label should be used only if a location needs to be addressed from the program.
12. The opcode field contains an instruction for the microcontroller or a directive for the assembler. The opcode instructs the microcontroller to perform some operation.
13. The operand field contains information used by the opcode to complete an instruction. The operand might be a register name, a data RAM address, a numeric value, and so forth.
14. The comment field is not processed by the assembler, but gives the programmer a method for commenting about blocks of code so a program is more readable.
15. A directive is a command to the assembler or linker that often generates no code for the program or data memory. Directives simplify the programming task.

2-6   QUESTIONS AND PROBLEMS

1. In Figure 2-1, the microcontroller-based system contains no memory. Where is the memory for the microcontroller located?
2. What is the Harvard architecture?
3. Compare and contrast the Harvard architecture with the von Neumann architecture.
4. What is pipelining and how does it increase the efficiency of program execution?
5. The program memory is _____ bits wide.
6. The data memory is _____ bits wide.
7. What is the first and last address of the program memory in a PIC?
8. What is the first and last address of the data memory?
9. The program memory uses a _____ bit address in the PIC18 family.
10. The data addresses for the PIC18 family are _____ bit addresses.
11. What is the difference between an 18F series PIC and an 18C series PIC?
12. If an EEPROM is available in the PIC18, where is it located and how is it addressed?
13. If a PIC18F6620 is used for system development, its program memory begins at address _____ and ends at address _____.
14. If a PIC18F442 is used for system development, its program memory begins at address _____ and ends at address _____.
15. A 12K memory contains how many hexadecimal memory locations?
16. If a 16K span of the memory begins at program memory address 0x004000, what is the last address of the memory span?
17. What are the locations (start and ending addresses) of the data memory in the 18F452 PIC microcontroller?
18. Where are the registers that program the I/O devices in the system found?
19. Where are the special function registers located in any data memory of the PIC microcontroller?
20. What is the purpose of the working register?
21. What is the accumulator?
22. Where is the W register located in the PIC microcontroller?
23. What part of the PIC memory system is often called the register file?
24. What is the purpose of the three register file select registers in a PIC?
25. What is the purpose of the bank select register and why is it needed?
26. Where would one look to find out if a carry occurred after an addition?
27. What is the purpose of the B status register bit?
28. What are the TRIS registers?
29. How is the TRISB register programmed to select the output operation for Port B bits 0 through 2 and input operation for Port B pins 3 through 7?
30. How is the number 0x33 sent to the Port A pins if Port A is programmed as an output port?
31. Why is there a set of registers for the product of a multiplication and where are these registers?
32. How many 8-bit registers comprise the program counter?
33. What is the purpose of the program counter and why is it called a counter?
34. What is the IDE?
35. The template file for a microcontroller contains what information?
36. The IDE is used for software development and what other tasks in the development process?
37. What is the project wizard and why do you think it's needed?
38. What is a simulator?
39. What are the four parts of an assembly language statement and what is the purpose of each?

40. Given the following assembly language statements, identify the field of each section of the statement.
    a. Start: GOTO Heaven
    b. ADDLW   0x29   ;add a 0x29
    c. Loopy1:   MOVFF   WREG, 0x145   ;a move in action
41. Which of the following labels are valid?
    a. 2FAR
    b. FAR A WAY
    c. FAR_A_WAY
    d. FarAWay
42. What is the assembly process?
43. What is a source file?
44. What is an object file?
45. What is the purpose of the linker?
46. What is a .lkr file?
47. What is the purpose of the UDATA directive?
48. What is the purpose of the CODE directive?
49. A DE directive is used to place data into the _____.
50. What is accomplished by the CODE 0x1000 directive?
51. What is a LIST directive used for in a project?
52. The DATA1 RES   2 statement will do what?
53. What is accomplished by a GOTO instruction?

CHAPTER 3

PIC18 Family Instruction Set

This is one of the most important chapters in this textbook. It explains the operation of the instructions in the instruction set of the PIC18 family of microcontrollers. Although much programming uses a high-level programming language such as C, it is important to understand the operation of each instruction so that the microcontroller is better understood. This chapter also demonstrates the microcontroller's limitations.

Each instruction or grouping of instructions is presented with small applications that can be entered and executed in the IDE. This allows the operation of an instruction to be investigated and better understood through the simulator in the IDE. This chapter refines many of the ideas and concepts discussed in the Chapters 1 and 2. Once the instruction set is understood, applications that access I/O devices are discussed in later chapters.

Upon completion of this chapter you will be able to:

1. Describe the operation of the addressing modes available to the microcontroller.
2. Detail the operation of each instruction in the microcontroller instruction set.
3. Develop and simulate short program examples using the IDE.
4. Describe how to program using programming constructs.
5. Use indirect addressing to access data memory.
6. Use table addressing to access program memory.
7. Generate and use macro sequences in programs.

3-1   LITERAL INSTRUCTIONS

Before an instruction is used in a program, the addressing modes the instruction can use must be understood. This section explains the various addressing modes available to the PIC18 family of microcontrollers and its instructions using assembly language.

Literal Instructions

Literal addressing is probably the easiest type of addressing to understand so it is explained first. A **literal** is a constant such as a number or ASCII character. Most of the literal addressing instructions operate with the working register or WREG. Table 3-1 lists all the available literal instructions in the PIC18 family instruction set. Most literal instructions use the second byte of the

TABLE 3-1 Literal instructions.

Opcode	Operand1	Operand2	Examples	Comments
ADDLW	Literal		ADDLW 0x20	Add 0x20 to W
			ADDLW .100	Add 100 to W
ANDLW	Literal		ANDLW 0x0F	AND 0x0F with W
			ANDLW .15	AND 15 decimal with W
			ANDLW 0b00001111	AND 00001111 binary with W
IORLW	Literal		IORLW 0x80	IOR 0x80 with W
			IORLW 1	IOR a 1 with W
LFSR	FSR register number	Literal	LFSR 0, 0x123	Load FSR0 with 0x123
			LSFR 2, 0x10	Load FSR2 with 0x010
MOVLB	Literal		MOVLB 2	Load BSR with 2
			MOVLB 0	Load BSR with 0
MOVLW	Literal		MOVLW 3	Load W with 3
			MOVLW 0x34	Load W with 0x34
MULLW	Literal		MULLW .100	Multiply W by 100
			MULLW 2	Multiply W by 2
RETLW	Literal		RETLW 2	Return with W = 2
			RETLW 0x2A	Return with W = 0x2A
SUBLW	Literal		SUBLW 5	Subtract W from 5
			SUBLW .19	Subtract W from 19
XORLW	Literal		XORLW 4	Exclusive-OR 4 with W
			XORLW 0xF0	Exclusive-OR 0xF0 with W

16-bit instruction to hold the literal data. If you need more detail about the numeric machine coding of the instructions, please refer to Appendix A, which lists all of the instructions in numeric machine code format along with other useful information about them.

Notice from Table 3-1 that there are not many literal instructions to learn. The first three letters of most of the literal instructions indicate the operation performed by the instruction. As mentioned in Chapter 1, the CPU is capable of only a few operations: ADD (addition), AND (logical AND), IOR (inclusive OR), MOV (copy), MUL (multiplication), SUB (subtraction), and XOR (exclusive OR). The last two letters of the opcode indicate something about the instruction. For example, the letter L indicates Literal and the letter W indicates the WREG register. This means that the ADDLW instruction adds the literal operand to the WREG register. The only instructions that do not follow this scheme are the LFSR and MOVLB instructions. The LFSR instruction loads a literal (usually register file address) into one of the three FSR file selection registers used to indirectly address the data memory. The MOVLB instruction loads the bank select register (BSR) with a number between 0, for bank 0, through 15, for bank 15.

The arithmetic and logic instructions are not complete in the PIC microcontroller. A divide instruction is not implemented in the PIC family, but multiplication is implemented. If division is needed in a system, software must be developed that divides.

In general, an arithmetic or logic instruction changes the status register bits. The exception is the multiply instruction, which does not affect any of the status register bits. (Recall that the status register bits are important because the microcontroller uses them to make decisions). To illustrate the changes to the status register bits, a simple program is executed as shown in Example 3-1. This program does not accomplish much, but the status register changes for the ADDLW instruction. This sample program moves a 0x7F into the WREG register with a MOVLW instruction and then adds a 1 to WREG with the ADDLW instruction to generate a sum of 0x80 in WREG. Note how the GOTO Stop instruction is used to end the program. Because the only thing that a microcontroller does is continually fetch and execute software, such a trap (GOTO Stop) is used at times to end a program. In actual system software, this type of trap is not needed, because most operating systems repeat a program continually with the infinite while construct. If the operating system stops, the program has crashed.

EXAMPLE 3-1

```
        MOVLW   0x7F    ;load W with 0x7F
        ADDLW   1       ;add 1 to W

Stop:   GOTO    Stop    ;stop here
```

Place this simple example into the IDE template file (refer to the file in Chapter 2, Example 2-6) at the section that is labeled Main. Once this program executes, display the special function registers under the View menu and notice that the WREG register (at address 0xFE8) contains a 0x80 and the status register contains a 0x1A. To view the special function registers, run the program and then stop the run before looking at the special function registers. If you look at the bottom of the screen, the contents of the W register are displayed along with other useful information. If you do not stop the run, you will not see the correct values in the registers when using the simulator. The simulator displays updated information only when it is stopped.

The 0x1A in the status register tells us that the result is negative because the negative bit (N) is a logic one. An 0x80 or 1000 0000$_2$ in WREG is indeed negative. Recall that the rightmost five bits of the status register contain indicators such as the negative bit. These bits are N, OV, C, DC, and Z from left to right beginning at bit position 4. With the outcome of 0x1A in the status register, N = 1, OV = 1, C = 0, DC = 1, and Z = 0. The result is negative, the result overflowed the W register, there was no carry, there was a half-carry or digit carry, and the result is not zero.

The overflow occurred in this example because a +1 (0x01) is added to a +127 (0x7F) and the result is −128 (0x80). The maximum positive number that fits into any 8-bit register is +127 (0x7F) and the maximum negative number is −128 (0x80). If either value is exceeded, the result is an overflow. Overflow conditions apply only to signed numbers. If the numbers in this example are considered unsigned, then the 0x80, which is an unsigned 128, is correct and the N and OV status register bits have no meaning for the result of an unsigned operation. Figure 3-1 illustrates the addition in Example 3-1 and the location of the two carry bits.

EXAMPLE 3-2

```
        MOVLW   6       ;move 6 into W
        SUBLW   5       ;subtract W (6) from 5

Stop:   GOTO    Stop
```

When a subtraction occurs, the carry (C) and half-carry (DC) status bits actually hold borrows (0 = borrow, 1 = no borrow). If the program in Example 3-2 is executed, the result is a 0xFF in WREG and the status register contains a 0x10. The SUBLW instruction subtracts the contents of the WREG register from the literal value. This seems backwards and it is: beware of the way that subtraction functions in this microcontroller. In this case, a 6 is subtracted from a 5 and the

FIGURE 3-1   Location of the C and DC bits for the addition in Example 3-1.

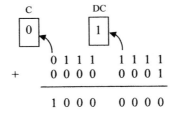

FIGURE 3-2 Location of the C and DC bits for the addition in Example 3-2.

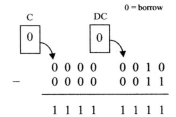

$$
\begin{array}{r}
0\ 0\ 0\ 0 \quad 0\ 0\ 1\ 0 \\
-\quad 0\ 0\ 0\ 0 \quad 0\ 0\ 1\ 1 \\
\hline
1\ 1\ 1\ 1 \quad 1\ 1\ 1\ 1
\end{array}
$$

result is negative 1 or 0xFF. The status register has a zero in both the C and DC bits. This means that to accomplish this subtraction, borrows were needed as shown in Figure 3-2.

Literal Logic Instructions. There are three logical literal instructions: AND, Inclusive-OR, and Exclusive-OR. The truth-tables for all three operations appear in Figure 3-3. These operations are performed in a bit-wise fashion (bit-by-bit) when executed in a program. The AND operation

FIGURE 3-3 AND, Inclusive-OR, and Exclusive-OR operations.

AND $(T = A \cdot B)$

A	B	T
0	0	0
0	1	0
1	0	0
1	1	1

$$
\begin{array}{r}
\text{X X X X}\ \ \text{X X X X} \\
\text{AND}\quad 0\ 0\ 0\ 0\ \ 1\ 1\ 1\ 1 \\
\hline
0\ 0\ 0\ 0\ \ \text{X X X X}
\end{array}
$$

Inclusive-OR $(T = A + B)$

A	B	T
0	0	0
0	1	1
1	0	1
1	1	1

$$
\begin{array}{r}
\text{X X X X}\ \ \text{X X X X} \\
\text{IOR}\quad 0\ 0\ 0\ 0\ \ 1\ 1\ 1\ 1 \\
\hline
\text{X X X X}\ \ 1\ 1\ 1\ 1
\end{array}
$$

Exclusive-OR $(T = A \oplus B)$

A	B	T
0	0	0
0	1	1
1	0	1
1	1	0

$$
\begin{array}{r}
\text{X X X X}\ \ \text{X X X X} \\
\text{XOR}\quad 0\ 0\ 0\ 0\ \ 1\ 1\ 1\ 1 \\
\hline
\text{X X X X}\ \ \overline{\text{X X X X}}
\end{array}
$$

is often used to selectively clear a single or multiple bits to zero, because when a zero is ANDed with anything the result is zero. The Inclusive-OR (IOR) operation is often used to selectively set a single or multiple bits to one, because when one is Inclusive-ORed with anything the result is a one. The Exclusive-OR (XOR) operation selectively inverts a single bit or multiple bits from zero to one or from one to zero, because when a one is Exclusive-ORed with anything the result is inverted. These three facts are also illustrated in Figure 3-3 using X to represent any number and show how these operations modify the number X.

Suppose a situation requires that the rightmost two bits (bits 0 and 1) of WREG must be set (1) and that bits 6 and 7 need to be cleared (0). This is accomplished by using an AND instruction to clear bits 0 and 1 and an IOR instruction to set bits 6 and 7. The sample program in Example 3-3 accomplishes this task. Here WREG is first loaded with a 0x1F as a test value. If the program is executed in the IDE, the WREG register changes to a 0xDC.

EXAMPLE 3-3

```
        MOVLW   0x1F
        ANDLW   0xFC    ;clear bits 0 and 1
        IORLW   0xC0    ;set bits 6 and 7

Stop:   GOTO    Stop
```

Another example requires that the leftmost three bits of the W register are inverted. This is accomplished by using XOR with a literal value of 0xE0 (1110 0000) to invert the leftmost three bits of W. Example 3-4 shows a test program that inverts the leftmost three bits of a 0x90 (1001 0000). The result is a 0x70 (0111 0000).

EXAMPLE 3-4

```
        MOVLW   0x90
        XORLW   0xE0    ;invert left 3 bits

Stop:   GOTO    Stop
```

Comparing the utility of all the logic instructions, the AND operation is used most often in programming to mask off bits in a register. For example, suppose the WREG contains the number 0x4A and only the rightmost four bits (A) are needed. How are the leftmost four bits masked off? The ANDLW 0x0F instruction erases the 4 in 0x4F, leaving the 0x0A as the result. Likewise, if only the leftmost 4 bits, are needed, the ANDLW 0xF0 instruction is used to erase the rightmost 4 bits leaving a 0x40 in the W register.

3-2   **BIT-ORIENTED INSTRUCTIONS**

Because logic operations are being discussed, this section describes the bit-oriented operations. The bit operations set, clear, and toggle test only a single bit, where the AND, IOR, and XOR byte-oriented instructions change a single bit or multiple bits. Table 3-2 lists the bit-oriented instructions available to the PIC family. Many applications for the microcontroller require the use of the bit-oriented instructions to control and test individual bits in a program. This is especially true when interfacing and controlling I/O devices.

The a-bit position, in Table 3-2, is needed only in the assembler when a numeric register file address is used in an instruction to specify either the access bank or the bank select register. If the instruction refers to a register by its name or the name of a memory location instead of its numeric address, the a-bit is not used in a program. Notice in Table 3-2 that the a-bit is not specified when the WREG register is used as operand 1. The assembler is intelligent enough to know

TABLE 3-2 Bit-oriented instructions.

Instruction	Operand1	Operand2	Operand3	Examples	Comment
BCF	Register	Bit #	a-bit	BCF WREG, 7	Clear bit 7 of W
				BCF 0x10, 1, 0	Clear bit 1 of access bank register 0x10
				BCF 0x10, 1, ACCESS	Clear bit 1 of access bank
BSF	Register	Bit #	a-bit	BSF WREG, 1	Set bit 1 of W
BTFSC	Register	Bit #	a-bit	BTFSC WREG, 2	If bit 2 of W is 0, then skip the next instruction
BTFSS	Register	Bit #	a-bit	BTFSS WREG, 7	If bit 7 of W is 1, then skip the next instruction
BTG	Register	Bit #	a-bit	BTG WREG, 4	Toggle (invert) bit 4 of W

that the access bank is where the WREG is located so the a-bit is not needed in instructions that name a specific register such as WREG.

To illustrate the BCF (bit clear F) and BSF (bit set F) instructions, Example 3-3 is rewritten in Example 3-5 using the bit-oriented instructions in place of the ANDLW and IORLW instructions. If these examples are compared, Example 3-5 seems easier to understand although there is no difference in the result. There are few differences between the examples; one difference is that the bit-oriented instructions do not modify the status register, but the ANDLW and IORLW instructions do modify the status register. Also, because pairs of bits are changed, two bit-oriented instructions are needed to clear two bits or set two bits. It is more efficient to use ANDLW and IORLW to modify multiple bits.

EXAMPLE 3-5

```
        MOVLW   0x1F
        BCF     WREG, 0    ;clear bit 0
        BCF     WREG, 1    ;clear bit 1
        BSF     WREG, 6    ;set bit 6
        BSF     WREG, 7    ;set bit 7

Stop:   GOTO    Stop
```

Example 3-6 uses the conditional bit-oriented instructions to set bit zero of WREG only if bit 7 is zero. The BTFSS instruction (bit test file register and skip if set) tests bit 7 of WREG and if it is a one, the next instruction (BCF) is skipped. If bit 7 of WREG is a zero, then the BCF instruction is executed clearing bit zero. To test this process, use the IDE and its simulator and use different values for the MOVLW instruction such as 0x7F and 0xFF.

EXAMPLE 3-6

```
        MOVLW   0x7F       ;load test data
        BTFSS   WREG, 7
        BCF     WREG, 0    ;clear bit 0

Stop:   GOTO    Stop
```

Suppose that a program is required for the PICDEM2 PLUS board that displays a 0x05 on the LEDs connected to Port B bits 0 through 3 when the pushbutton connected to Port A bit 4 is not pressed down. The program must display a 0x03 when the pushbutton is pressed down. The pushbutton produces a logic zero when pressed down, otherwise it produces a logic one. Example 3-7 lists a program that accomplished this task. If a different demonstration board is used, the I/O ports and bit positions may need to be adjusted for the board. The I/O ports are addressed as PORTA or PORTB for transferring data. The direction of a port pin is programmed as an output by placing a zero in the corresponding TRISA or TRISB register bit, and as an input by placing a in the corresponding TRISA or TRISB bit. The ADCON1 register is programmed for digital pin operation by placing a 0x7F in an ADCON1 register or as analog by placing a 0x00 in an ADCON1 register. Note that the function of ADCON1 and values used in the ADCON1 register differ in various versions of the PIC18 family members. This software is written for a PIC18F1320. If another version of the microcontroller is in use, the value may differ or the ADCON1 register may not be needed at all.

EXAMPLE 3-7

```
Main:
        MOVLW   0x7F        ;program all ports as digital
        MOVWF   ADCON1

        MOVLW   0x00
        MOVWF   TRISB       ;Port B is output

        MOVLW   0xFF
        MOVWF   TRISA       ;Port A is input
Main1:                      ;main program loop
        MOVLW   0x05

        BTFSS   PORTA, 0    ;test Pushbutton
        MOVLW   0x03        ;and skip this if Pushbutton is up

        MOVWF   PORTB       ;change Port B and the LEDs

        GOTO    Main1       ;repeat
```

If the status register bits are addressed in an instruction, the name of the bit is often used to address a particular bit. For example, to clear the carry bit the instruction BCF STATUS, C is used instead of a BCF STATUS, 2. Either example instruction functions.

3-3   BYTE-ORIENTED INSTRUCTIONS

The **byte-oriented instructions** are the most numerous in the instruction set and usually in programs. The byte-oriented instructions allow **variable** data to be used in a program, whereas the literal instructions allow **constant** data to be used in a program. The byte-oriented instructions typically use the W register and a location in the register file to perform some operation. Table 3-3 lists all of the byte-oriented instructions with an example or two per instruction. Most of these instructions have three operands: The first is the register file location, the second determines the destination, and the third selects the access bank or a register file bank as determined by the bank select register (BSR). If the second operand (d-bit) is a 0, then the destination is the WREG, and if the second operand is a 1, then the destination is the register file location. If the third operand (a-bit) is a zero, then the access bank is used for the register file location (usually 0x000–0x07F and 0xF80–0xFFF), and if it is a one, then the bank determined by the BSF is used for the register file. The a-bit is required only when referencing a file register by number; if referenced by a label, the a-bit is not used in an instruction. The default is a = 1. The d-bit is needed only to place the result of an operation into WREG; the default is d = 1. In assembly language, the letter W or F is used to indicate the d-bit if desired. Some examples of this appear in Table 3-3.

Suppose that two 16-bit numbers are added to form a 16-bit sum. Prior to this section, this operation was more difficult to perform, but because the byte-oriented instructions contain an **addition with carry** instruction it is possible to more easily add numbers that are wider than 8-bits. Example 3-8 illustrates a short program that adds the 16-bit number in access bank locations 0x10 and 0x11 to the 16-bit number in access bank locations 0x12 and 0x13. Both numbers

TABLE 3-3 Byte-oriented instructions.

Instruction	Operand1	Operand2	Operand3	Examples	Comment
ADDWF	reg	d-bit	a-bit	ADDWF 0x10, W, 0	Add W and access bank location 0x10, store result in W
				ADDWF 0x10, F, 0	Add W and access bank location 0x10, store result in access bank location 0x10
ADDWFC	reg	d-bit	a-bit	ADDWFC 0x10, 0, 0	Add W with carry and access bank location 0x10, store result in W
				ADDWFC 0x10, 1, 1	Add WREG with carry and BSR bank location 0x10, store result in BSR bank location 0x10
ANDWF	reg	d-bit	a-bit	ANDWF 0x10, 0, 1	AND W with BSR bank location 0x10, store result in W
CLRF	reg	a-bit		CLRF WREG	Clear W to 0x00
				CLRF 0x10, ACCESS	Clear access bank location 0x10 to 0x00
				CLRF BOB	Clear location BOB
COMF	reg	d-bit	a-bit	COMF WREG	One's complement W
				COMF 0x10, 0, 0	One's complement access bank location 0x10
				COMF 0x10, 1, 1	One's complement BSR bank location 0x10
CPFSEQ	reg	A-bit		CPFSEQ 0x10, 0	Compare W with access bank location 0x10 and skip the next instruction if they are equal
				CPFSEQ 0x11, 1	Compare W with BSR bank location 0x11 and skip the next instruction if they are equal
CPFSGT	reg	a-bit		CPFSGT 0x12, 0	Compare W with access bank location 0x12 and skip the next instruction if contents of access bank location 0x12 is greater than WREG
CPFSLT	reg	a-bit		CPFSLT 0x13, 1	Compare W with BSR bank location 0x13 and skip the next instruction if contents of BSR bank location 0x13 is less than W
DECF	reg	d-bit	a-bit	DECF WREG	Subtract 1 from W
				DECF 0x10, 0, 0	Subtract 1 from access bank location 0x10 and store result in access bank location 0x10
				DECF 0x10, 1, 0	Subtract 1 from access bank location 0x10 and store result in access bank location 0x10
DECFSZ	reg	d-bit	a-bit	DECFSZ WREG	Decrement W and skip the next instruction if the result is zero

(continued)

TABLE 3-3 (*continued*)

Instruction	Operand1	Operand2	Operand3	Examples	Comment
DCFSNZ	reg	d-bit	a-bit	DCFNSZ 0x10, 1, 0	Decrement access bank location 0x10 and skip the next instruction if the result is not zero
INCF	reg	d-bit	a-bit	INCF WREG	Increment W
				INCF 0x10, 0, 0	Increment access bank location 0x10
INCFSZ	reg	d-bit	a-bit	INCFSZ WREG	Increment W and if the result is zero skip the next instruction
INFSNZ	reg	d-bit	a-bit	INFSNZ WREG	Increment W and if the result is not zero skip the next instruction
IORWF	reg	d-bit	a-bit	IORWF 0x10, 0, 0	Inclusive-OR W with access bank location 0x10 and store result in access bank location 0x10
MOVF	reg	d-bit	a-bit	MOVF 0x10, 0, 0	Copy access bank location 0x10 into W
MOVFF	reg source	reg destination		MOVFF WREG, 0x130	Copy W into location 0x130
MOVWF	reg	a-bit		MOVWF 0x10, 0	Copy W into access bank location 0x10
MULWF	reg	a-bit		MULWF 0x10, 0	Multiply W with access bank location 0x10 and store product in PRODL and PRODH
NEGF	reg	a-bit		NEGF WREG	Two's complement (negate) W
RLCF	reg	d-bit	a-bit	RLCF WREG	Rotate left W through carry
RLNCF	reg	d-bit	a-bit	RLNCF WREG	Rotate left W without carry
RRCF	reg	d-bit	a-bit	RRCF WREG	Rotate right W through carry
RRNCF	reg	d-bit	a-bit	RRNCF WREG	Rotate right W without carry
SETF	reg	a-bit		SETF WREG	Place an 0xFF into W
				SETF 0x11, ACCESS	Place an 0xFF into access bank location 0x11
SUBFWB	reg	d-bit	a-bit	SUBFWB 0x10, 0, 0	Subtract with borrow access bank location 0x10 from W and store result in access bank location 0x10
SUBWF	reg	d-bit	a-bit	SUBWF 0x10, 0, 0	Subtract W from access bank location 0x10 and store result in access bank location 0x10
SUBWFB	reg	d-bit	a-bit	SUBWFB 0x10, 0, 0	Subtract W and borrow from access bank location 0x10 and store result in access bank location 0x10
SWAPF	reg	d-bit	a-bit	SWAPF WREG	Swap nibbles in W
TSTFSZ	reg	d-bit		TSTFSZ WREG	Skip the next instruction if W is zero
XORWF	reg	d-bit	a-bit	XORWF 0x10, 0, 0	Exclusive-OR W with access bank location 0x10 and store result in access bank location 0x10

FIGURE 3-4 Sample 16-bit addition.

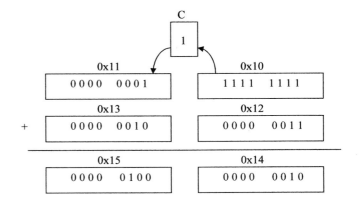

are stored using the little endian format where the least significant portion is stored in the lowest numbered data location. The result of this 16-bit addition is stored at access bank locations 0x14 and 0x15. The program looks long, but the part that performs the addition (last six instructions) is fairly short. Notice how the least significant part is added and the more significant parts are added with carry. Using this same technique, numbers of any width are added. Figure 3-4 shows the operation of the program and the how carry is used to form the 16-bit sum.

EXAMPLE 3-8

```
        MOVLW   0xFF                    ;store 0x01FF
        MOVWF   0x10                    ;into 0x10 and 0x11
        MOVLW   0x01
        MOVWF   0x11

        MOVLW   0x03                    ;store 0x0203
        MOVWF   0x12                    ;into 0x12 and 0x13
        MOVLW   0x02
        MOVWF   0x13

        MOVF    0x10, W, ACCESS         ;add 0x10 and 0x12
        ADDWF   0x12, W, ACCESS
        MOVWF   0x14                    ;save result at 0x14

        MOVF    0x11, W, ACCESS         ;add 0x11 and 0x13 with carry
        ADDWFC  0x13, W, ACCESS
        MOVWF   0x15                    ;save result at 0x15

Stop:   GOTO    Stop
```

A much cleaner looking and easier to write version of this program appears in Example 3-9. The difference between Example 3-8 and Example 3-9 is that the assembler is used to define the memory locations, which precludes the use of all the operands that appears in Example 3-8. The UDATA area is already defined in the program template and the section in Example 3-9 is added to the template without an additional UDATA directive (user data). If the data are to be placed into the access bank, then the UDATA_ACS directive is used before the RES (reserve memory) instructions. This gives a choice as to where the data are stored. Data stored in the access bank usually execute faster because the bank selection register does not need to be initialized with the bank address.

EXAMPLE 3-9

```
        UDATA
Num1L   RES     1                       ;variables reserved by name in UDATA
Num1H   RES     1
Num2L   RES     1
Num2H   RES     1
AnsL    RES     1
AnsH    RES     1
```

```
Main:
        MOVLW   0xff                        ;store 0x01ff
        MOVWF   Num1L                       ;into Num1
        MOVLW   0x01
        MOVWF   Num1H

        MOVLW   0x03                        ;store 0x0203
        MOVWF   Num2L                       ;into Num2
        MOVLW   0x02
        MOVWF   Num2H

        MOVF    Num1L                       ;add low parts
        ADDWF   Num2H, W
        MOVWF   AnsL                        ;save result AnsL

        MOVF    Num1H                       ;add high parts with carry
        ADDWFC  Num2H, W
        MOVWF   AnsH                        ;save result at AnsH

Stop:   GOTO    Stop
```

To perform 16-bit subtraction, the program in Example 3-9 is modified by changing the ADDFW and ADDWFC instructions to SUBFW and SUBWFB instructions. Also, the operands register file locations for the subtraction instructions need to be switched because of the way subtraction functions. A subtract instruction always subtracts WREG from the register file location appearing in the instruction. Example 3-10 shows a 16-bit subtraction. This example uses the same data definitions from Example 3-9.

EXAMPLE 3-10

```
        MOVLW   0xFF                        ;store 0x01FF
        MOVWF   Num1L                       ;into Num1
        MOVLW   0x01
        MOVWF   Num1H

        MOVLW   0x03                        ;Store 0x0203
        MOVWF   Num2L                       ;into Num2
        MOVLW   0x02
        MOVWF   Num2H

        MOVF    Num1L                       ;subtract low parts
        SUBWF   Num2H, W
        MOVWF   AnsL                        ;save result AnsL

        MOVF    Num1H                       ;subtract high parts with borrow
        SUBWFB  Num2H, W
        MOVWF   AnsH                        ;save result at AnsH

Stop:   GOTO    Stop
```

Rotates. The PIC18 family instruction set has four, byte-oriented rotate instructions. Bytes are rotated right or left either through the carry or without the carry. Figure 3-5 illustrates these four rotate instructions.

One method of multiplying or dividing is by rotating a number. If a number is rotated left, with a zero placed into the rightmost bit, the number is multiplied by a factor of 2. Likewise, if a number is rotated right with a zero placed in the leftmost bit, it is divided by two. For a signed division by two, the sign bit is rotated into the leftmost bit of the result. If a number can be multiplied or divided by two it can be multiplied or divided by any number.

For example, to multiply by a five (a five is a 4x plus a 1x or 101_2), multiply the number by 4 (two rotates left) and then add the original number. This is illustrated in Example 3-11. This example multiplies a test number of 4 in WREG by a factor of 5 and leaves the result of 20 in WREG. This program functions correctly, but the multiply instruction in the PIC is more efficient. See the same program using the MULWF instruction in Example 3-12. Also notice in the program how the carry bit is cleared by using the BCF instruction.

FIGURE 3-5 Four rotate instructions.

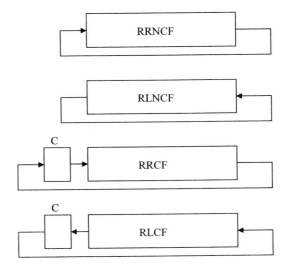

EXAMPLE 3-11

```
        UDATA_ACS

Num     RES     1               ;register in UDATA access area

Main:
        MOVLW   4               ;load test value
        MOVWF   Num             ;save 1x at 0x10
        BCF     STATUS, 0       ;clear carry
        RLCF    WREG            ;WREG x 2
        BCF     STATUS, 0
        RLCF    WREG            ;WREG x 4
        ADDWF   Num, 0          ;WREG x 5

Stop:   GOTO    Stop
```

EXAMPLE 3-12

```
        UDATA_ACS

Num     RES     1               ;register in UDATA access area

Main:
        MOVLW   5
        MOVWF   NUM             ;save the 5 in 0x10
        MOVLW   4               ;load test value
        MULWF   NUM             ;multiply by 5
        MOVFF   PRODL, WREG     ;get 8-bit product into WREG

Stop:   GOTO    Stop
```

If a division by a power of 2 is needed, it is accomplished by shifting a number to the right. By using the rotate instruction, a shift right is accomplished by first clearing the carry flag followed by a RRCF instruction. Suppose the content of WREG needs to be divided by 8. This is accomplished by three right shifts as illustrated in Example 3-13. Notice that the answer is rounded up if a carry is found after the last rotate right. The bit test and skip if zero instruction is used to skip the INCF (increment) if the carry was zero. Here a 100 decimal is divided by 8 and the result of 12.5 left in WREG as 13 (a rounded-up 12.5). More information and software for division is provided in Chapter 4 on assembly language programming examples.

EXAMPLE 3-13

```
        MOVLW   .100            ;load WREG with 100 decimal
        BCF     STATUS, 0       ;divide by 8
        RRCF    WREG
```

```
            BCF     STATUS, 0
            RRCF    WREG
            BCF     STATUS, 0
            RRCF    WREG

            BTFSC   STATUS, 0       ;round result
            INCF    WREG

Stop:       GOTO    Stop
```

To illustrate the rotate instructions on a demonstration board, such as the PICDEM2 PLUS, program Port B as an output port and program Port A as an input port. This sample program randomly selects one of the four LEDs connected to Port B when the pushbutton connected to Port A bit position 4 is pressed down. To accomplish this, the rotate without carry instruction is used in Example 3-14. In this example, when the pushbutton is pressed all the LEDs will light and when released, only one will light randomly.

EXAMPLE 3-14

```
Main:
            MOVLW   0x00            ;program Ports A & B
            MOVWF   TRISB
            MOVLW   0xFF
            MOVWF   TRISA
            MOVLW   0x11
            MOVWF   PORTB           ;start Port B at 0x11
Main1:
            BTFSS   PORTA, 4
            RLNCF   PORTB           ;rotate if Pushbutton down

            GOTO    Main1           ;repeat
```

3-4   PROGRAM CONTROL INSTRUCTIONS AND INDIRECT ADDRESSING

Before much worthwhile programming is accomplished, the program control instructions must be learned. This section presents the program control instructions and some additional examples that use them to perform simple tasks. The **program control instructions** modify the flow of a program through conditional branches and function calls. They also provide some control over the operation of a few internal features of the microcontroller. Also presented is indirect addressing through the file select registers (FSR0, FSR1, and FSR2). Table 3-4 illustrates the program control instructions available to the PIC18 family of microcontrollers.

The GOTO instruction has already been used in a program, but it has not been explained with enough detail to be properly understood. A GOTO is a 32-bit instruction that branches to any program memory location in the program memory. The BRA (**branch always**) instruction, which also does an unconditional branch, is a 16-bit instruction that has a limited branch range. A BRA instruction contains an 11-bit number stored with it that is not an address, but a distance. This distance or **displacement**, as it is often called, allows the BRA to jump ahead in a program by up to 1,024 bytes from the address of the next instruction in the program or back by up to −1024 bytes from the next instruction in the program. This type of branch is often called a **relative branch** because the instruction is moved to some other part of memory and the distance of the branch remains the same. All of the conditional jumps use this displacement form of addressing, but have an even shorter range of between +127 or −128 bytes from the next instruction in a program. The reason for these limits is the available number of bits in the instruction for addressing memory. The GOTO is often called an **absolute branch** because the address is fixed or absolute. A branch (BRA) or conditional branch is often called a **relative branch** because the branch location is determined by the relative location or position of the branch in the program memory.

TABLE 3-4 Program control instructions.

Instruction	Operand1	Operand2	Examples	Comment
BC	n1		BC AGAIN	Branch to AGAIN if carry is 1
BN	n1		BN WOW	Branch to WOW if negative
BNC	n1		BNC FLOP	Branch to FLOP if carry is 0
BNN	n1		BNN POSITIVE	Branch to POSITIVE if not negative
BNOV	n1		BNOV BIG	Branch to BIG if no overflow
BNZ	n1		BNZ MORE	Branch to MORE if not zero
BOV	n1		BOV NOW	Branch to NOW if an overflow
BRA	n2		BRA WINK	Branch to WINK unconditionally
BZ	n1		BZ BOSTON	Branch to BOSTON if zero
CALL	n3	S	CALL HOME	Invoke a function called HOME
CLRWDT			CLRWDT	Clear watchdog timer
DAW			DAW	Decimal adjust contents of WREG
GOTO	n3		GOTO ERIE	Unconditional jump to ERIE
NOP			NOP	No operation
POP			POP	Remove the return address at the top of the stack
PUSH			PUSH	Place the address of the next instruction on the top of the stack
RCALL	n2		RCALL 6	Relative CALL to PC + 2 + 2 * 6 ahead in the program
RESET			RESET	Software reset
RETFIE	S		RETFIE 1	Return from interrupt with shadow registers loaded into registers and interrupts enabled
RETLW	L		RETLW 6	Return from a function with 6 in WREG
RETURN	S		RETURN 1	Return from a function with shadow registers loaded into registers
SLEEP			SLEEP	Enter standby mode

Notes: n1 = 8-bit displacement, n2 = 11-bit displacement, n3 = 21-bit absolute address, s = shadow bit, and l = 8-bit literal.

Many of the control instructions are used to implement the program constructs that make up the building blocks of a program. These constructs are if-then-else, repeat-until (do-while or for), while, and functions.

The CALL and RETURN instructions are used to create functions or subroutines in a program. The CALL instruction pushes the return address onto the internal program stack and then does a GOTO in the memory location indicated in the operand of the CALL instruction. The RETURN instruction returns from a function just as it does in C language. What the RETURN actually does is remove the return address from the program stack and place it into the program counter. A CALL is used to CALL a function and a RETURN is used to return from a function. The RETURN always returns the instruction following the most recent CALL.

The PUSH and POP instructions also access the program stack within the microcontroller. A PUSH is used to preload a return address of PC + 2 onto the program stack and a POP is used to remove and discard a return address from the stack. Both instructions should be used with care.

The NOP instruction performs absolutely no operation except it does require one instruction cycle. NOP instructions are sometimes used in time delays to add an additional clock cycle to execution. A NOP is also added to some instructions by the assembler.

The RETLW instruction is used to return from a function with the WREG loaded with the operand of the RETLW instruction. This finds an application in some types of assembly language programs to return a value from a function.

FIGURE 3-6   If-then-else construct.

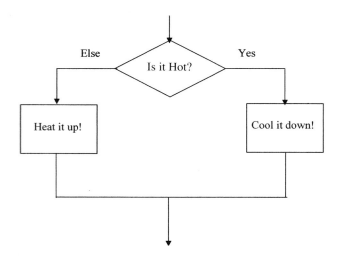

Program Constructs

Program constructs are the building blocks of programming and need to be discussed so that programs are written efficiently. The first construct described is the if-then-else construct. Figure 3-6 illustrates the general form of this construct.

Notice how the question in Figure 3-6 asks, It is hot? If it is hot, then the program branches to a block of code that cools it down, else (or otherwise) it heats it up. This construct is used to ask a question about something in a program. A code block that performs the task of asking a question is illustrated in Example 3-15. This example tests access bank register 0x10 for the value of 4. If it is 4, a 6 is placed into 0x10 and if it is not, a 9 is placed into 0x10.

EXAMPLE 3-15

```
        MOVLW   4
        SUBWF   0x10, 0
        BNZ     NOPE        ;if not a 4
        MOVLW   6           ;if a 4
        BRA     DONE1

NOPE:
        MOVLW   9

DONE1:
        MOVWF   0x10        ;save the 6 or 9
```

Suppose the numbers 0x00 through 0x0F in WREG are converted from hexadecimal code to ASCII code. The ASCII numbers are 0x30 through 0x39 and the letters A through F are 0x41 through 0x46. To accomplish this conversion, a 0x30 is added to a hexadecimal number (0x00-0x09) and a 0x37 is added to a hexadecimal letter (0x0A-0x0F). The if construct, as illustrated in Example 3-16, is used to perform this simple task. Here a 0x30 is added, then, if the result is less than 0x3A, seven more are added for converting the letters. This example uses two temporary locations in the data register file area to function properly. Although the definition of these locations does not appear here, they are defined using RES 1 in the UDATA areas of the source file. The CPFSGT instruction compares the contents of Temp2 (0x3A) with W and if Temp2 is greater than W, the ADDLW instruction is skipped.

EXAMPLE 3-16

```
ADDLW   0x30                ;add 0x30
MOVWF   Temp1
MOVLW   0x3A
MOVWF   Temp2
```

```
MOVFF    Temp1, WREG
CPFSGT   Temp2                          ;skip for 0 through 9

ADDLW    0x07                           ;only for 3A or greater
```

An even shorter method, although less clear, for performing the conversion is listed in Example 3-17. Here the DAW instruction is used in a slightly unconventional manner to accomplish the conversion. The DAW instruction is meant to adjust the number in WREG to BCD after a BCD addition. It does this by examining the value of the low nibble and if it is greater than 9 or the DC bit is set, it adds 6. It also examines the high nibble and if it is greater than 9 or the C bit is set, it adds 0x60. This causes the result to be 0x30 through 0x39 for the numbers and 0x40 through 0x45 for the letters. As can be seen, the letters are one short of the correct value needed so the BTFSC (bit test and skip if cleared) instruction is used to test bit 6 of the result. Bit 6 is cleared for hexadecimal numbers and set for hexadecimal letters. This test and the BTFSC instruction then increment WREG only for 0x40 through 0x45 to generate the correct values for the letters of 0x41 through 0x46.

EXAMPLE 3-17

```
ADDLW    0x30
DAW
BTFSC    WREG, 6

INCF     WREG
```

Although this is efficient, it would probably be more functional if placed into a function. Functions are often used to reduce the coding task. Example 3-18 shows how to place Example 3-17 into a function called Hex2ASCII, which is a good name for the function. Always try to name functions so the name denotes the operation of the function. This Hex2ASCII function converts the hex digit in WREG into an ASCII digit in WREG.

EXAMPLE 3-18

```
;
;*** Hexadecimal to ASCII function ***
;
; Converts right nibble of WREG to ASCII returned in WREG
;
Hex2ASCII:
        ANDLW    0x0F              ;clear high nibble

        ADDLW    0x30              ;convert to ASCII
        DAW
        BTFSC    WREG, 6

        INCF     WREG

        RETURN

;
;*** Main program ***
;
Main:
        MOVLW    0x5c              ;load test data

        MOVWF    ASCII_H
        CALL     Hex2ASCII         ;convert low nibble
        MOVWF    ASCII_L           ;store it at ASCII_L

        MOVFF    ASCII_H, WREG
        SWAPF    WREG              ;get high nibble

        CALL     Hex2ASCII         ;convert it to ASCII
        MOVWF    ASCII_H           ;store it at ASCII_H
Stop:   BRA Stop
```

The CALL instruction places the return address onto the internal stack (32x21) and then branches to the absolute location of the function, which in this case is Hex2ASCII. The **return address** always points to the instruction immediately following the CALL. If the first CALL in the program is examined, the return address is the location of instruction MOVWF ASCII_L. Whenever a RETURN instruction is encountered, the microcontroller retrieves the most recent return address from the stack and branches to that address. Because the stack contains 31 locations, functions can be nested to a depth of 31. Note that often only a few locations are needed on the stack for most programs.

There are two types of return instructions: an ordinary return as described (RETURN), and the other (RETURN 1) which is called a fast return because, although it is not faster than the RETURN, it does restore the values in WREG, STATUS, and BSR. To use this restore feature, the function must be called with a CALL SUBROUTINE, 1 instruction. The CALL, with a second parameter of 1, automatically saves WREG, STATUS, and BSR in internal, invisible shadow registers. The RETURN 1 extracts the data from these shadow registers. Another type of return is the RETLW instruction. The RETLW 6 instruction returns to the caller, but it also places a 6 into WREG on the return. Most functions use the CALL and RETURN pair for operation.

Example 3-19 illustrates how a new function called Hex2ASC is written using the nesting of the Hex2ASCII function. Function Hex2ASCII is called by Hex2ASC two times, which requires one stack location for the return address for Hex2ASC and a second for Hex2ASCII. The program is much cleaner because of the new function that converts an entire byte. The only registers that the functions use are WREG, ASCII_L, and ASCII_H. It also uses two stack levels when it executes. The only way this might be improved is to return the two ASCII characters in the PRODL and PRODH registers. That way, no general registers would be used for the conversion because general registers are often scarce.

EXAMPLE 3-19

```
;************* FUNCTION Hex2ASCII **********
;
; uses 1 stack level
; uses WREG
;
; converts right nibble of WREG from hex to ASCII
;
;*********************************************

Hex2ASCII:

        ANDLW   0x0F            ;isolate low nibble

        ADDLW   0x30
        DAW
        BTFSC   WREG, 6

        INCF    WREG

        RETURN

;*********** FUNCTION Hex2ASC *************
;
; uses 2 stack levels
; uses WREG, ASCII_L, and ASCII_H
; uses Hex2ASCII
;
; converts byte in WREG from hex to ASCII
;       where ASCII_L contains rightmost ASCII digit
;       and ASCII_H contains leftmost ASCII digit
;
;*********************************************
```

```
Hex2ASC:

        MOVWF    ASCII_H
        CALL     Hex2ASCII        ;convert low nibble
        MOVWF    ASCII_L

        MOVFF    ASCII_H, WREG
        SWAPF    WREG             ;swap nibbles

        CALL     Hex2ASCII        ;convert high nibble
        MOVWF    ASCII_H

        RETURN

;************** MAIN PROGRAM ******************
Main:
        MOVLW    0x5c             ;load test data

        CALL     Hex2ASC          ;convert entire byte to ASCII

Stop:  BRA Stop
```

The repeat-until construct repeats an operation a fixed number of times or until a certain condition occurs. Figure 3-7 illustrates the repeat-until construct.

Suppose that the data memory registers between locations 0x10 and 0x28 are cleared to 0x00 by a program. This can be accomplished with many MOVWF instructions. A better way is to use indirect addressing with a repeat-until. A loop is repeated 0x19 times to store 0x19 zeros into 0x19 different register file locations. There are 0x19 locations between 0x10 and 0x28 (the difference plus 1). Example 3-20 illustrates a program that accomplishes this task.

EXAMPLE 3-20

```
        LFSR     0, 0x10          ;address 0x10 with FSR0
        MOVLW    0x19             ;load WREG with count

Loop:

        CLRF     POSTINC0         ;clear a location and increment pointer

        DECFSZ   WREG             ;decrement counter, skip BRA if 0

        BRA      Loop             ;repeat loop until WREG is zero

Stop:  BRA       Stop
```

FIGURE 3-7 Repeat-until construct.

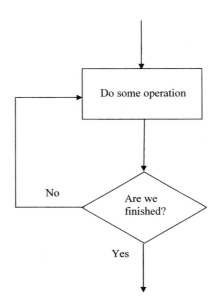

TABLE 3-5   Indirect addressing details.

Mnemonic Operands	Example	Comment
INDF0	MOVFF WREG, INDF0	Copy WREG into the memory location indirectly addressed by FSR0
INDF1	INCF INDF1	Increment the contents of the memory location addressed by FSR1
INDF2	MOVWF INDF2	Copy WREG into the memory location addressed by FSR2
POSTDEC	MOVWF POSTDEC0	Copy WREG into the memory location addressed by FSR0 and then subtract 1 from FSR0
	MOVFF POSTDEC0, POSTDEC1	Copy the contents of the memory location addressed by FSR0 into the memory location addressed by FSR1 and then decrement both FSR0 and FSR1
POSTINC	CLRF POSTINC2	Clear the memory location addressed by FSR2 and then increment FSR2
PREINC	DECF PREINC1	Increment FSR1 and then decrement the contents of the location now addressed by FSR1
PLUSW	INCF PLUSW0	Increment the contents of the memory location addressed by FSR0 + WREG

Indirect Addressing

Example 3-21 illustrates the first example of indirect addressing. The microcontroller has three registers that are used to indirectly access or address memory, called the file select registers (FSR0, FSR1, and FSR2). For some reason, the only instruction that uses FSR as an operand is the LFSR instruction, which is used to place a data file memory address into an FSR. All other addressing through the file select registers uses a different mnemonic. Table 3-5 lists the mnemonics and their functions as they apply to indirect addressing through the FSR registers.

To illustrate the power of these indirect operands, suppose that a block of data from data memory addresses 0x10 through 0x32 must be added to a block of data located at data addresses 0x40 through 0x62. Each block contains 0x23 locations. The program requires a programmed loop to execute 0x23 times and also requires that indirect addresses are used to access both areas of data memory using two file select registers (FSR0 and FSR1). The program to accomplish this task is listed in Example 3-22. Notice how register 5 is used for the programmed loop as a counter.

EXAMPLE 3-21

```
        LFSR    0, 0x10         ;address 0x10 with FSR0
        LFSR    1, 0x40         ;address 0x40 with FSR1
        MOVLW   0x23            ;get a 0x23 for the count
        MOVWF   5               ;save count in location 5
Loop:
        MOVF    POSTINC0, 0     ;get Block 1 into WREG
        ADDWF   POSTINC1        ;add WREG to Block 2

        DECFSZ  5               ;decrement the counter
        BRA     Loop            ;loop 0x23 times
Stop:   BRA     Stop
```

Suppose the memory contains information acquired from some external device and a program is needed to search through the data memory for the occurrence of the number 0x0D. The table is located at addresses 0x20 through 0x5A. This table is 0x3B locations in length. If the 0x0D is found, the program continues at label FoundIt and if it is not found, the program continues at label NotFound.

EXAMPLE 3-22

```
        LFSR    0, 0x20         ;address 0x20 with FSR0
        MOVLW   0x3B
        MOVWF   5
Loop:
        MOVLW   0x0D            ;load WREG with 0x0D
        SUBWF   POSTINC0, 0
        BZ      FoundIt          ;if found

        DECFSZ  5               ;check counter
        BRA     Loop            ;if not zero
NotFound:

;   **** code for 0x0D not found ****

FoundIt:

;   **** code for 0x0D found ****

Stop:   BRA     Stop
```

The final programming construct is the **while construct** illustrated in Figure 3-8. Suppose that Example 3-22 is rewritten using the while construct instead of the repeat-until construct. In this example, the search for a 0x0D is repeated while the counter is not zero. To accomplish this task for all locations, the counter value must be increased by 1. Example 3-23 illustrates the new software using the while construct.

EXAMPLE 3-23

```
        LFSR    0, 0x20         ;address 0x20 with FSR0
        MOVLW   0x3C
        MOVWF   5
Loop:
        DCFSNZ  5               ;check counter
        BRA     NotFound

        MOVLW   0x0D            ;while count is not zero
        SUBWF   POSTINC0, 0
        BNZ     Loop            ;if not yet found
```

FIGURE 3-8 While construct.

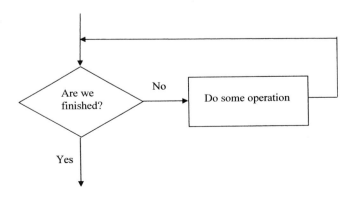

```
FoundIt:

;    **** code for 0x0D found ****

NotFound:

;    **** code for 0x0D not found ****

Stop:  BRA    Stop
```

If the two examples are compared, the placement of the decrement counter instruction is essentially the only thing that changes. In the repeat–until version of Example 3-22, the decrement is the last test in the loop and in Example 3-23, the decrement is the first test in the loop. Otherwise, both tasks are similar except the while version uses a count that is one greater than the repeat–until version because it is decremented first. Which is better is purely in the eyes of the beholder.

3-5   TABLE INSTRUCTIONS

There are two table instructions: TBLRD (table read) and TBLWT (table write). These **table instructions** transfer information between the program memory, which is persistent memory, and the data memory, which is volatile memory. Tables that are used with code can therefore be stored in the program memory and, when required, transferred to the data memory for use. The table read instruction reads data from the program memory and stores it into the data memory. The table write instruction reads data from the data memory and writes it to the program memory. A table read reads program memory and a table write writes program memory. Note that a table write to the program memory is not possible in the OTP version (18C family). The microcontroller must have a flash memory to accomplish any write to the program memory. The OTP version of the microcontroller does not have flash memory.

The table instructions use a pointer to access the program memory called a table pointer (TBLPTR). The table pointer is located in the SFR area and is three bytes in length to hold a 21-bit address. The TBLPTRL register holds the low-order 8-bits, the TBLPTRH register holds the next higher order 8-bits, and the TBLPTRU holds the most significant part of the program memory address. Data transfers occur through the TABLAT register to or from the program memory.

The variations of the table read and write instructions are illustrated in Table 3-6. These variations allow the table pointer to be incremented before or after, or decremented after, a table read or write. No decrement before a table read or write is allowed. One variation does not

TABLE 3-6   Table instructions.

Instruction	Comment
TBLRD*	Data transferred from the program memory to the TABLAT, table pointer is not changed
TBLWT*	Data transferred from the TABLAT to the program memory, table pointer is not changed
TBLRD*+	Data transferred from the program memory to the TABLAT, table pointer is then incremented
TBLWT*+	Data transferred from the TABLAT to the program memory, table pointer is then incremented
TBLRD*-	Data transferred from the program memory to the TABLAT, table pointer is then decremented
TBLWT*-	Data transferred from the TABLAT to the program memory, table pointer is then decremented
TBLRD+*	Table pointer is incremented, then data transferred from the program memory to the TABLAT
TBLWT+*	Table pointer is incremented, then data transferred from the TABLAT to the program memory

increment or decrement the table pointer. Table reads or writes transfer an 8-bit number between the program memory as addressed by the table pointer register (TBLPTR) and a SFR called the table latch register (TABLAT).

Table Reads. The **table pointer (TBLPTR)** (a 21-bit address and a 22nd bit for addressing a configuration area) and the **table latch (TABLAT)** (an 8-bit register) are located in the SFR section of the data memory. The program memory area is from addresses 0x000000 through 0x1FFFFF, and the configuration memory is at location 0x300000 through 0x30FFFF. Suppose that a number between 0 and 9 needs to be converted from BCD code into 7-segment code for a segmented numeric LED display. To accomplish the conversion it is customary to use a lookup table. Because the data memory is relatively small and dynamic, the lookup table is stored in the program memory which is large and static. By storing the lookup table into the program memory it is persistent. When the lookup table is needed, a location is copied into the data memory from the program memory using a table read instruction to access the lookup table. How is this accomplished? First, the lookup table is stored in the program memory using a block of code that appears in Example 3-24.

EXAMPLE 3-24

```
;************** 7-segment lookup table ***************
;
TABLE7        CODE_PACK       0x1FF0
Look7:
         db          0x06     ;0
         db          0x5b     ;1
         db          0x79     ;2
         db          0x4F     ;3
         db          0x66     ;4
         db          0x6D     ;5
         db          0x7D     ;6
         db          0x07     ;7
         db          0x7F     ;8
         db          0x3F     ;9
```

The lookup table in Example 3-24 is placed before the main program just before the RESET_VECTOR1 CODE 0x0000 statement in the template file. This is after the user data and EEPROM data are defined. It is important to use the directive CODE_PACK to store the bytes as two bytes per word in the program memory. When the TBLRD instruction accesses program memory, it addresses bytes of program memory so the table must be packed into bytes for a byte-sized table. The numbers in the table correspond to the bits needed to implement the 7-segment code as illustrated in Figure 3-9.

Segments are lettered from a through g and a logic 1, in this example, illuminates a given segment. If, for example, a 0x66 is sent to the LED display, segments b, c, f, and g will light, displaying the number 4 on the LED display. The table contains all of the 7-segment codes in numeric order as illustrated in Example 3-24. Software to access this table is listed in Example 3-25.

FIGURE 3-9 7-segment numeric display.

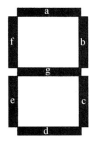

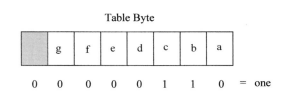

The most effective way to obtain a code from the table is to use a single TBLRD instruction that accesses the proper location in the lookup table for the desired code. First address Look7 by loading the TBLPTR with Look7. Here a function is written to access the table and convert from BCD to 7-segment code. The processor under test has a 4K byte program memory so the table is placed at program memory locations 0x1FF0 through 0x1FF9 near the top of the program memory.

EXAMPLE 3-25

```
;********** To7seg **********************
;
;   uses 1 level of stack data
;   uses WREG, TABPTR, and TABLAT
;
;   accesses table at 0x1FF0 through 0x1FF9
;
;   converts WREG from BCD to 7-segment code
;   7-segment code is returned in WREG
;
;

To7seg:
        ANDLW   0x0F              ;clear high nibble

        ADDLW   0xF0              ;generate table address
        MOVWF   TBLPTRL
        MOVLW   0x1F
        MOVWF   TBLPTRH
        MOVLW   0
        MOVWF   TBLPTRU

        TBLRD*                    ;get 7-segment code
        MOVFF   TABLAT, WREG

        RETURN
```

This same technique can be used for accessing any type of fixed table in the program memory, provided the table is stored in a location that is easily modified by the number that is being converted. To illustrate this, suppose that a table is used to convert from hex to ASCII code instead of the technique used in Example 3-16. Undoubtedly, the method of Example 3-16 is much shorter and much more efficient, but nonetheless a table can be used for the same conversion. See Example 3-26 for the lookup table and the software for the conversion. This function has been called Hex2ASCII just like the one in Example 3-18 because it performs exactly the same task.

EXAMPLE 3-26

```
;************** ASCII lookup table ****************
;
;
TABLE7          CODE_PACK       0x1FF0
HexASC          DB      0x30, 0x31, 0x32, 0x33
                DB      0x34, 0x35, 0x36, 0x37
                DB      0x38, 0x39, 0x41, 0x42
                DB      0x43, 0x44, 0x45, 0x46

;********** Hex2ASCII **********************
;
;   uses 1 level of stack data
;   uses WREG, TABPTR, and TABLAT
;
;   accesses table at 0x1FF0 through 0x1FFF
;
;   converts WREG from hex to ASCII code
;   ASCII is returned in WREG
;
;
```

```
Hex2ASCII:
      ANDLW  0x0F              ;clear high nibble

      ADDLW  0xF0              ;generate table address
      MOVWF  TBLPTRL
      MOVLW  0x1F
      MOVWF  TBLPTRH
      MOVLW  0
      MOVWF  TBLPTRU

      TBLRD*                   ;get ASCII code
      MOVFF  TABLAT, WREG

      RETURN
```

Table Writes. **Table write** operations can be risky. For example, if a reset occurs during a write to the program memory, the data can be corrupted. If possible, some method of verification should be used, but may not always be possible. Other possible problems with a table write can occur if the power fails or if interrupts are not disabled during a table write. For these reasons, table writes are discussed in a later section of the text when these issues can be properly addressed.

One method of verification is to write the data into two separate tables. Each table also contains a count byte. When a table is written, the count byte is incremented. To store the tables, one table is written with its count byte incremented, and then the next table is written with its count byte incremented. If a crash occurs, these two tables will not match and the count bytes will not match. The error recovery is a rollback to the table that contains the lowest number in its count byte. Although this is not a complete error recovery, it does detect an error and perform a rollback.

The table–write instruction is used in Chapter 10 in a bootstrap loader program to write information into the program memory. The only application for a table write is to write to the program memory or to the configuration registers.

3-6   MACRO SEQUENCES

Macro sequences are groups of instructions that function in the same way as a function, but there is no linkage to or return from a macro sequence. Another difference is that the instructions in the macro are placed in the program each time that the macro is invoked.

Defining the Macro

A macro is given a name followed by the word MACRO, followed by any parameters associated with the macro. Example 3-27 illustrates a macro sequence defined for the assembler that shifts "what" to the left. The label "what" is called a replaceable parameter. Each time the instructions in the macro are used in a program, the label "what" is replaced by the parameter to the right of the macro name. In this example, the macro called ShiftL is used twice and each time the parameter "what" is replaced by WREG.

EXAMPLE 3-27

```
ShiftL   MACRO    what        ;macro called ShiftL

         ADDWF    what

         endm

;use the macro ShiftL 2 times

         MOVLW 1
         ShiftL   WREG        ;use macro for WREG
         ShiftL   WREG        ;use macro for WREG
```

Macro sequences can make it a little easier to write a program. If the same sequence of instructions appears in a program many times, they are written as a function or as a macro. The choice of whether to use a macro or a function is determined by observing the number of instructions in the sequence. If there are more than about 10 instructions in the sequence, it is usually written as a function and if there are fewer than 10, it is written as a macro. Each time a macro is used in a program the instructions in the macro are placed in the program. Each time a function is used in a program the instructions are CALLed, which takes time, but because the instructions are stored in the memory only once it saves memory. Macros execute faster, but consume more memory space. Functions save memory, but the CALL and RETURN instructions require time to execute.

Example 3-18 listed a program that used a function for conversion from hexadecimal to ASCII. Example 3-28 lists the same program, but instead of using a function for Hex2ASCII a macro is used. If the two examples are compared, there is not much difference between them. The function ends with a return and the macro ends with an ENDM. The ENDM inserts no code into a program and the RETURN insert one instruction. The macro is one instruction shorter than the function. In the main program the function is called twice and the macro is invoked twice. When a function is called, the CALL instruction requires two program words for storage and the macro requires none. What does appear in the program listing in Example 3-28 is that the instructions in the macro are placed in the program two times, once for each macro in the program.

EXAMPLE 3-28

```
;
;*** Hexadecimal to ASCII macro ***
;
;     Converts right nibble of WREG to ASCII returned in WREG
;

Hex2ASCII MACRO

        ANDLW   0x0F            ;clear high nibble

        ADDLW   0x30            ;convert it to ASCII
        DAW
        BTFSC   WREG, 6

        INCF    WREG

        ENDM

;
;*** Main program ***
;

Main:
        MOVLW   0x5c            ;load test data

        MOVWF   ASCII_H

        Hex2ASCII               ;convert low nibble

        MOVWF   ASCII_L         ;store it at ASCII_L

        MOVFF   ASCII_H, WREG
        SWAPF   WREG            ;get high nibble

        Hex2ASCII               ;convert it to ASCII

        MOVWF   ASCII_H         ;store it at ASCII_H

Stop:   GOTO    STOP
```

If you compile this example, the way that the macro is listed in the program listing file is hard to decipher. Each instruction in the macro appears twice with different nonsequential addresses. If each instruction is mentally moved to it's proper place in the program, they are actually in the program at the correct place. Why the instructions in the macro are listed in this fashion is a puzzle.

3-7   SUMMARY

1. Literal instructions allow literal or constant data to be used mainly with the WREG. Literal instructions use opcodes such as MOVLW or ADDLW, where the first three letters are the operation and the last two indicate what is operated upon. (L = literal, W = WREG). The ADDLW instruction adds literal data to WREG. For example, the ADDLW 6 instruction adds a 6 to WREG.

2. Other literal instructions are available for subtraction (SUBLW), include-OR (IORLW), exclusive-OR (XORLW), AND (ANDLW), and multiplication (MULLW). There is also a literal return from a procedure (RETLW) that allows a literal to be returned in WREG.

3. Two of the literal instructions do not use WREG. The MOVLB and the LFSR load the bank select register (BSR) or the file selection register (FSR0, FSR1, or FSR2) with literal data.

4. The bit-oriented instructions allow individual bits to be cleared (BCF), set (BSF), or toggled (BTG). This allows complete control of the state of any bit anywhere in the data memory, including the special function registers.

5. Other bit-oriented instructions are BTFSC and BTFSS, which test a bit and either skip the next instruction if the bit is cleared (BTFSC) or if it is set (BTFSS).

6. There are more byte-oriented instructions than any other type of instruction. All of them work with byte-sized data. The way that the instruction appears indicates the operation of the instruction. For example, an ADDWF adds a byte from the WREG to a register file location or from the register file location to the WREG. The direction of the operation is selectable.

7. Data are addressed by using a register number or indirectly by using a pointer in a file select register (FSR). If an FSR is used, the operands for indirect addressing are INDF0, INDF1, and INDF2. Also available for indirectly addressing memory are POSTINC0, POSTINC1, and POSTINC2, as well as PREINC0, PREINC1, and PREINC2. PLUSW0, PLUSW1, and PLUSW2 are also available to allow WREG to be added to an FSR register.

8. The control instructions allow conditions to be tested and then, based on the outcome, can branch to another part of the program. These instruction often implement an if statement.

9. Program constructs are the building blocks of programming. The program constructs most often used are if–then–else, repeat–until, while, and function.

10. Table instructions allow access to the program memory. The program memory is read using a table read instruction (TBLRD) or written using a table write instruction (TBLWR).

11. Data is addressed in the program memory using the TBLPTR register and all data are transferred through the TABLAT register.

12. Lookup tables are often used to convert data from one form to another.

13. Macro sequences are used in programs to increase the execution speed, but usually require more memory space than an equivalent function.

3-8   QUESTIONS AND PROBLEMS

1. What is literal data?
2. The literal instructions usually function with which microcontroller register?
3. What is an ADDLW instruction?
4. Select a literal instruction to perform each of the following tasks:
   a. Place a 0x34 into the WREG register
   b. Add a 3 to the WREG register
   c. Inclusively-OR a 6 with the WREG register

    d. Load the bank select register with a 2

    e. Subtract the WREG register from a 0

5. Explain what each of the following literal instructions accomplish:

    a. MOVLW 0x6A

    b. XORLW 4

    c. LFSR 0x123

    d. MULLW .10

    e. MOVLB 2

6. Develop a short sequence of instructions that load WREG with a 3, then add 9 to it, and finally inclusively-OR a 5 to it.

7. Write a short sequence of instructions that place a 0x0F into WREG and then clear the rightmost two bits, set the leftmost bit, and invert bit position 5 using only literal instructions to accomplish the task.

8. Select bit-oriented instructions that:

    a. Clear bit position 2 of WREG

    b. Set bit position 3 of access bank register 0x11

    c. Toggle bit position 5 of access bank register 0x1A

    d. Set the carry status register bit

    e. Clear the N status register bit

9. What is accomplished by the BTFSC instruction?

10. What is accomplished by the BTFSS instruction?

11. Develop a sequence of instructions using only the bit-oriented and literal instructions that will add a 1 to WREG if the carry status register bit is a 1.

12. The access bank contains what register file locations?

13. Explain how register file location 0x200 is accessed in a program.

14. What is the difference between a byte-oriented addition and a literal addition?

15. Describe what each of the following instructions accomplish:

    a. MOVFF WREG, 0x10

    b. ADDWF 0x20, 0

    c. ADDWF 0x20, 1, 1

    d. DECF 0x34

    e. INCF 0x34, 1

16. Develop a short program that stores a 0x5A into access bank locations 0x10, 0x20, and 0x30.

17. Develop a short program that doubles the contents of access bank location 0x12.

18. Develop a short sequence of instructions that squares the contents of the WREG.

19. Which byte-oriented instruction adds with the carry status bit?

20. Where is the borrow found after a subtraction?

21. What does a SETF instruction accomplish?

22. Develop a short sequence of instructions that adds the 24-bit number in locations 0x10, 0x11, and 0x12 to the 24-bit number in locations 0x20, 0x21, and 0x22. The numbers are stored in the little endian format. The result is stored in 0x20, 0x21, and 0x22, and any carries between bytes must be included.

23. Explain how FSR0 is used to indirectly address a memory location in any instruction.

24. What is accomplished by the DCFSNZ instruction?

25. Develop a sequence of instructions that multiply the contents of the WREG register by ten. You may not use the multiply instruction.

26. What is accomplished by the operand POSTINC1?

27. How is the content of WREG 2's complemented?

28. How is the content of WREG 1's complemented?

29. What is accomplished by a BRA instruction?

30. Compare and contrast a BRA with a GOTO.

31. Is a BNZ instruction a relative or an absolute branch?
32. What is the range or distance that a BZ instruction can branch in a program?
33. Write a short program that places a 0x12 into memory from location 0x00 to location 0x22 in the access bank.
34. A checksum is the sum (with any carries dropped) of all the numbers in a block of memory. Write a program that generates an 8-bit checksum for the contents of memory bank 0 (0x00 through 0xFF) and stores it in WREG.
35. What are CALL and RETURN instructions?
36. How is the stack used by the CALL and RETURN functions?
37. Develop a function that develops a 16-bit checksum for all the locations in memory bank 0. The checksum should be stored in PRODL and PRODH in little endian format because we are not multiplying.
38. Explain how Example 3-17 converts from hex to ASCII.
39. What are the table instructions?
40. Write a function that accesses a table of prime numbers to test whether a number is prime or not. The 8-bit number under test is input to the function in WREG. The table must contain the prime numbers up to 251. If a number is a prime number, the RETURN should occur with the carry bit set and if it is not a prime number, the RETURN should occur with the carry bit cleared.
41. How is a macro sequence defined?
42. How much memory is required when the macro is invoked?
43. Develop a macro that shifts the 8-bit contents of any register to the right one place. (A shift right places a zero in the leftmost bit of the result).

CHAPTER 4

Assembly Language Programming

Now that many of the PIC instructions are becoming second nature, they can be used fairly effectively in short programs. Yet, there may be a need to learn more about assembly language programming. However, if you decide that programming in assembly language looks difficult, C language can be chosen for program development for the PIC. If the programming environment is primarily C for project development, then move ahead to Chapter 5, skipping this chapter. The material learned in Chapter 3 is sufficient to handle most inline assembly code needed inside of a C program.

This chapter presents more software and algorithms using assembly language for the assembly language programmer. Many of these algorithms are also covered in Chapter 5, when needed, with C language programming. The best of both worlds is to complete both Chapters 4 and 5. What is learned in this chapter is worthwhile and the techniques just reinforce what already is known about the assembler.

Upon completion of this chapter you will be able to:

1. Develop the software for a data stack.
2. Write a program to use a queue.
3. Develop software to divide and multiply numbers.
4. Manipulate binary-coded decimal (BCD) numbers using addition and subtraction.
5. Convert between binary and BCD.
6. Develop time delay software.

4-1 STACK AND QUEUE STRUCTURES

Queues and stacks are two data structures that appear in many programs to perform a variety of tasks. A **queue** stores data on a first-in, first-out basis often called a **FIFO** (*first-in, first-out*), whereas a **stack** stores data in a first-in, last-out basis often called a **LIFO**. A queue is a buffer often used between external I/O devices and the microcontroller. A stack is often used to store data temporarily and efficiently in the memory. For example, the program counter is stored in a stack as a return address from a function. Functions may be nested because of the stack. The stack discussed here enables data storage on a separate stack stored in the data memory.

Data Stack

Microchip has not implemented a data stack in the hardware. However, a data stack may have been contemplated because there are a PUSH and POP instructions. The purpose of the PUSH and POP instructions are not to store and retrieve data from a data stack; they are used to PUSH the PC register content + 2 onto the program stack, and the POP instructions discard a return address from the program stack.

A data stack is a good place to store data temporarily inside of functions or for other applications. The data stack is an integral part of the C language compiler. If the contents of all the registers are saved on the stack inside of a function, the function becomes transparent to the program using the function. This transparency provides local variables with local scoping in a C language function. Saving the registers used within a function prevents errors. To create a data stack, the software requires a pointer to a section of data memory and a method to store and retrieve information from the data stack. The FSR2 register, which can function as a stack pointer, is used in Example 4-1 to address an area of memory such as the data stack. To place data onto the stack, use the POSTDEC2 form of indirect addressing and to retrieve data from the stack, use the PREINC2 form of indirect addressing.

Sixteen bytes of memory reserved for the data stack should be a large enough stack for most programs. The data stack efficiently reuses the same area of data memory. As data are placed onto the data stack, the stack grows downward in memory and as data are removed from the data stack, the stack shrinks upward. Placing information on a stack is called pushing, and removing data from a stack is called popping. Example 4-1 illustrates stack pointer initialization and how to form instructions to place data onto the stack and remove data from the stack. The stack is defined in the UDATA_ACS or UDATA area as Stack RES 16. Either area is usable with the same amount of access time because of the use of FSR2. The order of the data in a data attack is that is reversed; a data stack is a FILO (first-in, last-out) storage array. In Example 4-1, a 2 is placed on the stack followed by a 3. When data are removed from the data stack, the 3 is retrieved first followed by the 2.

EXAMPLE 4-1

```
        UDATA_ACS

Stack   RES     .16

        LFSR    2, Stack + .15      ;initialize stack pointer
                                    ;(address the top of stack)
;
;code to place two values on the stack
;
        MOVLW   2                   ;place a 2 onto the stack
        MOVWF   POSTDEC2

        MOVLW   3                   ;place a 3 onto the stack
        MOVWF   POSTDEC2
;
;       **** do other things ******
;
;
;code to retrieve the two values from the stack
;
        MOVF    PREINC2, 0          ;retrieve the 3 from the stack

        MOVF    PREINC2, 0          ;retrieve the 2 from the stack
```

Now that the stack is proven as a valuable asset, it should be included in most programs for transitory data storage. Many examples in the remainder of this chapter, assume that FSR2 is used as a data stack pointer and the data stack is 16 bytes deep.

Suppose that Example 4-1 is rewritten so that it converts a byte in WREG into a two-digit ASCII number returned in ASCII_L and ASCII_H. Example 4-2 illustrates this reworked function (Hex2ASC). The only thing that is changed when Hex2ASCII is called is memory locations ASCII_L and ASCII_H. The stack is used to save the contents of WREG before the conversion and restored from the stack at the end of the conversion. The only registers that are changed from outside of the Hex2ASC function are register file locations ASCII_L and ASCII_H.

EXAMPLE 4-2

```
;
;data memory data
;
MyData UDATA_ACS

Stack         RES .16              ;define stack space
ASCII_L       RES 1                ;place for ASCII low
ASCII_H       RES 1                ;place for ASCII high

;************* FUNCTION Hex2ASCII ***********
;
; uses 1 program stack level
; uses 2 data stack levels
; uses WREG
;
; converts right nibble of WREG from hex
;      to ASCII returned on the data stack
;
;*********************************************

Hex2ASCII:
      ANDLW  0x0F               ;isolate low nibble
      ADDLW  0x30
      DAW
      BTFSC  WREG, 6
      INCF   WREG
      RETURN

;************ FUNCTION Hex2ASC **************
;
; uses 2 program stack levels
; uses 1 data stack location
; uses Hex2ASCII
; uses ASCII_L and ASCII_H
;
; converts byte in WREG from hex to ASCII
;      results are in ASCII_L and ASCII_H
;
;*********************************************

Hex2ASC:
      MOVFF  WREG, POSTDEC2      ;stack WREG (save WREG)

      CALL   Hex2ASCII          ;convert low nibble
      MOVFF  WREG, ASCII_L
      MOVFF  PREINC2, WREG      ;get WREG
      MOVFF  WREG, POSTDEC2     ;restack WREG

      SWAPF  WREG               ;swap nibbles
      CALL   Hex2ASCII          ;convert high nibble
      MOVFF  WREG, ASCII_H

      MOVFF  PREINC2, WREG      ;unstack WREG
      RETURN

;************** MAIN PROGRAM *****************
;
; Illustrates a function call to Hex2ASC
```

```
;       Converts 0x5c to ASCII in ASCII_L (0x43)
;       and in ASCII_H (0x35)
:

Main:
        ;
        ;if a stack is used in a program it should be
        ;initialized first and only once
        ;

        LFSR   2, Stack+.15        ;initialize data stack pointer
        MOVLW  0x5c                ;load test data

        CALL   Hex2ASC             ;convert entire byte to ASCII
                                   ;results on data stack

        ;********* Other software here **********

Stop: BRA Stop
```

Queue

A **queue** is often called a buffer or spooler memory. The queue maintains the order of the data placed into it. Queues are often placed between I/O devices and the program. The queue acts as a buffer so that data are not lost. Unlike the data stack, the queue requires two pointers to operate. One pointer addresses the location where the data are stored into the queue, and the other pointer addresses the location where the data are removed from the queue. Two conditions about the queue must be detected: the *full* **condition** and the **empty condition**. When a queue is empty, both pointers address the same memory location. When a queue is full, the exit pointer addresses one location less than or lower than the entry pointer. The queue is cyclic memory that reuses the same area of memory just as a data stack reuses the same area of memory. One location above the top of the queue is the first location in the queue. The pointers must wrap around from the top of the queue to the bottom when a pointer is incremented past the end of the queue.

EXAMPLE 4-3

```
                UDATA_ACS           ;fast memory

Queue           RES .16             ;the queue
QEntry          RES 1               ;entry pointer
QExit           RES 1               ;exit pointer

                UDATA

Stack           RES .16             ;the stack

;
;The following is placed in the CODE section of a program to
;    initialize the queue
;
                MOVLW           Queue           ;initialization
                MOVFF           WREG, QEntry
                MOVFF           WREG, QExit
```

Example 4-3 shows how to reserve 16 bytes of data memory for a queue and 2 bytes for the pointers to address the queue. It is assumed that the queue will be in a section of the memory that does not span a 256-byte boundary and is located below 0x07F in the data memory access bank, if possible. To do this, place the queue declarations at the beginning of the UDATA_ACS section of the program. The data access section of data memory is used for the queue because it is addressed by a single 8-bit address and does not need the bank selection register. If a stack is also used, place it at the start of the UDATA section. To make a queue operational, two functions are required: one to store data in the queue and the other to remove data from the queue. These functions are called PutQ and GetQ in Example 4-4. In order to make them as transparent as possible, the stack structure is also used with FSR2 functioning as a stack pointer. The PutQ function saves the WREG register into the queue, and the GetQ function retrieves data from the queue and returns it in WREG. In both functions, the carry status register bit indicates an error if C = 1 on the return and no error if C = 0 on the return. In Example 4-4, the PutQ function error is full and the GetQ function error is empty. To initialize the queue, make sure that QEntry and QExit are loaded with the address of the queue as shown.

EXAMPLE 4-4

```
;**************** PutQ *********************
;
; uses 1 program stack level
; uses 3 data stack levels addressed by FSR2
;
; Stores WREG into the queue
;
; C = 1 on return for a full queue
;
PutQ:
        MOVFF   FSR0L, POSTDEC2        ;stack FSR0 and WREG
        MOVFF   FSR0H, POSTDEC2
        MOVFF   WREG, POSTDEC2
        MOVFF   QEntry, FSR0L
        MOVLW   0
        MOVWF   FSR0H
        MOVF    POSTINC0, 0
        MOVLW   Queue+.16
        SUBWF   FSR0L, 0
        BNZ     PutQ1
        MOVLW   Queue
        MOVFF   WREG, FSR0L
PutQ1:
        MOVF    QEntry, 0             ;check if full
        SUBWF   FSR0L, 0
        BNZ     PutQ2
        SETF    STATUS, 0            ;if full
        MOVFF   PREINC2, WREG        ;cleanup stack
        BRA     PutQ3
PutQ2:
        MOVF    QEntry, 0
        MOVFF   FSR0L, QEntry
        MOVFF   WREG, FSR0L
        MOVFF   PREINC2, INDF0       ;save queue data
        CLRF    STATUS, 0
PutQ3:
        MOVFF   PREINC2, FSR0H       ;unload stack
        MOVFF   PREINC2, FSR0L
        RETURN

;**************** GetQ *********************
;
; uses 1 program stack level
; uses 3 data stack levels addressed by FSR2
;
; Retrieves WREG from the queue
;
; C = 1 on return for a empty queue
;
GetQ:
        MOVF    QEntry, 0            ;check if empty
        SUBWF   QExit, 0
        BNZ     GetQ1
        SETF    STATUS, 0           ;if empty
        RETURN
```

```
GetQ1:
        MOVFF   FSR0L, POSTDEC2         ;stack FSR0
        MOVFF   FSR0H, POSTDEC2

        MOVFF   QExit, FSR0L
        MOVLW   0
        MOVWF   FSR0H
        MOVFF   POSTINC0, POSTDEC2      ;stack queue data
        MOVFF   FSR0L, QExit
        MOVLW   Queue+.16               ;check boundary
        SUBWF   FSR0L, 0
        BNZ     GetQ2
        MOVLW   Queue
        MOVFF   WREG, QExit
GetQ2:
        MOVFF   PREINC2, WREG           ;unstack queue data and FSR0
        MOVFF   PREINC2, FSR0H
        MOVFF   PREINC2, FSR0L
        CLRF    STATUS, 0
        RETURN
```

A queue is most useful when interfacing an external device to the microcontroller. Data are entered and stored in the queue from the I/O device. When the microcontroller is ready for the data, the microcontroller retrieves the data from the queue. The queue provides a method of buffering an external device from the microcontroller. For example, a keypad might be connected to the microcontroller through its I/O ports. Some typists can enter data quickly enough that the microcontroller may miss something because it's off doing another task. The queue prevents any data from being missed, especially if the input data are interrupt-processed as explained in later chapters.

4-2   ## COMPLEX ARITHMETIC

This section does not cover complex arithmetic; however, the algorithms and methods explained are certainly complex for a microcontroller that cannot divide. Therefore, this section explains how to develop software for division and multiplication of numbers that are larger than 8 bits, BCD arithmetic, and various other arithmetic operations that are complex to the PIC microcontroller.

BCD Arithmetic

Embedded systems often use BCD numbers because of the time required to convert between binary and BCD. The data required for display devices is often in either BCD or ASCII format. The only difference between BCD and ASCII is a bias of 0x30. The difference between binary and BCD is much more complicated, requiring an algorithm that is more significant for conversion. For example, take the binary number 11 1110 1000 (0x3E8). Is there a simple way to convert 0x3E8 to 1000 decimal or 0001 0000 0000 0000 in packed BCD? There is a way, but it requires considerable computer time and a divide instruction, neither of which is often available to an embedded application written in assembly language, so numbers often remain in BCD format.

Whenever data are entered from a keypad or keyboard, the number is in BCD or ASCII format. These BCD or ASCII numbers can be converted to binary, but again it requires time and a multiply instruction. Because the PIC microcontroller provides only one of the two instructions (multiply and divide) needed for the conversion of numbers between binary and BCD, the numbers often remain in the BCD format.

BCD Addition and Subtraction.   To perform BCD addition, an instruction called **decimal adjust WREG** (DAW) is provided for the conversion. The DAW instruction corrects the result of a

BCD-packed (two digits per byte) addition. This instruction is required because the microcontroller adds only binary numbers. The DAW instruction performs the following tests after an addition of packed BCD numbers in WREG to correct the sum to a valid BCD number in WREG. (Make sure to check the errata for the microcontroller in use, because one version of the microcontroller has a problem with the DAW instruction. It seems that in one microcontroller, the carry must be cleared before the DAW instruction functions properly. In this section, DAW will function correctly even in the errant microcontroller.)

1. If the rightmost half-byte is greater than 9 or DC = 1, add 0x06
2. If the leftmost half-byte is greater than 9 or C = 1, add 0x60

The DAW instruction adds 0x00, 0x06, 0x60, or 0x66 to WREG to correct the result after a packed BCD addition. If the result is greater than 100 BCD, the carry bit (C) is set to indicate the 100-bit position in the result. Therefore, if 99 BCD and 05 BCD are added, the result is 04 BCD with C set to a one to indicate the 100 after the DAW instruction executes.

Suppose that the 4-digit packed BCD number in access file registers 0x10 and 0x11 are added to the 4-digit packed BCD number in access file registers 0x20 and 0x21. The program illustrated in Example 4-5 performs this task. Notice that the carry from the addition of byte 0x10 and 0x20 is added to the addition of 0x11 to 0x21 using the add with carry instruction. Enter the program into the IDE and execute it with a few different BCD numbers in the registers to learn how this program functions. Also notice that if the final carry is a logic one, a 0x01 is placed into location 0x22 for the 10,000 position of the result.

EXAMPLE 4-5

```
Main:
                MOVF    0x10, 0      ;add 0x10 to 0x20
                ADDWF   0x20, 0
                DAW                  ;make it BCD
                MOVWF   0x20

                MOVF    0x11, 0      ;add with carry 0x11 to 0x21
                ADDWFC  0x21, 0
                DAW                  ;make it BCD
                MOVWF   0x21

                CLRF    0x22
                BTFSC   STATUS, 0
                INCF    0x22         ;on the final carry

Stop:   BRA     Stop
```

Suppose that two 4-digit unpacked BCD numbers are added. The first number is stored at locations 0x10 through 0x13 and the second is at 0x20 through 0x23, with 0x24 containing the 10,000 position of the result. An unpacked BCD number is stored as one digit per byte. Because four additions are required to form the sum, indirect addressing is the best choice for developing this program, as shown in Example 4-6. This program can add any size unpacked BCD numbers by changing the count. The program adds two singled unpacked numbers at a time and if the result is 0x10 or greater after the DAW instruction, the program strips the 0x10 with an ANDLW 0x0F instruction that preserves the carry status bit. The ANDLW instruction changes only the zero and negative flag bits.

EXAMPLE 4-6

```
Main:
                LFSR    0, 0x1C      ;address numbers
                LFSR    1, 0x2C
                MOVLW   4            ;load counter
                MOVWF   0x30
                BCF     STATUS, 0    ;clear C
```

```
Loop:
            MOVF    POSTINC0, 0
            ADDWFC  INDF1, 0
            DAW

            BTFSC   WREG, 4
            BSF     STATUS, 0      ;if sum 10 through 18
            ANDLW   0x0F           ;strip 10

            MOVWF   POSTINC1

            DECFSZ  0x30           ;do loop
            BRA     Loop

            CLRF    INDF1
            BTFSC   STATUS, 0
            INCF    INDF1

Stop:   BRA     Stop
```

How are BCD numbers subtracted, since there is no DAW instruction for subtract? The DAW instruction functions correctly only for a BCD addition. BCD subtraction is accomplished by using complement addition. For example, if a 123 is subtracted from a 391, the correct difference is obtained by adding the ten's complement of 123 to the 391. The ten's complement of a 123 is an 877. An 877 plus a 391 is a 1268. The carry (shown as 1,000) is dropped and the result is 268, the correct difference. Figure 4-1 illustrates the ten's complement of 123.

It is a good idea to write a function that adds or subtracts any size packed BCD numbers. Is this possible? Yes, Example 4-7 illustrates such a function. This function has four calling parameters that are loaded before it functions correctly. The calling parameters are the addresses for number1 (FSR0), number2 (FSR1), and the result is stored on top of number 1; the width of the number is in bytes (PRODL), either a 0x00 (for an addition) or a 0x9A (for a subtraction) in PRODH. Note that for a subtraction, number1 is subtracted from number2.

FIGURE 4-1   Ten's complement.

$$
\begin{array}{r}
999 \\
-\ \underline{123} \\
876 \quad \longleftarrow 9\text{'s complement} \\
+\ \underline{1} \\
877 \quad \longleftarrow 10\text{'s complement}
\end{array}
$$

OR

$$
\begin{array}{r}
99A \\
-\ \underline{123} \\
877 \quad \longleftarrow 10\text{'s complement}
\end{array}
$$

EXAMPLE 4-7

```
;************** SubAdd Function **************
;
; uses 1 program stack level
; uses 2 data stack levels
; uses FSR0, FSR1, PRODL, and PRODH
;
; Adds/subtracts the packed BCD number
;       addressed by FSR0 to or from the number addressed by
;       FSR1. Result replaces number at FSR0.
;
; CALLING parameters:
;       FSR0 = address of number1 and result
;       FSR1 = address of number2
```

```
;        PRODL = number of bytes in each number
;        PRODH = 0x00 (addition) or 0x9A (subtraction)
;
SubAdd:
        MOVFF   WREG, POSTDEC2      ;stack WREG
        MOVFF   TABLAT, POSTDEC2    ;stack TABLAT

        BCF     STATUS, 0           ;clear carry
SubAdd1:
        MOVFF   STATUS, TABLAT
        MOVF    POSTINC0, 0

        TSTFSZ PRODH
        SUBWF   PRODH, 0

        MOVFF   TABLAT, STATUS
        ADDWFC POSTINC1, 0
        DAW
        MOVWF   POSTINC2
        MOVLW   0x99

        TSTFSZ PRODH
        MOVWF   PRODH

        DECFSZ PRODL
        BRA     SubAdd1

        MOVFF   PREINC2, TABLAT     ;unstack TABLAT
        MOVFF   PREINC2, WREG       ;unstack WREG

        RETURN
```

Multiplication

The instruction set includes a multiplication instruction, but it is limited to 8-bit multiplication. There are times when wider multiplication is required. Once the algorithm for multiplication is understood, software to perform multiplication on any size numbers can be written. Multiplication is a series of shifts and additions in the binary number system. The algorithm requires a product that is twice the width of the multiplier and multiplicand. For instance, a 16-bit multiplication requires a 32-bit product. The software algorithm for a 16-bit unsigned multiplication is:

1. Clear the 32-bit product.
2. Shift the multiplier right.
3. If the carry is set, add the multiplicand to the product.
4. If the multiplier is zero, end the function.
5. Shift the multiplicand left one place.
6. Repeat steps 2 through 5.

This multiplication algorithm functions only with unsigned numbers and is illustrated in the software in Example 4-8. In this example, the 16-bit multiplier is transferred to the function in the data memory location addressed by FSR0, the 16-bit multiplicand is at the location addressed by FSR1, and the location for storing the 32-bit product is in FSR2. This function is modifiable to multiply any numbers of any widths using the same algorithm.

EXAMPLE 4-8

```
;****************** MUL16 Function ******************
;
; uses 2 program stack levels
; uses WREG, FSR0, FSR1, FSR2, TABLAT, PRODL, and PRODH
; uses functions Add32 and Shift
;
```

```
; Multiplies contents of FSR1 location times contents of
;       FSR0 location, stores result at location FSR2
;       16-bit times 16-bits -> 32-bit result
;

MUL16:
        MOVLW   3
MUL16a:
        CLRF    PLUSW2          ;clear product
        DECF    WREG
        BNN     MUL16a
        MOVLW   3
        CLRF    PLUSW1          ;clear left 2 bytes of multiplicand
        MOVLW   2
        CLRF    PLUSW1

MUL16b:
        MOVLW   1
        RRCF    PLUSW0          ;shift multiplier right
        RRCF    INDF0
        BNC     MUL16c          ;if no carry
        CALL    Add32           ;add multiplicand to product

MUL16c:
        CALL    Shift           ;shift multiplicand left
        MOVF    PLUSW0, 0
        IORWF   INDF0, 0
        BNZ     MUL16b          ;if multiplier is not zero
        RETURN

;************** Add32 Function *******************

Add32:
        MOVFF   FSR2L, PRODH
        MOVFF   FSR1L, PRODL
        BCF     STATUS, 0
        MOVLW   4
        MOVWF   TABLAT

Add32a:
        MOVF    POSTINC1, 0
        ADDWFC  POSTINC2
        DECFSZ  TABLAT
        BRA     Add32a
        MOVFF   PRODH, FSR2L
        MOVFF   PRODL, FSR1L
        RETURN

;*************** Shift Function *************

Shift:
        MOVFF   FSR1L, PRODH
        BCF     STATUS, 0
        MOVLW   4

Shift1:
        RLCF    POSTINC1
        DECFSZ  WREG
        BRA     Shift1
        MOVFF   PRODH, FSR1L
        RETURN
```

This function is easily modified to multiply any size numbers, but if the task is 16-bit multiplication, it is accomplished by using the MULWF instruction four times to generate a 16-bit product in ans3-ans0 with the following algorithm:

1. Multiply low bytes and place result in ans0, ans1.
2. Multiply high bytes and place result in ans2, ans3.

3. Cross multiply low1 times high2; add product to ans1, ans2, ans3.
4. Cross multiply low2 times high1; add product to ans1, ans2, ans3.

This process is illustrated in Example 4-9 and is much shorter than the function illustrated in Example 4-8. This function uses the multiplier with FSR0, the multiplicand with FSR1, and the product with FSR2. Example 4-9 also executes faster and requires less memory than Example 4-8.

EXAMPLE 4-9

```
;****************** MUL16f Function *****************
;
; uses 1 stack level
; uses WREG, PRODL, PRODH, TABLAT, FSR0, FSR1, FSR2
;
; 16-bit unsigned multiplication. Contents of location
; addressed by FSR0 is multiplied times contents of location
; addressed by FSR1, result is at location addressed by FSR2
;
MUL16f:
        MOVF    INDF0, 0
        MULWF   INDF1
        MOVFF   PRODL, INDF2
        MOVLW   1
        MOVFF   PRODH, PLUSW2

        MOVFF   FSR1L, TABLAT
        MOVF    PLUSW0,0
        MULWF   PREINC1
        MOVLW   2
        MOVFF   PRODL, PLUSW2
        MOVLW   3
        MOVFF   PRODH, PLUSW2
        MOVFF   TABLAT, FSR1L

        MOVFF   FSR1L, TABLAT
        MOVF    INDF0, 0
        MULWF   PREINC1
        MOVFF   TABLAT, FSR1L
        MOVFF   FSR2L, TABLAT
        MOVF    PRODL, 0
        ADDWF   PREINC2
        MOVF    PRODH, 0
        ADDWFC  PREINC2
        CLRF    WREG
        ADDWFC  PREINC2
        MOVFF   TABLAT, FSR2L

        MOVF    PREINC0, 0
        MULWF   INDF1
        MOVF    PRODL, 0
        ADDWF   PREINC2
        MOVF    PRODH, 0
        ADDWFC  PREINC2
        CLRF    WREG
        ADDWFC  PREINC2
        RETURN
```

Division

Unfortunately, there is no instruction for division. If a division is required, an algorithm is available to accomplish it. Division is the opposite of multiplication, so instead of shifting right, shift left is used and instead of addition, subtraction is used to generate a quotient. It is customary to divide a double-wide dividend and generate a whole number remainder and a whole number quotient. For example, if a 22 is divided by a 5, the quotient is a 4 and the remainder is a 2. The following algorithm divides an unsigned 16-bit number by an unsigned 8-bit number and both the quotient and the remainder are unsigned 8-bit numbers.

1. Shift the dividend and quotient left one place.
2. Compare the most significant 8 bits of the dividend with the divisor and if it is greater than the divisor, subtract the divisor for the most significant 8 bits of the dividend and increment the quotient
3. Repeat step 2 eight times.
4. The remainder is the most significant 8 bits of the dividend.

To illustrate division, the function in Example 4-10 divides the contents of PRODL and PRODH by TABLAT with the quotient in PRODL and the remainder in PRODH upon a return.

EXAMPLE 4-10

```
;**************** Div function ******************
;
; uses 1 program stack level
; uses WREG, PRODL, PRODH, TABLAT, TBLPTRL, and TBLPTRH
;
; Quotient returned in PRODL and remainder is returned in PRODH
;       TABLAT is preloaded with divisor
;       PRODL and PRODH are preloaded with dividend
Div:
        MOVLW   8                       ;load counter with 8
        MOVFF   WREG, TBLPTRL

Div1:
        BCF     STATUS, 0               ;shift quotient left
        RLCF    TBLPTRH
        BCF     STATUS, 0               ;shift dividend left
        RLCF    PRODL
        RLCF    PRODH

        MOVF    TABLAT, 0               ;compare divisor with dividend
        SUBWF   PRODH, 0
        BNC     Div2                    ;if divisor greater than dividend

        MOVFF   WREG, PRODH
        INCF    TBLPTRH

Div2:
        DECFSZ  TBLPTRL                 ;decrement count
        BRA     Div1

        MOVFF   TBLPTRH, PRODL          ;quotient to PRODL
        RETURN
```

The function in Example 4-10 is modifiable for any size division, and can also perform signed division, if the numbers are first made positive before using the Div function. If signed division is attempted, the quotient and remainder are both negative if the numbers are divided with different signs, otherwise both are positive. An exclusive OR is used to compare the signs of the dividend and the divisor to determine if the result must be made negative or positive.

Example 4-11 shows how a 32-bit number is divided by a 16-bit number to produce a 16-bit quotient and a 16-bit remainder. Compare this with Example 4-10 to see how to change to a higher precision division. Instead of trying to pass the numbers to the functions in the registers, the numbers are passed using the file select registers to address them and also PRODL and PRODH. Here FSR0 points to the 32-bit dividend and PRODL and PRODH are the 16-bit divisors. Upon completing the division, the quotient is returned in the location addressed by FSR2 and the remainder is in the location addressed by FSR0, which is the most significant two bytes of the dividend. The counter has been changed to 16 from 8, because this is a 16-bit division.

EXAMPLE 4-11

```
;***************** Div16 function ******************
;
; uses 1 program stack level
; uses WREG, PRODL, PRODH, TABLAT, TBLPTRL, TBLPTRH,
;      FSR0, and FSR1
;
; Quotient returned in location addressed by FSR2
;      PRODL and PRODH are preloaded with divisor
;      FSR0 is preloaded with address of dividend
;      FSR0 address is remainder on return

Div16:
        MOVLW   .16                     ;load counter with 16
        MOVFF   WREG, TABLAT
Div16a:
        MOVLW   1                       ;shift quotient left
        BCF     STATUS, 0
        RLCF    INDF2
        RLCF    PLUSW2

        BCF     STATUS, 0               ;shift dividend left
        RLCF    POSTINC0
        RLCF    POSTINC0
        RLCF    POSTINC0
        RLCF    POSTDEC0

        MOVF    PRODL, 0                ;compare divisor and dividend
        SUBWF   POSTINC0, 0
        MOVWF   TBLPTRL
        MOVF    PRODH, 0
        SUBWFB  POSTDEC0, 0
        MOVWF   TBLPTRH
        BNC     Div16b

        MOVFF   TBLPTRL, POSTINC0       ;subtract
        MOVFF   TBLPTRH, POSTDEC0

        INCF    INDF2                   ;increment quotient
        BNC     Div16b
        MOVLW   1
        INCF    PLUSW2

Div16b:
        MOVLW   0
        IORWF   POSTDEC0
        IORWF   POSTDEC0
        DECFSZ  TABLAT                  ;decrement counter
        BRA     Div16a

        IORWF   POSTINC0
        IORWF   POSTINC0

        RETURN
```

4-3   CONVERTING BETWEEN DECIMAL AND BINARY

Numbers often remain in BCD, but there are occasions where BCD numbers must be converted to binary. If numbers are required in BCD form, but are in binary form, a conversion must also be performed. This section explains conversions between binary and BCD. ASCII is also needed at times, but the conversion between ASCII and BCD is the simple task of adding or subtracting 0x30 as explained in earlier chapters.

Binary to BCD

This conversion is fairly easy to accomplish if division is available. The algorithm (called **Horner's Rule**) for binary-to-BCD conversion uses a division by 10 and is as follows:

1. Divide by 10.
2. Save the remainder as a digit in the answer.
3. Repeat steps 1 and 2 until the quotient is zero.

Suppose that a $0110\ 0111_2$ must be converted to a series of BCD digits. If $0110\ 0111_2$ is divided by 1010_2, a quotient of 1010_2 and a remainder of 0011_2 is obtained. The 0011_2 is the least significant digit of the result. Because the quotient of 1010_2 is not zero, divide by 1010_2 again. This time we obtain a quotient of 0001_2 and a remainder of 0000_2. Again, 0001_2 is divided by 1010_2 and a quotient of 0000_2 with a remainder of 0001_2 is obtained. The numbers 0011_2, 0000_2, and 0001_2 are the order of the remainders obtained in this example. The result is 103, but the order in which the number is returned is reversed.

Example 4-12 shows a function that uses the 8-bit division from Example 4-10 to convert the number in PRODL to a BCD number stored in memory at the location addressed by FSR0. This number is stored with its least significant digit first, and is terminated in the memory with the number 0x10. A 103 is stored as a 3, 0, 1, and 0x10.

EXAMPLE 4-12

```
;************* BinBCD *************
; uses 2 program stack levels
; uses Div as its registers as well as FSR0
;
; Converts binary from PRODL to the BCD string in memory at FSR0 terminated
;       with a 0x10
;

BinBCD:

        MOVLW   .10
        MOVFF   WREG, TABLAT

BinBCD1:
        CALL    Div                     ;divide by 10
        MOVFF   PRODH, POSTINC0         ;save remainder
        CLRF    PRODH
        MOVF    PRODL, 0
        BNZ     BinBCD1                 ;while quotient is not zero

        MOVLW   0x10                    ;save end mark
        MOVWF   POSTINC0
        RETURN
```

Suppose the string of BCD digits stored in the memory is converted to an ASCII NULL string, which makes it compatible with C language. This is accomplished by the function listed in Example 4-13.

EXAMPLE 4-13

```
;************* ASCIInull *************
;
; uses 1 program stack level
; uses FSR0, FSR1, WREG, TABLAT
;
; Reverses the order of a string of characters
; and converts them to an ASCII null string
;
```

```
ASCIInull:
        MOVFF   FSR0L, FSR1L
        MOVFF   FSR0H, FSR1H
ASCIInulla:
        MOVLW   0x30                    ;to ASCII
        ADDWF   INDF0
        MOVLW   0x40
        CPFSEQ  POSTINC0
        BRA     ASCIInulla              ;if not end

        DECF    FSR0L
        MOVLW   0
        SUBWFB  FSR0H
        MOVWF   POSTDEC0

ASCIInullb:
        MOVFF   INDF0, TABLAT           ;swap data
        MOVF    INDF1, 0
        MOVWF   POSTDEC0
        MOVFF   TABLAT, POSTINC1
        MOVF    FSR0H, 0
        CPFSEQ  FSR1H, 0
        BRA     ASCIInullb              ;if not end

        MOVF    FSR0L, 0
        CPFSEQ  FSR1L, 0
        BRA     ASCIInullb              ;if not end

        RETURN
```

The binary-to-decimal conversion algorithm also works with other number bases if the divisor is changed to the radix value. For example, to convert from a binary number to an octal string of digits, divide by 8. This algorithm is used to convert to any number base by changing the divisor to the radix number.

Recall that to convert from binary to BCD (Examples 4-9 and 4-10) required that the remainders are saved in the memory, but the order of the remainders is reversed. One way to change the order of the remainders is by placing them onto the stack before storing them in the ASCII string. Example 4-14 shows how this is done using a stack to replace the two functions in Examples 4-12 and 4-13. Look at how much shorter the task becomes when using a data stack to reverse the data.

EXAMPLE 4-14

```
;************* BinBCDs **************
; uses 2 program stack levels
; uses Div as its registers as well as FSR0 and WREG
; uses TABLAT and PRODH
;
; Converts binary from PRODL to the BCD string in memory at FSR0 terminated
;       with a 0x00
;
BinBCDs:

        MOVLW   .10                     ;get divisor of 10
        MOVFF   WREG, TABLAT
        MOVLW   0                       ;save end mark on stack
        MOVWF   POSTDEC2
BinBCDs1:
        CALL    Div                     ;divide by 10
        MOVLW   0x30
        ADDWF   PRODH                   ;convert to ASCII
        MOVFF   PRODH, POSTDEC2         ;save remainder on stack
        CLRF    PRODH
        MOVF    PRODL, 0
        BNZ     BinBCDs1                ;while quotient is not zero
```

```
BinBCDs2:
        MOVF    PREINC2,0       ;retrieve stack data
        MOVWF   POSTINC0
        BNZ     BinBCDs2
        RETURN
```

Example 4-14 uses PRODH and TABLAT for the conversion. In order that they are not used elsewhere, it is worthwhile to modify the function so these registers are saved on a data stack before the function uses them and retrieved from the data stack before the function returns, as illustrated in Example 4-15.

EXAMPLE 4-15

```
;************* BinBCDs **************
; uses 2 program stack levels
; uses 6 program levels (max) of a data stack addressed by FSR2
; uses Div as its registers as well as FSR0 and WREG
;
; Stores the BCD string in memory at FSR0 terminated
;    with a 0x00 (C-style null string)
;
BinBCDs:
        MOVFF   TABLAT, POSTDEC2    ;stack TABLAT and PRODH
        MOVFF   PRODH, POSTDEC2

        MOVLW   .10
        MOVFF   WREG, TABLAT
        MOVLW   0                   ;save end mark on stack
        MOVWF   POSTDEC2
BinBCDs1:
        CALL    Div                 ;divide by 10
        MOVLW   0x30
        ADDWF   PRODH               ;convert to ASCII
        MOVFF   PRODH, POSTDEC2     ;save remainder on stack
        CLRF    PRODH
        MOVF    PRODL, 0
        BNZ     BinBCDs1            ;while quotient is not zero
BinBCDs2:
        MOVF    PREINC2,0
        MOVWF   POSTINC0            ;retrieve stack data
        BNZ     BinBCDs2

        MOVFF   PREINC2, PRODH      ;unstack PRODH and TABLAT
        MOVFF   PREINC2, TABLAT
        RETURN
```

Notice how the TABLAT and PRODH registers are placed onto the stack before the function begins and retrieved from the stack just before the return instruction. This function now uses only FSR0, WREG, and the data stack. Normally the working register is not saved. How much data stack space is needed? Two bytes are needed for TABLAT and PRODH and up to three more bytes for the remainders that are placed on the stack. The end mark requires one byte of data stack space for a total of up to six bytes of data stack space for this example.

BCD to Binary

Converting from a BCD number to binary requires multiplication instead of division. The algorithm for the conversion requires that the binary number is started at zero and as each digit is encountered, the binary number is multiplied by 10 and the new digit is added to the binary result. The algorithm is:

1. Start with a result of zero.
2. Multiply the result by 10.

3. Add the new BCD digit to the result.

4. Repeat steps 2 and 3 until the final digit is reached.

To illustrate this algorithm, the function in Example 4-16 converts an ASCII null string from the memory into a binary number returned as an 8-bit unsigned number in the TABLAT register. The ASCII null string is addressed by the file selection register FSR0 on entrance to the function.

EXAMPLE 4-16

```
;************** ASCIIBin ***************
;
; uses 1 stack level
; uses WREG, TABLAT, FSR0
;
; Converts the ASCII null string address by FSR0
;       into an 8-bit binary number in TABLAT
;

ASCIIBin:
        CLRF    TABLAT          ;clear result
ASCIIBin1:
        MOVLW   0
        CPFSEQ  INDF0
        BRA     ASCIIBin2
        RETURN                  ;return on null
ASCIIBin2:
        MOVLW   .10             ;x10
        MULWF   TABLAT
        MOVFF   PRODL, TABLAT
        MOVLW   0x30            ;ASCII to BCD
        SUBWF   POSTINC0, 0
        ADDWF   TABLAT          ;add digit
        BRA     ASCIIBin1
```

As with the binary-to-decimal conversion algorithm, this algorithm also works for any number base by changing the multiplication number of 10 to the radix number. For example, if a string of octal digits exists, multiply by 8. This function uses 8-bit multiplication, but if 16-bit multiplication is used, a larger binary result can be obtained.

4-4   TIME DELAYS

Many applications for a microcontroller require time delays. This section explains how to create precise time delays using software to generate them. For example, suppose a solenoid must be activated for 100 ms to properly move some mechanical device. To accomplish this delay of 100 ms requires some method of developing a time delay.

Time Delay Software

To start with, examine the simple time delay function shown in Example 4-17. The contents of WREG are decremented and if the result is zero, a return occurs. If the result is not zero, WREG is again decremented. This is simple and also quite short, and has a predictable execution time determined by how many times the WREG is decremented.

EXAMPLE 4-17

```
Delay:
        DECFSZ WREG             ;decrement WREG
        BRA    Delay            ;branch if WREG is not zero

        RETURN
```

Assuming that WREG is loaded before the function is called, the amount of time required to execute this procedure is the number of instruction cycles required for the CALL and RETURN plus the number of instruction cycles required for WREG to decrement to zero. The decrement of WREG is accomplished by the DECFSZ followed by a BRA instruction. Refer to Appendix A and locate the number of instruction cycles required for each instruction. The CALL and RETURN instructions each require two instruction cycles for a total of four instruction cycles. The DECFSZ instruction requires one instruction cycle and the BRA instruction requires two instruction cycles. Example 4-18 illustrates an equation for the number of instruction cycles required for this function.

EXAMPLE 4-18

```
Clocks = 4 + WREG x 3
or
```

$$WREG = \frac{Clocks - 4}{3}.$$

From this equation, if WREG is a 1, the amount of time required to execute the Delay function is 7 cycles. If the maximum number of 0 (a 0 causes 256 iterations) is used for WREG, then the number of instruction cycles is $4 + (256 \times 3)$, or 772 cycles. An instruction cycle is four external clock pulses in the PIC18 family. If the external clock frequency is 4MHz (250 ns), then the instruction cycle time is 1MHz (1.0 μs). This means that the Delay function of Example 4-17 causes between 7.0 μs and 772 μs of time delay, dependent on the number chosen for WREG. This is not nearly enough time, for the time delay mentioned at the beginning of this section of 100 ms.

Example 4-19 shows one method for accomplishing a longer time delay. Here a count of 165 is chosen for WREG. At a 4-MHz external clock, this gives us a delay time of $165 \times 3 + 4$ or 499 μs (round that to 500 μs for this example). If exactly 500 μs is needed, a NOP (no operation) instruction (a NOP requires 1 instruction cycle to perform no operation) is added to the Delay function just before the RETURN, as shown in Example 4-20. For the functions shown in Examples 4-19 and 4-20, the outer count is 200. If we execute the 500 μs delay 200 times, a time delay of exactly 100 ms is obtained. Of course, the accuracy of this delay depends on the accuracy of the clocking source. This time is not exact because we did not count the time required to decrement the outer counter in TABLAT, but for a solenoid this time is well within a millisecond of 100 ms, so it is accurate enough for controlling a solenoid.

EXAMPLE 4-19

```
Delaym:
        MOVLW   .200            ;200 times
        MOVWF   TABLAT
Delaym1:
        CALL    Delay
        DECFSZ  TABLAT
        BRA             Delaym1

        RETURN
Delay:                          ;165 times
        MOVLW   .165
Delay1:
        DECFSZ  WREG
        BRA     Delay1

        RETURN
```

EXAMPLE 4-20

```
Delaym:
        MOVLW   .200            ;200 times
        MOVWF   TABLAT
```

```
Delaym1:
        CALL    Delay
        DECFSZ TABLAT
        BRA             Delaym1

        RETURN
Delay:                          ;165 times
        MOVLW  .165
Delay1:
        DECFSZ WREG
        BRA    Delay1

        NOP                     ;1 more instruction cycle
        RETURN
```

Even longer time delays are obtained by using the 100-ms delay as a basis. For example, if a 5-second delay is required, it is obtained by CALLing the Delay function 50 times. (See Example 4-21.) Later in the text other methods of time delays are explained using the internal timers. Still another method is explained using a real-time clock (RTC) interrupt when interrupts are explained and used for various applications. The main problem with a software timer is that the microcontroller can do nothing else while the delay is running. In some cases this is acceptable, but in many cases it is not. If a system uses an interrupt, for example, a software time delay can be interrupted by an interrupt, causing the time delay to be incorrect.

EXAMPLE 4-21

```
D5sec:
        MOVLW  .50              ;5-second delay uses Delaym
        MOVWF  TBLPTRL
D5sec1:
        CALL    Delaym
        DECFSZ TBLPTRL
        BRA    D5sec1

        RETURN
```

4-5   PROGRAMMING EXAMPLES

This section contains several programming examples that illustrate various programming techniques. As with all examples in this textbook, they should be entered into the IDE and executed at full speed and single-stepped so the programs are thoroughly understood.

Program Example One

This first programming example illustrates a use of I/O for the first time. This example uses bit positions 0 (RA0) and 1 (RA1) of I/O port A on the PIC microcontroller to read data from a PC-style keyboard. The PC-style keyboard is selected because it is easy to connect to the microcontroller because it is TTL compatible using zero and five-volt logic. The 6-pin DIN connector on a PC-style keyboard is illustrated in Figure 4-2.

The connector uses only four of the six connections for power and two signal lines. The signal lines are data and clock. The keyboard is supplied power from the +5.0 V and ground inputs and requires a maximum of 250 mA of current, which must be supplied from the microcontroller power supply. Figure 4-3 illustrates the data and clock signals that commonly appear on the interface.

One frame of keyboard data consists of 11 clock pulses that are sampled at the positive edge to obtain the data from the keyboard. Clock pulses are transmitted at a nominal rate of 10 KHz and

FIGURE 4-2 6-pin DIN
connector of a PS-style
keyboard.

Plug on keyboard
cable (male)

Pins

1 = Data
2 = No connection
3 = Ground
4 = +5 V
5 = Clock
6 = No connection

must be between 10 KHz and 10.7 KHz for the interface to function properly. A bit time (time between clocks) is therefore nominally 100 μs. To intercept keyboard data, the clock pulse is examined and each time the clock pulse transitions from a logic zero to a logic one that data are sampled by the microcontroller. This means a program or function is written to watch the clock, and each time the clock changes to a one, the data are input to the microcontroller and assembled into a character from the keyboard.

A framing error occurs when the start bit it not a zero and the stop bit is not a one in their respective places. Framing errors most commonly occur when the data are received at an incorrect data rate. Thus, the microcontroller program could contain software to hunt for the correct framing rate. The parity bit is added to the data to cause an odd number of ones to be sent. In the sample data waveform of Figure 4-3, a parity bit of one is sent to form an odd parity. The data has two one-bits and the parity bit has one parity bit, for a total of three bits. Because three is an odd

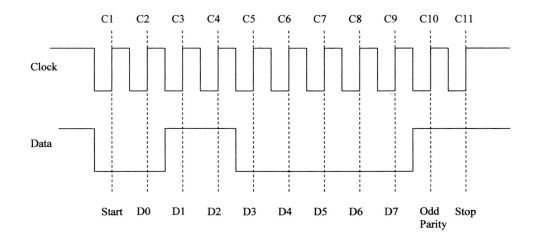

Start bit is always zero

Parity is always odd

Stop is always one

FIGURE 4-3 Keyboard clock and data signals.

number, the message has odd parity. If the wrong parity is detected (even parity), then the message is invalid. Parity errors occur if noise is a problem on a connection. In some cases, a question mark is inserted in a message if a parity error occurs, and in other cases flow control is used to request the retransmission of the data.

Because bits 0 and 1 of Port A are used in this example, the instruction used to read Port A must be known before software is written to read a key from the keyboard. Port A is read by reading the Port A register file at location 0xF80. In a program, PORTA refers to 0xF0. Port A is programmed by programming the Port A TRISA register at location 0xF92. A one in a bit position of TRISA selects the input mode for a corresponding pin and a zero selects the output mode. In this example, bits 0 and 1 are used to input data so TRISA is programmed with a 0x03. Example 4-22 shows the software required to program Port A bits 0 and 1 so they function as input pins. The port pins must also be programmed as digital pins in some versions of the PIC by sending a 0x7F to the ADCON1 register.

EXAMPLE 4-22

```
MOVLW   0x7F            ;program the ports as digital
MOVWF   ADCON1

MOVLW   3
MOVWF   TRISA           ;bits 0 and 1 are inputs
```

A function that receives a single character from the keyboard is illustrated in Example 4-23. This software assumes that RA0 (bit 0 of Port A) is connected to the data pin from the keyboard and RA1 (bit 1 of Port A) is connected to the clock signal. The function obtains a character from the keyboard and returns it in WREG. If an error is detected (framing or parity), carry is set on a return and if a valid character exists in WREG, carry is cleared. To make the procedure more efficient and easier to read, a function called Wait4Clock is included that waits for the clock to go to zero, and then waits for the clock to become a one before returning. The Wait4Clock function appears first in the listing. The bit test and skip instruction are used to wait for the clock input to change states in the function.

EXAMPLE 4-23

```
;
;************** Wait4Clock ****************
;
;  Waits for the clock to go low and then high
;      before returning
;
Wait4Clock:
        BTFSC   PORTA, 1        ;test RA1
        BRA     Wait4Clock      ;if clock = 1
Wait4Clock1:
        BTFSS   PORTA, 1        ;test RA1
        BRA     Wait4Clock1     ;if clock = 0

        RETURN

;************** GetData ***************
;
; uses 2 program stack levels
;
; uses PRODL, PRODH, and TABLAT
;
; Reads character from the keyboard and returns it in WREG
;
;       error = carry set
```

```
GetData:
        CALL    Wait4Clock      ;wait for initial clock

        BTFSS   PORTA, 0        ;test RA0
        BRA     GetData1        ;if data is a start bit, GOTO GetData1

        BSF     STATUS, 0       ;if not a valid start, set CARRY

        RETURN                  ;and return
;
; program continues at this point if a valid start bit is detected
;       PRODL is used as a data bit counter
;       PRODH is used to count the ones in the data stream for parity
;
GetData1:
        MOVLW   8
        MOVWF   PRODL           ;set count of 8
        CLRF    PRODH           ;clear parity count

GetData2:
        CALL    Wait4Clock      ;wait for a data bit clock

        BTFSC   PORTA, 0        ;test RA0
        INCF    PRODH           ;count a one-parity bit
                                ;and clear CARRY

        BTFSC   PORTA, 0        ;test RA0
        SETF    STATUS, 0       ;set carry if data = 1

        RRCF    TABLAT          ;rotate carry (data bit) into TABLAT

        DECF    PRODL           ;decrement count
        BNZ     GetData2        ;repeat 8 times for the 8 data bits

        CALL    Wait4Clock      wait for parity bit clock

        MOVF    PORTA, W        ;get data bit to WREG

        XORWF   PRODH           ;exclusive or with parity count
        BTFSS   PRODH, 0
        BRA     GetData3        ;if parity good

        BSF     STATUS,0        ;if parity error, set CARRY

        RETURN

GetData3:                       ;if no parity error
        CALL    Wait4Clock

        BTFSC   PORTA, 0        ;test RA0 for stop bit
        BRA     GetData4        ;if good, stop bit

        BSF     STATUS, 0       ;if framing error, set CARRY

        RETURN

GetData4:                       ;if no errors
        MOVF    TABLAT, W       ;get data byte to WREG
        BCF     STATUS, 0       ;clear CARRy

        RETURN
```

The function illustrated in Example 4-23 retrieves a byte from the keyboard, but it does not obtain an ASCII key code until the way that a keyboard operates is understood and handled. Whenever a key is typed, the keyboard sends the **scan code** for the key. The scan code is not ASCII code; it is a number that represents a key on the keyboard. Table 4-1 lists the scan codes for a keyboard.

Notice that some scan codes are two or more characters in length. All scan codes that are longer than a single character begin with the 0xE0 code, which is detected by the software to receive the additional bytes for these multiple byte codes.

When a key is pressed on the keyboard, the keyboard interface sends the scan code and when a key is released, the keyboard interface sends the break code followed by the scan code for

TABLE 4-1   Keyboard scan codes.

Key	Code	Key	Code	Key	Code
0	0x45	Z	0x1A	.	0x49
1	0x16	F1	0x05	/	0x4A
2	0x1E	F2	0x06	Enter	0x5A
3	0x26	F3	0x04	Escape	0x76
4	0x25	F4	0x0C	Print Screen	0xE0, 0x12, 0xE0, 0x7C
5	0x2E	F5	0x03	Scroll Lock	0x7E
6	0x36	F6	0x0B	Pause	0xE1, 0x14, 0x77
7	0x3D	F7	0x83	Insert	0xE0, 0x70
8	0x3E	F8	0x0A	Home	0xE0, 0x6C
9	0x46	F9	0x01	Page Up	0xE0, 0x7D
A	0x1C	F10	0x09	Delete	0xE0, 0x71
B	0x32	F11	0x78	End	0xE0, 0x69
C	0x21	F12	0x07	Page Down	0xE0, 0x7A
D	0x23	Backspace	0x66	Up Arrow	0xE0, 0x75
E	0x24	Space	0x29	Left Arrow	0xE0, 0x6B
F	0x2B	Tab	0x0D	Down Arrow	0xE0, 0x72
G	0x34	Caps Lock	0x58	Right Arrow	0xE0, 0x74
H	0x33	Left Shift	0x12	Num Lock	0x77
I	0x43	Left Control	0x14	Keypad /	0xE0, 0x4A
J	0x3B	Left Alt Option	0xE0, 0x1F	Keypad *	0x7C
K	0x42	Left Alternate	0x11	Keypad -	0x7B
L	0x4B	Right Shift	0x59	Keypad +	0x79
M	0x3A	Right Control	0xE0, 0x13	Keypad Enter	0xE0, 0x5A
N	0x31	Right Alt Option	0xE0, 0x27	Keypad .	0x71
O	0x44	Right Alternate	0xE0, 0x11	Keypad 0	0x70
P	0x4D	Applications	0xE0, 0x2F	Keypad 1	0x69
Q	0x15	'	0x0E	Keypad 2	0x72
R	0x2D	-	0x4E	Keypad 3	0x7A
S	0x1B	=	0x55	Keypad 4	0x6B
T	0x2C	\	0x5D	Keypad 5	0x73
U	0x3C	[	0x54	Keypad 6	0x74
V	0x2A	]	0x5B	Keypad 7	0x6C
W	0x1D	;	0x4C	Keypad 8	0x75
X	0x22	`	0x52	Keypad 9	0x7D
Y	0x35	,	0x41		

the key. For example, when the A key is pressed, a 0x1C is sent. When the A key is released, a 0xF0 followed by a 0x1C is sent. This allows multiple keys to be pressed and tracked by the software. A left shift followed by an A key generates the following codes when the keys are pressed with the shift key first: 0x12, 0x1C. If the keys are released with the A released first followed by the shift, the following code is sent: 0xF0, 0x1C, 0xF0, 0x12. When a key is released that generates more than a byte, such as the Insert key, the 0xF0 code follows the 0xE0 code as in 0xE0, 0xF0, 0x70. The only exception to these rules is the Pause key, which generates a code beginning with 0xE1 and a break of 0xE1, 0xF0, 0x14. The print screen key break sequence is the other exception, which is 0xE0, 0xF0, 0x7C, 0xE0, 0xF0, 0x12.

Software can be written to complete the keyboard interface, but in most cases this entire process is handled with an interrupt. The interrupt is needed because of the speed of the keyboard interface. If the software listed here is used to read a keyboard, not much else can be accomplished by the microcontroller. The remainder of the keyboard software is not explained until later in the text after interrupts are understood. Here we learn how to detect a clock pulse and receive data through a serial connection using a few I/O pins.

Program Example Two

The first programming example illustrates how to test for a clock signal, a table to look up ASCII code (even though no software is written), and how to operate an I/O port. Some ham (amateur) radio operators use the Morse code to send information in order to conserve bandwidth. A signal using CW (continuous wave) to send Morse code uses a bandwidth of only 150 Hz. This Morse code was originally used by telegraph operators and developed for communicating between railroad stations in the nineteenth century and continued to be used well into the twentieth century.

Table 4-2 shows the Morse code for letters and numbers. The table contains the character, the Morse code pattern of dits (.) and dahs (_), and a hexadecimal number that contains a count of the dits and dahs (right three bits) and the pattern (left 5 bits). A dit is a logic zero and a dah is a logic one.

Suppose a system is needed that sends Morse code. Because not many people know the Morse code, we decide to produce a simple microcontroller-based system that generates Morse code. Because hardware has not yet been covered in this book, our system takes the ASCII character from WREG and generates Morse code from it. Once we know how to read a keyboard, such as a standard PC keyboard, characters from the keyboard are transferred to the software for the conversion to Morse code. This allows us to send Morse code by simply typing letters and numbers on the keyboard.

In addition to the Morse code table, which is stored in the program memory as a lookup table, some other facts must be known about Morse code. A dah bit time is three times longer than a dit bit time. The spacing between letters in a word is three dit times. The spacing between words is seven dit times. To send data at 50 words per minute (WPM), the dit time is 24 milliseconds. This (50 WPM) is called the "Paris" standard because it's derived from sending the word "PARIS" 50 times in one minute. The word "PARIS" has exactly 50 dit times with spacing between words.

The software requires time delays of 1 dit (24 ms) and 3 dits (72 ms). Because we do not yet know how to energize relay (keyer), a function called KeyOn is used to activate the relay and a function called KeyOff is used to deactivate a relay. KeyOn sets bit RB0 and KeyOff clears bit RB0. The space for ASCII code does not appear in Table 4-2, but it is 7 dit times between words. Example 4-24 illustrates the lookup table in the program memory for storing the Morse code. The table has two sections: Morse_L contains the letters and Morse_N contains the numbers.

TABLE 4-2 Morse code.

Character	Code	Byte	Character	Code	Byte	Character	Code	Byte
A	._	0x42	M	_ _	0xC2	Y	_.__	0xB4
B	_...	0x84	N	_.	0x80	Z	__..	0xC4
C	_._.	0xA4	O	_ _ _	0xE3	0	_____	0xFD
D	_..	0x83	P	.__.	0x64	1	.____	0x7D
E	.	0x01	Q	__._	0xD4	2	..___	0x3D
F	.._.	0x24	R	._.	0x43	3	...__	0x1D
G	__.	0xC3	S	...	0x03	4	...._	0x0D
H	....	0x04	T	_	0x81	5	.....	0x05
I	..	0x02	U	.._	0x23	6	_....	0x85
J	.___	0x74	V	..._	0x14	7	__...	0xC5
K	_._	0xA3	W	.__	0x63	8	___..	0xE5
L	._..	0x44	X	_.._	0x94	9	____.	0xF5

EXAMPLE 4-24

```
MORSE               CODE_PACK

Morse_L:            ;letters A through Z

        db                  0x42, 0x84, 0xa4, 0x83, 0x01
        db                  0x24, 0xc3, 0x04, 0x02, 0x74
        db                  0xa3, 0x44, 0xc2, 0x80, 0xe3
        db                  0x64, 0xd4, 0x43, 0x03, 0x81
        db                  0x23, 0x14, 0x63, 0x94, 0xb4
        db                  0xc4

Morse_N:            ;numbers 0 through 9

        db                  0xfd, 0x7d, 0x3d, 0x1d, 0x0d
        db                  0x05, 0x85, 0xc5, 0xe5, 0xf5
```

The procedure that takes a single ASCII character from WREG and sends the Morse code is listed in Example 4-25. Although I/O is yet to be explained, the procedure uses Port B, one of the I/O ports, and bit position zero to activate the key. If an oscilloscope is connected to this bit, the Morse code appears as ones and zeros that are used to turn a relay on and off. The relay and its interface are discussed in later chapters. The software is written for a 4-MHz clock; if the clock is a different frequency, the time delay counts must be adjusted as explained in the prior section of this chapter.

EXAMPLE 4-25

```
;************** SendChar Function **************
;
; uses program 3 stack levels
; uses LookUp, KeyOn, KeyOff, Delay24, Delay72, and Delay
; uses WREG, PRODL, PRODH, TABLAT, TBLPTR
;
; Sends the character from WREG to
; the keyer.
;
SendChar:
        MOVWF   PRODL           ;save WREG
        MOVLW   0x61
        SUBWF   PRODL, 0
        BN      SendChar1       ;if it is uppercase
        MOVLW   0x20
        SUBWF   PRODL           ;make it uppercase

SendChar1:
        MOVLW   0x20
        SUBWF   PRODL, 0
        BNZ     SendChar2       ;if not a space
        CALL    Delay72         ;4 more dits
        CALL    Delay24
        RETURN

SendChar2:
        MOVLW   0x30
        SUBWF   PRODL, 0
        BC      SendChar3       ;if "0" or greater
        RETURN

SendChar3:
        MOVLW   0x3a
        SUBWF   PRODL, 0
        BC      SendChar4       ;if not "0" — "9"
        CALL    LookUp
        BRA     SendCharSend

SendChar4:
        MOVLW   0x41
        SUBWF   PRODL, 0
```

```
        BC       SendChar5
        RETURN                   ;if less than "A"

SendChar5:
        MOVLW    0x5b
        SUBWF    PRODL, 0
        BNC      SendChar6
        RETURN                   ;if greater than "Z"

SendChar6:
        CALL     LookUp1

SendCharSend:
        MOVWF    PRODL           ;code to PRODH
        MOVWF    PRODH           ;count to PRODL
        MOVLW    7
        ANDWF    PRODL

SendCharSend1:
        CALL     KeyOn           ;keyer on
        RLCF     PRODH           ;get dit or dah
        BNC      SendCharSend2   ;if dit
        CALL     Delay72         ;waste 3 dit times
        BRA      SendCharSend3

SendCharSend2:
        CALL     Delay24         ;waste 1 dit time

SendCharSend3:
        CALL     KeyOff          ;keyer off
        CALL     Delay24         ;waste 1 dit time
        DECF     PRODL
        BNZ      SendCharSend1   ;if not done
        CALL     Delay24         ;waste 2 more dit times
        CALL     Delay24
        RETURN
;
; Lookup code from Morse code table
;
; Code returned in WREG
;
LookUp:
        MOVLW    0x30
        SUBWF    PRODL           ;convert from ASCII
        MOVLW    UPPER(Morse_N)
        MOVWF    TBLPTRU
        MOVLW    HIGH(Morse_N)
        MOVWF    TBLPTRH
        MOVLW    LOW(Morse_N)
        MOVWF    TBLPTRL
        BRA      LookUp2

LookUp1:
        MOVLW    0x41
        SUBWF    PRODL
        MOVLW    UPPER(Morse_L)
        MOVWF    TBLPTRU
        MOVLW    HIGH(Morse_L)
        MOVWF    TBLPTRH
        MOVLW    LOW(Morse_L)
        MOVWF    TBLPTRL

LookUp2:
        MOVF     PRODL,0
        ADDWF    TBLPTRL
        MOVLW    0
        ADDWFC   TBLPTRH
        ADDWFC   TBLPTRU
        TBLRD*                   ;lookup code
```

```
                MOVF    TABLAT,0
                RETURN
;
; One-dit time delay
;
Delay24:
                MOVLW   .48             ;48 times
                MOVWF   TABLAT

Delay24a:
                CALL    Delay
                DECFSZ TABLAT
                BRA     Delay24a
                RETURN
;
; Three-dit time delay
;
Delay72:
                CALL    Delay24
                CALL    Delay24
                CALL    Delay24
                RETURN
;
; 500-microsecond time delay
;
Delay:                          ;165 times
                MOVLW   .165            ;500 microseconds
Delay1:
                DECFSZ WREG
                BRA     Delay1
                RETURN
;
; Bit 0 of Port B is set
;
KeyOn:
                BSF     PORTB, 0        ;set bit 0 of Port B
                RETURN
;
; Bot 0 of Port B is cleared
;
KeyOff:
                BCF     PORTB, 0        ;clear bit 0 of Port B
                RETURN
```

The SendChar procedure is fairly simple. The bulk of it filters the ASCII character so that only numbers, letters, and spaces are detected; any other ASCII character is ignored. Once a valid ASCII character is detected, the Morse code pattern and count are looked up from the two tables in the program memory using the TBLRD* instruction. Notice how the address is loaded into the TBLPTR register in the LookUp function. The assembler directives UPPER, HIGH, and LOW are used to access different portions of the address for storage into TBLPTR. Finally, the Morse coded pattern is separated into a count and a bit pattern for PortB, bit 0. The bit pattern is then shifted a bit at a time to determine how long bit zero remains a logic one. The dit is 24 ms and the dah is 72 ms.

4-6   SUMMARY

1. A data stack is a place in the data memory where information is stored on a first-in, last-out basis. A file select register functions as a stack pointer using the POSTDEC form of addressing to store data on the stack and the PREINC form of addressing to retrieve data from the stack.

2. A queue is a data buffer that stores data on a first-in, first-out basis. A queue requires two pointers for operation, one addresses the entry point and the other addresses the exit point. Error conditions detected by queue software are full and empty.

3. Binary-coded decimal (BCD) addition and subtraction use the DAW instruction to correct the result after a BCD addition. For a BCD subtraction, ten's complement addition is used to form the difference and use the DAW instruction.

4. The DAW instruction corrects the result of a BCD addition by examining the carry and digit carry status register bits as well as the values in both digits of the result. The result determines whether a 0x00, 0x06, 0x60, or 0x66 is added for the correction.

5. Binary multiplication uses a shift and add algorithm to generate a product. Binary multiplication is also performed by the MULLW or MULWF instruction. The algorithm is explained so that numbers of any size can be multiplied.

6. Binary division requires a shift and subtract algorithm to generate the quotient and remainder. The quotient and remainder are always unsigned integers when using this algorithm.

7. Binary-to-decimal conversion requires a division by ten. The algorithm used for the conversion is called the Horner's Rule. This algorithm also functions for converting a binary number to any number base by changing the divide by number.

8. Decimal-to-binary conversion is accomplished by multiplication by ten. This conversion also converts from any number base to binary by changing the multiplication number to the number base.

9. Time delays are often produced by software. This is accomplished because the amount of time required to execute an instruction is known. Therefore, accurate time delays are constructed with programmed loops.

4-7 QUESTIONS AND PROBLEMS

1. What is a data stack?
2. What is a program stack?
3. Compare the data stack with the program stack.
4. If the FSR1 register is used as a stack pointer, select an instruction or instructions that:
   a. Places the contents of WREG onto the stack.
   b. Removes data from the stack and places it into WREG.
   c. Places the product of a multiplication onto the stack.
5. Which instruction is used to initialize FSR2 as a stack pointer?
6. If WREG is placed on the stack followed by PRODL, which of these registers is the first to be removed from the stack?
7. What is a queue?
8. Describe the error conditions that must be detected by queue software.
9. If a queue is 16 bytes in size, how many bytes of data can be stored in the queue?
10. What is meant when a queue is said to be cyclic memory?
11. When BCD numbers are added, what instruction must be used to correct the result?
12. Describe the difference between packed and unpacked BCD numbers.
13. Determine the number added by a DAW instruction for the following packed BCD additions:
    a. 03 + 08
    b. 10 + 28
    c. 92 + 99
    d. 92 + 90
14. How are BCD numbers subtracted?
15. What is the ten's complement and how is it generated?

16. Why would BCD numbers ever be used in a program?
17. Multiplication can be performed using a series of _____ shifts and additions.
18. The multiply instruction functions only with 8-bit numbers. How can it be used for 16-bit multiplication?
19. How long does a multiply instruction require for execution?
20. Develop a function that multiplies the content of WREG by the content of TABLAT. The function must return the result in WREG as an 8-bit product, ignoring the most significant part of the product.
21. Develop a function that squares the number in WREG, returning the result in WREG as an 8-bit result.
22. Describe the changes required to Example 4-8 so that it multiplies two signed 16-bit numbers and generates a 32-bit signed product.
23. Division is accomplished by a series of _____ shifts and subtractions.
24. If a number is divided by 4, what would be the best method?
25. Could the algorithm for division given in this chapter be modified for 24-bit division? If so, what change is made?
26. Is it possible to write software that divides a 256-bit number by a 128-bit number?
27. What is Horner's Rule?
28. Modify Example 4-15 so it converts to octal instead of decimal.
29. Suppose that Example 4-15 must convert a number to hexadecimal. Describe the changes that must be made to accomplish this and produce both numbers and letters in the resulting ASCII character string.
30. Describe how a number is converted from decimal to binary using software.
31. Is it possible to convert a number from octal to binary? If so how?
32. How long does it take to execute the function listed in Example 4-15 if the microcontroller clock is 8 MHz?
33. What is an instruction cycle and how many system clocks are required to execute an instruction cycle?
34. Given the following function, how long does it take to execute if the system clock is 40 MHz? (Include the time to CALL it and RETURN from it).

```
Wait:
        MOVLW  .10
Wait1:
        DECFSZ WREG
        BRA    Wait1
        RETURN
```

35. Develop a series of functions for a 100-ms time delay called D100 and 1-second time delay called D1 if the microcontroller clock is 20 MHz.
36. Use the function in Example 4-23 to read a key from the keyboard. Your new function must place a 0x00 into data memory location 0x010 whenever the Right Shift key is pressed down and a 0xFF into data memory location 0x010 whenever the Right Shift key is released. All other keys and keystrokes must be ignored.
37. Repeat question 36 for both shift keys.
38. Locate a Morse code chart on the Internet and obtain the characters used for punctuation. Add the punctuation codes to the Morse code table in the memory of Example 4-24 in a third section called Morse_P.
39. What are the purposes of the LOW, HIGH, and UPPER directives in the program listed in Example 4-25?

CHAPTER 5

Programming the PIC18 with C Language

Now that you are familiar with the PIC assembler, you need to know how to program the PIC18 family using C language. Many programming environments use C language instead of assembly language because the program development cycle is much shorter and the software is much easier to write and maintain. This chapter uses the C language compiler that is available from Microchip as a free download. This chapter does not teach C language, because it is assumed that you know some programming. The C18 compiler is a C language compiler for the PIC18 family of microcontrollers. The free version is a limited version of the full compiler and is sufficient for learning how to program a PIC and develop applications. The differences between the free version and the full version are that optimization is disabled in the free version, which means that the code generated may be a little longer, and the extended instructions available to the PIC18 family are disabled.

This chapter presents more software and algorithms using C language as a platform. Some assembly language appears in some of the sample programs. As mentioned in the introduction to Chapter 4, the best of both worlds is to complete both Chapters 4 and 5.

Upon completion of this chapter you will be able to:

1. Use the C18 C language compiler in the PIC integrated development system for programming the PIC18 family of microcontrollers.
2. Learn some of the functions provided in the companion libraries for the C18 compiler.
3. Use and generate random numbers.
4. Store tables and other information in program memory.
5. Develop C programs using time delays and the time delay library.
6. Use states from a state transition diagram to implement a program.
7. Develop software using the math library.

5-1　　　**C18 C LANGUAGE COMPILER**

The program that Microchip provides for C language program development in the IDE is called the C18 compiler. The C18 compiler is available as a free download from the Microchip website (http://www.microship.com). Although this free version is limited to 60 days, it has enough functionality to accomplish any programming task, even though the code generated may be slightly longer than generated by the full version. The limited version does not include the extended instructions; however, it does not seem to present any problems except the resulting programs are slightly larger.

When developing software for the PIC18 family, unsigned char variables must be the first choice in any program, because the microprocessor architecture is 8 bits. If larger numbers are used, the program requires more time to execute, so use larger numbers sparingly. The greater the deviation from the native hardware, the longer the program will require for execution and the greater its length. A 16-bit integer addition requires more time and program storage space than an 8-bit integer addition; a 32-bit floating-point addition requires a lot more time and program storage space than an 8-bit addition. Choose the variables in a program wisely; if a number is unsigned, make sure that it is declared as unsigned. The PIC is not a PC, so the amount of program memory is extremely small in comparison. Because the memory is limited, use care when developing the software and choosing the variables.

Mixed language programming using the assembler and C together is allowed and often used in a system. Assembly language is placed in blocks starting with the _asm directive and ending with the _endasm directive. The assembly code blocks are not surrounded with { } as are C-programming blocks. Comments placed in an assembly block must use C-style comments and not the semicolon comments common in assembly language programs. Data defines in C language are accessible to software written using the inline assembler. Another difference between assembly language and inline assembly language within C language is that nothing is assumed. If the data memory accessed is in the access bank, then use the ACCESS keyword, and so forth.

C Language and the IDE

Before much else can be learned about C18, its use in the IDE must be understood. The IDE is also used for C program development just as it is for assembly language program development, so a new interface does not need to be learned. There are few differences between setting up the IDE for C language development and for assembly language development.

To use the C18 compiler in the IDE, install the compiler following the onscreen prompts. Make sure that all items are selected when installing the compiler. The steps required to set up the IDE are similar to the steps to set up the assembler.

1. Under the "Configure" label on the menu bar, select "Select Device" from the drop-down menu and choose the microcontroller for the project.
2. Under the "Project" label on the menu bar, select "Project Wizard" from the drop-down menu and again select the microcontroller for the project.
3. In step 2 of the project wizard, select the "Microchip C18 Toolsuite" and click Next. The paths are all correct if the C18 compiler is installed properly.
4. Enter a name for the project and a directory, and then click Next.
5. In step 4 you might want to start with Example1 from the mcc18/example/GettingStarted directory. Add the Example1.c file.
6. Under the "Project" label on the menu bar, select "Build Options," then browse and select "h" in the mcc18 directory as an include path. Click on "Apply" and then "Ok."
7. Right-click on "Linker files" in the list of project files on the screen and select "add file" and then add the 18fxxxx.lkr file. (Note that the xxxx is the number of the microcontroller used for the project).
8. Under "Debugger," select the tool "MPLAB SIM." This adds the simulator tool to the project. If an in-circuit emulator (ICE) is available, add the correct emulator for development.

The steps listed are no more difficult than setting up a project with the assembler. Example 5-1 shows the Example1.c program listing. This program listing is much shorter than the template program for use with the assembler because C language is much more powerful and less verbose than assembly language.

EXAMPLE 5-1

```
/*
 * This is Example 1 from "Getting Started with MPLAB C18."
 */

#include <p18cxxx.h>    /* for TRISB and PORTB declarations */

/* Set configuration bits for use with ICD2 / PICDEM2 PLUS Demo Board:
 *   - set HS oscillator
 *   - disable watchdog timer
 *   - disable low-voltage programming
 */

#pragma config OSC = HS
#pragma config WDT = OFF
#pragma config LVP = OFF

int counter;

void main (void)
{
    counter = 1;
    TRISB = 0;              /* configure PORTB for output */
    while (counter <= 15)
    {
        PORTB = counter;    /* display value of "counter" on the LEDs */
        counter++;
    }
}
```

This program is written for use with the PICDEM2 plus board. It is in standard C form using a main function to contain the program. This program uses a single variable called counter. The program configures Port B as an output port and then displays the numbers 0 through 15 on the LEDs connected to Port B. If a PICDEM2 plus board is available, this program can be executed on it. If the board is not available, the program can run in the simulator. The simulator displays the numbers as they change on Port B. To simulate this program, go to the View menu and display the special function registers and find Port B. Single-step through the program and watch Port B change as the program is single-stepped. This example uses the PIC18F1220 so to get it to function correctly, one instruction is added to the program in Example 5-1 that selects digital signals for all the I/O pins. Add a ADCON1 = 0x7F; statement to the program listing before counter = 1. Build and execute the program and then look at the simulator—you should see the screen displayed in Figure 5-1. To obtain the same image that appears in this figure, use the zoom tool (the square block) to zoom in and obtain the same screen as shown. What is wrong with the display? Nothing, although it seems as if there is a mistake because the count of 15 is displayed for more than one clock period, but if you look at the software, the program displays the counts 0000 through 1111 then restarts, which takes additional time. To correct the software example from Microchip, modify Example 5-1 so it looks like Example 5-2. Then build it, rerun it, and look at the output in the logic analyzer. The output now progresses from 0000 to 1111 and repeats without any gaps in the count.

EXAMPLE 5-2

```
/*
 * This is a modified Example 5-1
 *    written for the PIC18F1220
 */

#include <p18cxxx.h>    /* for TRISB and PORTB declarations */

/* Set configuration bits for use with ICD2 / PICDEM2 PLUS Demo Board:
 *   - set HS oscillator
 *   - disable watchdog timer
```

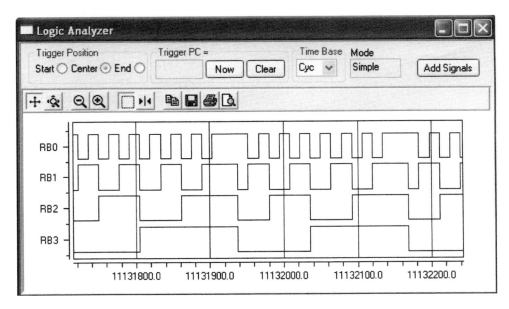

FIGURE 5-1   Screen shot of the simulator logic analyzer.

```
*   - disable low-voltage programming
*/

#pragma config OSC = HS
#pragma config WDT = OFF
#pragma config LVP = OFF

int counter;

void main (void)
{
     ADCON1 = 0x7F;    // make ports digital

     TRISB = 0;          /* configure PORTB for output */
     while  (1)
     {
          counter = 0;
          while (counter <= 15)
          {
               PORTB = counter;    // counter sent to LEDs
               counter++;
          }
     }
}
```

The #include statement includes a header file that defines all the I/O registers of the PIC18 family members selected for the project. This allows I/O operations to be performed for the microcontroller. This header must be included in all projects.

The #pragma config statements program the configuration bits of the microcontroller. The term **pragma** probably comes from pragmatic, which means a practical way of doing something. The pragma commands relieve the programmer of performing a long series of instructions or steps to accomplish some task. Hence it represents a practical or pragmatic way of accomplishing the task. The #pragma directive is used to give instructions to the compiler program. The configuration bits control the primary operation of the PIC18 microcontroller. These bits are explained in detail in Chapter 6 along with the hardware for the microcontroller. Each version of the PIC18 microcontroller has different configuration bits that are listed in the document "MPLAB C18 Config Settings," available at the Microchip website.

A Sample Program

Suppose that a program is needed that reads a switch on Port A and increments the number displayed on Port B one time per second only when the pushbutton is held down. This program is similar to the one in Example 5-1, but to accomplish the time delay a function from the standard C library is used. The C library provides time delays as listed in Table 5-1. These functions are important because just about any program used to control a peripheral requires a time delay. These functions make generating an accurate time delay a much simpler task than discussed in Chapter 4. Each of these functions provides a delay in instruction cycles from 1 to just about any number for any time delay. To uses these functions, the #include <delays.h> is used at the start of a program to include them. The standard library descriptions are available at the Microchip website in the document "MPLAB C18 Libraries." As with the time delays discussed in Chapter 4, the delays.h functions will be accurate only in a system that does not interrupt them.

Example 5-3 lists the program that increments the LEDs connected to Port B when the pushbutton is held down for at least one second. If it is held down for slightly more than 2 seconds, the LEDs increment twice, and so forth. Again this program uses the PIC18F1220. This microcontroller is chosen because of its cost (less than $3) and chip size. In this example, RA4 is connected to a pushbutton. The pushbutton signal is a logic one when not pressed and a logic zero when it is pressed down.

EXAMPLE 5-3

```
/*
 * This is Example 5-3 for a 4-MHz clock
 *    written for the PIC18F1220
 */

#include <p18cxxx.h>          //include port specifications
#include <delays.h>           //include time delays

/* Set configuration bits
 * - set HS oscillator
 * - disable watchdog timer
 * - disable low-voltage programming
 */

#pragma config OSC = HS
#pragma config WDT = OFF
#pragma config LVP = OFF
```

TABLE 5-1 Time delay functions from delays.h.

Function	Example	Note
Delay1TCY	Delay1TCY();	Inserts a single NOP instruction into the program
Delay10TCYx	Delay10TCYx(10);	Inserts 100 instruction cycles (number must be between 0 and 255) (0 causes a delay of 2560)
Delay100TCYx	Delay100TCYx(10);	Inserts 1000 instruction cycles (number must be between 0 and 255) (0 causes a delay of 25,600)
Delay1KTCYx	Delay1KTCYx(3);	Inserts 3000 instruction cycles (number must be between 0 and 255) (0 causes a delay of 256,000)
Delay10KTCYx	Delay10KTCYx(20);	Inserts 20,000 instruction cycles (number must be between 0 and 255) (0 causes a delay of 2,560,000)

```
// ******************** DATA MEMORY VARIABLES ************************

int counter;

// ********************* CONSTANTS *******************************
#define MSEC Delay1KTCYx(1)      // MSEC = 1 millisecond

// ******************** MAIN PROGRAM ******************************
void main (void)
{
      ADCON1 = 0x7F;           // configure PORTS A and B as digital
                               //    this might need to be changed depending
                               //    on the microcontroller version

      TRISB = 0;               // configure PORTB for output
      TRISA = 0xFF;            // configure PORTA for input
      PORTB = 0;               // LEDs off

      while (1)                // program infinite loop
      {
            counter = 0;                    // initialize counter

            while ((PORTA & 16) == 0)    // while pushbutton is down
            {
                  MSEC;                     // wait 1 msec
                  counter++;
                  if (counter == 1000)   // 1 second
                  {
                          PORTB++;
                          counter = 0;
                  }
            }
      }
}
```

In this program, a single variable is used to count in increments of 1 millisecond. Example 5-3 assumes that the microcontroller clock is 4 MHz $\left(\frac{1}{4\text{MHz}} = 250\,\text{ns}\right)$, which means that an instruction cycle time is 1 μs because 4 clocks are required for an instruction cycle. The Delay1KTCYx(1) function causes a delay of 1000 instruction cycles or 1 ms (1 μs × 1000). This time delay is defined as MSEC to make the program easier to read. Define statements should be used in this manner to clarify and make the program less cryptic. The pushbutton is tested by a while statement. Bit position 4 of Port A connects to the pushbutton so a 16 (0x10) is ANDed with Port A to find the state of the pushbutton. If the pushbutton is not pressed, bit position 4 is a logic one, which after ANDing with a 16, produces a 16. If the pushbutton is pressed, a zero is produced by PORTA ANDed with a 16. As long as the pushbutton is pressed, the while statement is true, which executes the instructions in the body of the while. Here, after one second, the contents of Port B increment and reset the counter to zero. This restarts the one-second countdown.

Of particular interest in this program is the way that a bit in Port A (pushbutton) is tested. As illustrated, the AND operation is used to isolate a bit in Port A. The number used with the AND operation singles out a bit, 1 selects bit zero, 2 selects bit one, 4 selects bit two, 8 selects bit three, and so forth (see Figure 5-2). Of secondary interest is the counter used to cause a one-second time delay in increments of a millisecond.

Why does this program contain an infinite loop in the form of the while(1) statement? The framework provided by the C language compiler repeatedly executes main(). The problem here is that main(), which initializes the program, needs to be executed only once. After initializing the program, the while(1) statement continuously executes the instructions in the loop, hence the infinite while loop. For an embedded system, or any system for that matter, the main program or operating system is usually an infinite loop. Figure 5-3 illustrates the structure of all operating systems, embedded or otherwise. The **cold start** is the hardware reset. The program

FIGURE 5-2 Extracting bits
of a number.

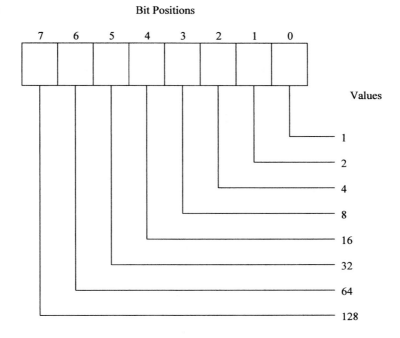

in Example 5-2 follows this model directly. In the PIC C18 compiler, a cold start is preformed by the main() function.

Another way to test a bit on a port is to use the PORTAbits C language directive. The PORTAbits.RA4 statement is used to test Port A bit position 4. The linker file for the microcontroller defines all of the pins of Port A so that the PORTAbits.RA4 instruction tests Port A bit position RA4. Example 5-4 illustrates Example 5-3 reworked using this bit test instruction. This method of testing a bit is not needed, but is a convenient way to test a bit. The assembly code produced for both methods of testing a bit is identical and there is no advantage of using one technique over the other. If a bit in a TRIS register needs to be changed, the TRISBbits.TRISB3 = 1

FIGURE 5-3 Structure of an
operating system.

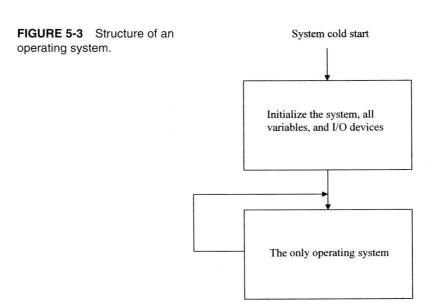

instruction is used as an example to set TRISB bit position 3 to a 1. This is sometimes needed to modify only a single bit in a TRIS register.

EXAMPLE 5-4

```
/*
 * This is Example 5-4 for a 4-MHz clock
 *   written for the PIC18F1220
 */

#include <p18cxxx.h>          // include port specifications
#include <delays.h>           // include time delays

/* Set configuration bits
 *   - set HS oscillator
 *   - disable watchdog timer
 *   - disable low-voltage programming
 */

#pragma config OSC = HS
#pragma config WDT = OFF
#pragma config LVP = OFF

int counter;

#define mSec Delay1KTCYx(1)              // mSec = 1 millisecond

void main (void)
{
      ADCON1 = 0x7F;      // configure PORTS A and B as digital
                          //    this might need to be changed depending
                          //    on the microcontroller version

      TRISB = 0;          // configure PORTB for output
      TRISA = 0xFF;       // configure PORTA for input
      PORTB = 0;          // LEDs off

      while (1)           // program infinite loop
      {
          counter = 0;                    // initialize counter

          while (PORTAbits.RA4 == 0)      // while pushbutton is down
          {
              mSec;                       // wait 1 msec
              counter++;
              if (counter == 1000)        // 1 second
              {
                  PORTB++;
                  counter = 0;
              }
          }
      }
}
```

5-2   USING C LANGUAGE INCLUDE FILES

This section provides more examples of C language programs using some of the other include files provided in the C18 library. Section 5-1 introduced the delays.h include file that contains time delays, because time delays are used in most programs written for a microcontroller. This section introduces some other useful include files used for programming the microcontroller.

Conversions

As mentioned in Chapter 4 with assembly language, data are seldom in the proper form. The conversion include files (in the stdlib.h) are used to convert data from one numeric form to

TABLE 5-2 Common conversion functions of stdlib.h.

Function	Example	Note
atob	atob(buffer)	Converts the number from string form in buffer; returned as a byte signed number (+127 to −128)
atof	atof(buffer)	Converts the number from string form in buffer; returned as a floating point number
atoi	atoi(buffer)	Converts the number from string form in buffer; returned as a 16-bit signed integer (+ 32,767 to −32, 768)
atol	atol(buffer)	Converts the number from string format in buffer; returned as a 32-bit signed integer (+ 2, 147, 483, 647 to −2, 417, 483, 648)
btoa	btoa(num, buffer)	Converts the signed byte into a string stored at buffer
itoa	itoa(num, buffer)	Converts the signed 16-bit integer to a string stored at buffer
itol	itol(num, buffer)	Converts the 32-bit signed integer to a string stored at buffer
rand	rand()	Returns a 16-bit random number (0 to 32, 767)
srand	srand(seed)	Sets the seed values to 16-bit integer seed
tolower	tolower(letter)	Converts byte-sized character letter from uppercase; returned as lowercase
toupper	toupper(letter)	Converts byte-sized character letter from lowercase, returns as uppercase
ultoa	ultoa(num, buffer)	Same as itol, except num is unsigned

another. Table 5-2 lists many of the commonly used conversion functions available in the library included with the C18 compiler in the standard library header file stdlib.h. The #include <stdlib.h> must be in a program to use the stdlib functions.

Suppose that a random number must be displayed on the Port B LEDs every half second. A program that accomplishes this using the rand() function appears in Example 5-5. Here a seed value of one is used to generate random numbers.

EXAMPLE 5-5

```
/*
 * This is Example 5-5 for a 4-MHz clock
 */

#include <p18cxxx.h>
#include <delays.h>
#include <stdlib.h>

/* Set configuration bits
 *  - set HS oscillator
 *  - disable watchdog timer
 *  - disable low-voltage programming
 */

#pragma config OSC = HS
#pragma config WDT = OFF
#pragma config LVP = OFF

void main (void)
{
        ADCON1 = 0x7F;        // configure PORTS A and B as digital
                              //    this might need to be changed depending
                              //    on the microcontroller version

        TRISB = 0;            // configure PORTB for output
        PORTB = 0;            // LEDs off
        srand(1);             // seed the random number
```

```
        while (1)              // repeat forever
        {
                Delay10KTCYx(50);    // wait 1/2 second
                PORTB = rand();      // display a random number
        }
}
```

Example 5-6 shows another method of displaying a random number, but generates a random number every time that a pushbutton is pressed. It actually increments a counter at full speed and generates a random number between 0 and 9. One use might be to select numbers for a daily lottery. This example assumes that Port A bit position 4 is a pushbutton switch. This program does not use any include files except the microcontroller include file p18cxxxx.h. The length of the program generated by this example is therefore much shorter, about half the size. The fewer include files and functions used in a program, the better. If there is a way to avod using a library file, do so.

EXAMPLE 5-6

```
/*
 * This is Example 5-6
 */

#include <p18cxxx.h>

/* Set configuration bits
 *   - set HS oscillator
 *   - disable watchdog timer
 *   - disable low-voltage programming
 */

#pragma config OSC = HS
#pragma config WDT = OFF
#pragma config LVP = OFF

int seed;

void main (void)
{
        ADCON1 = 0x7F;       // configure PORTS A and B as digital
                             //    this might need to be changed depending
                             //    on the microcontroller version

        TRISB = 0;           // configure PORTB for output
        TRISA = 0xFF;        // configure PORTA for input
        PORTB = 0;           // LEDs off
        seed = 1;            // self-generated random number

        while (1)            // repeat forever
        {
                while ( PORTAbits.RA4 == 0 )      // while pushbutton is down
                {
                        seed++;
                        if (seed == 10)    // if seed hits 10
                            seed = 1;
                        PORTB = seed;
                }
        }
}
```

Memory and String Functions

The next program uses a number in a character string to illustrate another function from the standard library. This program does not use any I/O and is strictly an exercise to illustrate the string functions. A character string called string1 contains the message "The time is 22 hours.". The program extracts the number 22 after finding it in the string and stores it in memory location hour. In Example 5-7 the character string is searched for the first occurrence of a number digit. The number in the string is then extracted from it using the atob function, which converts a string number into a byte or character size data.

EXAMPLE 5-7

```
/*
 * This is Example 5-7
 */

#include <p18cxxx.h>
#include <string.h>
#include <stdlib.h>

/* Set configuration bits
 *  - set HS oscillator
 *  - disable watchdog timer
 *  - disable low-voltage programming
 */

#pragma config OSC = HS
#pragma config WDT = OFF
#pragma config LVP = OFF

char buffer[] = "The time is 22 hours.";
char hour;
int a;

void main (void)
{
    for (a = 0; a < strlen(buffer); a++)
        if (buffer[a] >= '0' && buffer[a] <= '9')
            break;
    hour = atob (buffer + a);
}
```

This example uses the strlen function to determine the length of the string and a for loop with an if and a break to scan through the string in search of a numeric character. When the break occurs, variable a contains the relative position of the numeric character in the array. The atob function then converts the string number into an integer and stores it in hour. The string functions are listed in Table 5-3 along with an example of each. The string functions are located in the string.h file that must be included in a program if string functions are used in the program. There are many string functions that provide solutions to many string applications. Some of these are from standard C language (ANSI) and some are unique to the PIC microcontroller. The PIC has unique functions because of the program memory, which is not normally a feature in most computer systems.

To illustrate the string instructions, suppose that a number of character strings are stored in the program memory. Program memory is persistent so it is used to store static program data. To store information in the program memory, the directives rom and near or far are used in a program. The **rom** directive informs the compiler to place data into the program memory, and the **near** or **far** directives determine the size of the address. A near address is a 16-bit address and a far address is a 21-bit address. If the microcontroller has less than 64K of program memory, use near; if it has more that 64K of program memory, use far.

Example 5-8 shows how to store several character strings in an array in the program memory. The program also relegates 20 bytes of memory in the data memory so a string from the program memory can be copied into the data memory for manipulation by a program. Here the strcpypgm2ram (string copy program memory to RAM) is used to copy the second string (element 1) into the buffer space in the RAM. If the program memory and data memory are viewed in the IDE after executing this short program, the buffer area will contain the second string.

TABLE 5-3   String function located in string.h.

Function	Example	Note
memchr	memchr (area51, 'a' , 23)	Search the first 23 bytes of area51 for an 'a'
memchrpgm	memchrpgm (area1, 65, 5)	Search the first 5 bytes of area1 for a 65 (if found, a pointer is returned to the character; if not, a null is returned)
memcmp	memcmp (area1, area2, 4)	Compare area1 with area2 for 4 bytes
memcmppgm	memcmppgm (area3, area4, 2)	Compare program memory area3 with program memory area4 for 2 bytes
memcmppgm2ram	memcmppgm2ram (a1, a2, 5)	Compare a1 with program memory a2 for 5 bytes
memcmpram2pgm	memcmpram2pgm (a3, a4, 6)	Compare program memory a3 with a4 for 6 bytes (returns <0 is first less than second returns ==0 if strings are equal returns >0 if first string is greater than second string)
memcpy	memcpy (a1, a2, 4)	Copies from a2 to a1 for 4 bytes
memcpypgm	memcpypgm (a3, a4, 5)	Copies program memory a4 to program memory a3 for 5 bytes
memcpypgm2ram	memcpypgm2ram (a5, a6, 7)	Copies program memory a6 to a5 for 7 bytes
memcpyram2pgm	memcpyram2pgm (a7, a8, 2)	Copies a8 to program memory a7 for 2 bytes
memmove	memmove (a1, a2 , 3)	Same as memcpy except overlapping regions are allowed
memmovepgm	memmovepgm (a3, a4, 3)	
memmovepgm2ram	memmovepgm2ram (d, e, 3)	
memmoveram2pgm	memmoveram2pgm (f, g, 45)	
strcat	strcat (str1, str2)	Append str1 with str2
strcatpgm	strcatpgm (str3, str4)	Append str3 in the program memory with str4
strcatpgm2ram	strcatpgmram (str5, str6)	Append str5 with program memory string str6
strcatram2pgm	strcatpgmram (str3, str4)	Same as strcatpgm
strchr	strchr (str1, 'a')	Find the first letter a in str1
strchrpgm	strchrpgm (str2, '0')	Find the first zero in str2
strcmp	strcmp (str1, str2)	Compares str1 to str2
strcmppgm	strcmppgm (str3, str4)	Compares str3 in program memory to program memory str4
strcmppgm2ram	strcmppgmram (str5, str6)	Compares str5 to program memory str6
strcmpram2pgm	strcmprampgm (str3, str4)	Compares program memory str3 to str4 (returns >0 if first string is less than second string returns == 0 if strings are equal returns <0 if first string is greater then second string)
strcpy	strcpy (str1, str2)	Copies str2 into str1
strcpypgm	strcpypgm (str3, str4)	Copies str4 in the program memory to str3
strpypgm2ram	strcpypgm2ram (str5, str6)	Copies str6 to program memory location str5
strcpyram2pgm	strcpyram2pgm (str3, str4)	Same as strcpypgm
strcspn	strcspn (str1, str2)	Searches str1 for str2; match count returned
strcspnpgm	strcspnpgm (str3, str4)	Searches program memory str3 for program memory str4
strcspnpgmram	strcspnpgmram (str5, str6)	Searches program memory str5 for str6
strcspnrampgm	strcspnrampgm (str3, str4)	Searches str3 for program memory str4
strlen	strlen (str1)	Returns the length of str1
strlenpgm	strlenpgm (str2)	Returns the length of program memory str2
strlwr	strlwr (str1)	Converts uppercase of str1 to lowercase
strlwrpgm	strlwrpgm (str2)	Converts uppercase of program memory str2 to lowercase

TABLE 5-3   (*continued*)

Function	Example	Note
strncat	strncat (str1, str2, 3)	Appends str1 with str2 for 3 characters
strncatpgm	strncatpgm (str3, str4, 2)	Appends str3 in the program memory with str4 for 2 characters
strncatpgm2ram	strncatpgm2ram (str5, str6, 1)	Appends str5 with program memory string str6 for 1 characters
strncatram2pgm	strncatram2pgm (str3, str4, 2)	Same as strcatpgm
strncmp	strncmp (str1, str2, 2)	Compares str1 to str2 for 2 characters
strncmppgm	strncmppgm (str3, str4, 4)	Compares str3 in program memory to program memory str4 for 4 characters
strncmppgm2ram	strncmppgmram (str5, str6, 3)	Compares str5 to program memory str6 for 3 characters
strncmpram2pgm	strncmprampgm (str3, str4, 1)	Compares program memory str3 to str4 for 1 character (returns >0 if first string is less than second string returns == 0 if strings are equal returns <0 if first string is greater then second string)
strncpy	strncpy (str1, str2)	Copies str2 into str1
strncpypgm	strncpypgm (str3, str4)	Copies str4 in the program memory to str3
strnpypgm2ram	strncpypgm2ram (str5, str6)	Copies str6 to program memory location str5
strncpyram2pgm	strncpyram2pgm (str3, str4)	Same as strcpypgm
strpbrk	strpbrk (str1, str2)	Searches str1 for the characters in str2
strpbrkpgm	strpbrkpgm (str3, str4)	Searches program memory str1 for characters in program memory str2
strpbrkpgmram	strpbrkpgmram (str5, str6)	Searches program memory str5 for characters in str6
strpbrkrampgm	strpbrkrampgm (str3, str4)	Searches str3 for characters in program memory str4
Strrchr	strrchr (str, '0')	Returns a pointer to the last occurrence of 0
strspn	strspn (str1, str2)	Returns the number of consecutive characters in str1 from str2
strspnpgm	strspnpgm (str3, str4)	Returns the number of consecutive characters in program memory str3 from program memory str4
strspnpgmram	strspnpgmram (str5, str6)	Returns the number of consecutive characters in program memory str5 from str6
strspnrampgm	strspnrampgm (str7, str8)	Returns the number of consecutive characters in str7 from program memory str8
strstr	strstr (str1, str2)	Locates the first occurrence in str1 of str2
strstrpgm	strstrpgm (str3, str4)	Locates the first occurrence in program memory str3 of program memory str4
strstrpgmram	strstrpgmram (str5, str6)	Locates the first occurrence in program memory str5 of str6
strstrrampgm	strstrrampgm (str7, str8)	Locates the first occurrence in str7 of program memory str8 (pointer is returned if found, otherwise a null is returned)
strtok	strtok (str1, str2)	Inserts a null in str1 in place of tokens specified by str2
strtokpgm	strtokpgm (str3, str4)	Inserts a null in program memory str3 in place of tokens in str4
strtokpgmram	strtokpgmram (str5, str6)	Inserts a null in str5 in place pf program memory tokens in str6
strtokrampgm	strtokrampgm (str3, str4)	Same as strtokpgm (returns a pointer of a null if none found)
strupr	strupr (str1)	Converts lowercase to uppercase in str1
struprpgm	strupr (str2)	Converts lowercase to uppercase in program memory str2

EXAMPLE 5-8

```
/*
 * This is Example 5-8
 */

#include <p18cxxx.h>
#include <string.h>

/* Set configuration bits
 *   - set HS oscillator
 *   - disable watchdog timer
 *   - disable low-voltage programming
 */

#pragma config OSC = HS
#pragma config WDT = OFF
#pragma config LVP = OFF

// program memory data

rom near char lookUpTable[][20] =
{       "my first message",
        "my second message",
        "my third message",
        "my fourth message",
        "my fifth message"
};

// data memory data

char buffer[20];

// main program

void main (void)
{
        strcpypgm2ram (buffer, lookUpTable[1]);
}
```

Example 5-9 shows yet another example of accessing the program memory. This time, program memory is accessed to find the 7-segment code in a lookup table for an LED, 7-segment display. A function is used to access the table. The function called Get7Seg converts a BCD number to 7-segment code only if the number is 0 through 9. If any other number is passed to the function, a 0x00 is returned.

EXAMPLE 5-9

```
/*
 * This is Example 5-9
 */

#include <p18cxxx.h>

/* Set configuration bits
 *   - set HS oscillator
 *   - disable watchdog timer
 *   - disable low-voltage programming
 */

#pragma config OSC = HS
#pragma config WDT = OFF
#pragma config LVP = OFF

// program memory data

rom near char lookUp7Seg[] =        // 7-segment lookup table
{
      0x3f, 0x06, 0x5b, 0x4f, 0x66,
      0x6d, 0x7d, 7, 0x7f, 0x6f
};
```

```
// data memory
char data1;
// function Get7Seg
char Get7Seg (char bcd)
{
      if (bcd <= 9)
            return lookUp7Seg[bcd];
      else
            return 0;
}
// main program
void main (void)
{
      data1 = Get7Seg(3);
}
```

The program in Example 5-10 illustrates the use of a memory function from the string library. The memset function is used to place a 0x20 into all of the locations in a data memory buffer.

EXAMPLE 5-10

```
/*
 * This is Example 5-10
 */

#include <p18cxxx.h>
#include <string.h>

/* Set configuration bits
 *   - set HS oscillator
 *   - disable watchdog timer
 *   - disable low-voltage programming
 */

#pragma config OSC = HS
#pragma config WDT = OFF
#pragma config LVP = OFF

//data memory
char buffer[20];

void main (void)
{
      memset(buffer, 0x20, 20);
}
```

In Chapter 4, an assembly language program (Example 4-25) is used to access a table of Morse code characters with a function called SendChar. Review the example and compare it to the C language version in Example 5-11. Here, as in Chapter 4, program memory is used to contain the Morse code patterns for the function. The test program sends the letters a through z through bit position zero of Port B, just as in Example 4-25. The C language version is much clearer and easier to understand when compared to the assembly language version. The only thing that might be added is another lookup table for the Morse code punctuation characters. The best test, of course, is to use an oscilloscope to view the output signal on Port B bit position zero.

EXAMPLE 5-11

```
/*
 * This is Example 5-11 for 4-MHz clock
 */
```

```c
#include <p18cxxx.h>
#include <delays.h>
/* Set configuration bits
 *  - set HS oscillator
 *  - disable watchdog timer
 *  - disable low-voltage programming
 */

#pragma config OSC = HS
#pragma config WDT = OFF
#pragma config LVP = OFF

// program memory lookup tables

rom near char Morse_L[] =
{
      0x42, 0x84, 0xa4, 0x83, 0x01,
      0x24, 0xc3, 0x04, 0x02, 0x74,
      0xa3, 0x44, 0xc2, 0x80, 0xe3,
      0x64, 0xd4, 0x43, 0x03, 0x81,
      0x23, 0x14, 0x63, 0x94, 0xb4,
      0xc4
};

rom near char Morse_N[] =
{
      0xfd, 0x7d, 0x3d, 0x1d, 0x0d,
      0x05, 0x85, 0xc5, 0xe5, 0xf5
};

// functions

void sendDitDah (char pattern)         // send Morse character
{
      int a;
      int count = pattern & 7;
      for (a = 0; a < count & 7; a ++)
      {
            PORTB = 1;                   // key on

            if ((pattern & 0x80))        // check bit
                  Delay1KTCYx (72);      // dah
            else
                  Delay1KTCYx (24);      // dit

            PORTB = 0;                   // key off
            Delay1KTCYx (24);            // waste a dit
            pattern <<= 1;               // get next bit
      }
      Delay1KTCYx (48);                  // waste 2 more dits
}

void sendChar (char sendData)
{
      if (sendData >= 'a' && sendData <= 'z')
            sendData -= 0x20;          // make uppercase

      if (sendData == ' ')             // if space
            Delay1KTCYx (96);          // wait 96 ms

      else if (sendData >= 'A' && sendData <= 'Z')
            sendDitDah (Morse_L [sendData - 0x41]);

      else if (sendData >= '0' && sendData <= '9')
            sendDitDah (Morse_N [sendData - 0x30]);
}

// main program to test the sendChar function

void main (void)
{
      int a;

      ADCON1 = 0x7F;        // configure PORTS A and B as digital
```

FIGURE 5-4 Screen shot of the Morse code as sent by Example 5-11.

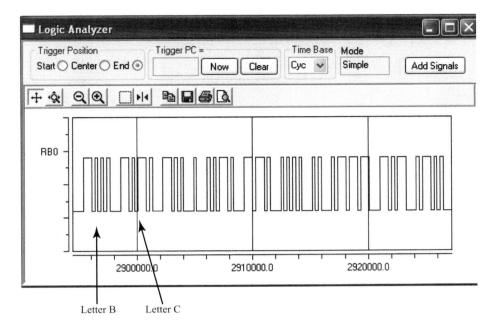

```
                                  //    this might need to be changed depending
                                  //    on the microcontroller version
        TRISB = 0x00;             // configure Port B as output
        for (a = 0; a < 26; a++)
            sendChar('a' + a);         // test sendChar
    }
```

Figure 5-4 shows the program of Example 5-11 in action. The logic analyzer is used to show the data output at pin RB0. Captured in Figure 5-4 are some Morse-coded characters beginning with the letter B. Because the program sends the letters in alphabetical order, C follows B, and so forth. Refer to Table 4-2 for a list of the Morse codes. The wide high pulse is a dah and the narrow high pulse is a dit.

The only way to capture this screen shot was to change the time delay functions from Delay1KTCYx to Delay10KTCYx, just to cause the signal to appear on the analyzer screen. Microchip, in version 7.11 of the IDE, did not include a way to modify the sampling rate of the logic analyzer. It is hoped that this will be corrected in future editions of the IDE.

5-3   SAMPLE C LANGUAGE PROGRAMS

Now that C language for the PIC18 family has been introduced, this section provides some sample programs that illustrate additional C language programming for the PIC microcontroller. As has been already realized, C for the PIC is much like C for any other system, although some of the libraries are expanded to provide functions that deal with the program memory space.

Example One

This first sample program (Example 5-12) reworks the assembly language example in Chapter 4 for a PC-style keyboard. This new example returns ASCII-coded characters for most of the keys on the keyboard. The control and alternate keys have not been implemented, but the shift key is

TABLE 5-4   Special function key codes generated by getChar in Example 5-12.

Key	Code	Key	Code	Key	Code
F9	0x80	**Left Alt**	0x8B	**Keypad 8**	0x96
F5	0x81	**Left Ctrl**	0x8C	**Num Lock**	0x97
F3	0x82	**Caps Lock**	0x8D	F11	0x98
F1	0x83	**Keypad 1**	0x8E	**Keypad +**	0x99
F2	0x84	**Keypad 4**	0x8F	**Keypad +**	0x99
F12	0x85	**Keypad 7**	0x90	**Keypad 3**	0x9A
F10	0x86	**Keypad 0**	0x91	**Keypad −**	0x9B
F8	0x87	**Keypad .**	0x92	**Keypad ***	0x9C
F6	0x88	**Keypad 2**	0x93	**Keypad 9**	0x9D
F4	0x89	**Keypad 5**	0x94	**Scroll Lock**	0x9E
Tab	0x8A	**Keypad 6**	0x95	F7	0x9F

implemented. Two program memory tables are used to return the ASCII code for both unshifted and shifted keystrokes on the keyboard. The getChar function is implemented using recursion to get additional characters from the keyboard. If the control key is implemented, additional tables for control = 0 and control = 1 are added. The control key codes are normally defined as 0x00 through 0x1A for the keys A through Z.

The ASCII code returned by the getChar function is standard ASCII code for all but the special keys such as page up, page down, and so forth. The special keys generate codes as illustrated in Table 5-4. Note that the two-digit codes have not been implemented in this example, but can be added by modifying the software and adding to the table.

EXAMPLE 5-12

```
/*
 * This is Example 5-12
 */

#include <p18cxxx.h>

/* Set configuration bits
 *  - set HS oscillator
 *  - disable watchdog timer
 *  - disable low-voltage programming
 */

#pragma config OSC = HS
#pragma config WDT = OFF
#pragma config LVP = OFF

// program memory data

// lookup table for the uppercase letters

rom near char upperCaseTable[] =
{
        0x00, 0x80, 0x00, 0x81, 0x82, 0x83, 0x84, 0x85,
        0x00, 0x86, 0x87, 0x88, 0x89, 0x8A, 0x7E, 0x00,
        0x00, 0x8B, 0x00, 0x00, 0x8C, 0x51, 0x21, 0x00,
        0x00, 0x00, 0x5A, 0x53, 0x45, 0x57, 0x40, 0x00,
        0x00, 0x43, 0x58, 0x44, 0x45, 0x24, 0x23, 0x00,
        0x00, 0x20, 0x56, 0x46, 0x54, 0x52, 0x25, 0x00,
        0x00, 0x4E, 0x42, 0x48, 0x47, 0x59, 0x5E, 0x00,
        0x00, 0x00, 0x4D, 0x4A, 0x55, 0x26, 0x2A, 0x00,
        0x00, 0x3C, 0x4B, 0x49, 0x4F, 0x29, 0x28, 0x00,
        0x4E, 0x3F, 0x4C, 0x3A, 0x00, 0x50, 0x5F, 0x00,
        0x00, 0x00, 0x22, 0x00, 0x7B, 0x2B, 0x00, 0x00,
        0x8D, 0x00, 0x0D, 0x7D, 0x00, 0x7C, 0x00, 0x00,
```

```
        0x00, 0x00, 0x00, 0x00, 0x00, 0x00, 0x08, 0x00,
        0x00, 0x8E, 0x00, 0x8F, 0x90, 0x00, 0x00, 0x00,
        0x91, 0x92, 0x93, 0x94, 0x95, 0x96, 0x1B, 0x97,
        0x98, 0x99, 0x9A, 0x9B, 0x9C, 0x9D, 0x9E, 0x00,
        0x00, 0x00, 0x00, 0x9F, 0x00, 0x00, 0x00, 0x00,
        0x00, 0x00, 0x00, 0x00, 0x00, 0x00, 0x00, 0x00
};
// lookup table for the lowercase letters

rom near char lowerCaseTable[] =
{
        0x00, 0x80, 0x00, 0x81, 0x82, 0x83, 0x84, 0x85,
        0x00, 0x86, 0x87, 0x88, 0x89, 0x8A, 0x60, 0x00,
        0x00, 0x8B, 0x00, 0x00, 0x8C, 0x71, 0x31, 0x00,
        0x00, 0x00, 0x7A, 0x73, 0x61, 0x77, 0x32, 0x00,
        0x00, 0x63, 0x78, 0x64, 0x65, 0x34, 0x33, 0x00,
        0x00, 0x20, 0x76, 0x66, 0x74, 0x73, 0x35, 0x00,
        0x00, 0x6E, 0x62, 0x68, 0x67, 0x79, 0x36, 0x00,
        0x00, 0x00, 0x6D, 0x6A, 0x75, 0x37, 0x38, 0x00,
        0x00, 0x2C, 0x6B, 0x69, 0x6F, 0x30, 0x39, 0x00,
        0x2E, 0x2F, 0x6C, 0x3B, 0x00, 0x60, 0x2D, 0x00,
        0x00, 0x00, 0x27, 0x00, 0x5B, 0x3D, 0x00, 0x00,
        0x8D, 0x00, 0x0D, 0x5D, 0x00, 0x5C, 0x00, 0x00,
        0x00, 0x00, 0x00, 0x00, 0x00, 0x00, 0x08, 0x00,
        0x00, 0x8E, 0x00, 0x8F, 0x90, 0x00, 0x00, 0x00,
        0x91, 0x92, 0x93, 0x94, 0x95, 0x96, 0x1B, 0x97,
        0x98, 0x99, 0x9A, 0x9B, 0x9C, 0x9D, 0x9E, 0x00,
        0x00, 0x00, 0x00, 0x9F, 0x00, 0x00, 0x00, 0x00,
        0x00, 0x00, 0x00, 0x00, 0x00, 0x00, 0x00, 0x00
};

// data memory data

int shift = 0;

// functions

// wait for one complete alternation of the clock input
//
// returns when clock goes high

//
void Wait4Clock (void)
{
        while (PORTAbits.RA1 == 1);        // while clock = 1
        while (PORTAnits.RA1 == 0);        // while clock = 0
}

int readChar (void)
{
        int parity = 0;
        int keyData = 0;
        int a;
        Wait4Clock();
        if (PORTAbits.RA0 == 1)
                keyData = 0x100;                    // if framing error

        else    // good start bit
        {
                for (a = 0; a < 8; a++)
                {
                        keyData >>= 1;          // assemble keyData

                        Wait4Clock ();

                        keyData |= (PORTAbits.RA0 == 1);
                        parity += PORTAbits.RA0;
                }
                Wait4Clock();                   // get parity bit

                parity ^= PORTAbits.RA0;
```

```
                if ((parity & 1) == 0)
                    keyData = 0x101;       // if parity error

                Wait4Clock();

                if (PORTAbits.RA0 == 0)
                    keyData = 0x100;       // if no stop framing error,
            }
        return keyData;
}

int     getChar (void)
{
        int step = 0;
        int scanCode = readChar();
        while (step == 0 && scanCode < 0x100)
        {
                switch (scanCode)
                {
                    case 0x12:               // shift
                    case 0x59:
                    {
                        shift = 1;
                        scanCode = readChar();
                        break;
                    }
                    case 0xF0:               // release
                    {
                        scanCode = readChar();
                        switch (scanCode)
                        {
                            case 0x12:    // shift
                            case 0x59:
                            {
                                shift = 0;
                                break;
                            }
                        }
                        scanCode = readChar();
                        break;
                    }
                    default:
                    {
                        step = 1;
                        break;
                    }
                }
        }
        if (step = 1 && scanCode < 0x100)
        {
                if (shift)
                    scanCode = upperCaseTable [scanCode];
                else
                    scanCode = lowerCaseTable [scanCode];
        }
        return scanCode;
}
```

This function reads the keyboard and returns an ASCII character, but a lot of time is needed from the microcontroller. To reduce this, the software is best relegated to interrupt techniques discussed in Chapter 8.

Example Two

A system exists for model railroading called DCC (digital command controls). A DCC system sends digital signals down the rails of the layout that controls engines and various other components of a

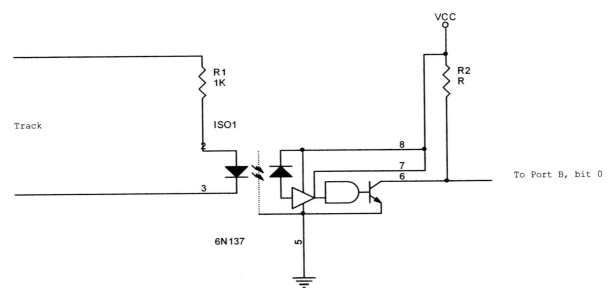

FIGURE 5-5   Detecting the DCC signal.

model railroad. One of the functions provided on the controller activates accessory devices on the layout by a programmable address. The address is sent out and intercepted by a microcontroller-based system that activates the device. The circuit used to extract the serial track signal from the layout appears in Figure 5-5. Here an optical isolator is used to isolate the system from the track voltage and generate a TTL-compatible signal for the microcontroller. The polarity of the connection to the track does not matter because the track signal is an AC signal and the information is carried in the width of the pulse applied to the track.

The signal is sent on the track as an AC signal that swings between $\pm$ 14 V and contains digital information as illustrated in Figure 5-6. A logic one data bit is a signal that has a high and low pulse width of between 52 μs and 64 μs each. A logic zero data bit is much wider and has a high and low time pulse width of between 90 μs and 10,000 μs. The data packet sent for an accessory contains 3 bytes. A preamble is a series of 12 bits of logic 1 data followed by an address start bit of a logic zero. The first packet byte contains primary address 0x80 (128) to 0xBF (191). The second byte contains three more address bits, a data control bit, and a secondary 3-bit address. The third and final byte of the packet contains an Exclusive-OR checksum for the packet. This accessory type 1 packet is illustrated in Figure 5-6. The address bits are indicated using the standard convention of using the one's complement of the three most significant address bits.

Example 5-13 lists a program that decodes all the bits in the packet and tests for a checksum error. The program does not program the address of other features found in a typical accessory controller nor does it control an accessory. Before this can be done, some of the hardware circuitry needs to be explained.

EXAMPLE 5-13

```
/*
 * This is Example 5-13 for a 4-MHz clock
 */

#include <p18cxxx.h>
#include <delays.h>
```

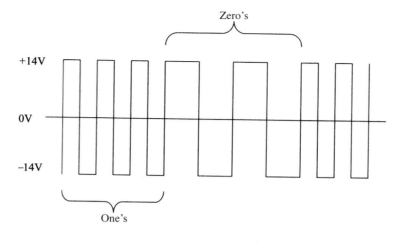

(a)

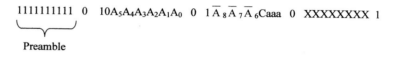

Preamble

AAAAAAAAA = 9-bit primary address (0–511)
C = control bit (ON/$\overline{\text{OFF}}$)
aaa = accessory address (0–7)
XXXXXXXX = Exclusive-OR checksum

(b) NMRA type 1 accessory packet

FIGURE 5-6   (a) Signal found on the tracks in a DCC system and (b) the type 1 packet.

```
/* Set configuration bits
 *   - set HS oscillator
 *   - disable watchdog timer
 *   - disable low-voltage programming
 */

#pragma config OSC = HS
#pragma config WDT = OFF
#pragma config LVP = OFF

// program memory data

int count;
int state;
int dataBit;
int preambleCount;
int firstByte;
int secondByte;
int checkSum;

// functions
```

```c
void assembleByte (int* locale, int nextState)
{
      if (preambleCount != 9)
      {
            *locale >>= 1;
            *locale += dataBit;
            preambleCount++;
      }
      else
      {
            preambleCount = 0;
            state = nextState;
      }
}
void state0 (void)
{
      if (dataBit == 1)
      {
            preambleCount--;
            if (preambleCount == 0 && dataBit == 0)
            {
                  state = 1;                 // good preamble received
                  preambleCount = 0;
            }
            else
                  preambleCount++;    // more 1's in preamble
      }
      else
            preambleCount = 10;        //restart wait
}
void state1 (void)
{
      assembleByte (&firstByte, 2);
      if (dataBit == 1)
            state = 0;                 // on error
}
void state2 (void)
{
      assembleByte (&secondByte, 3);
      if (dataBit == 1);
            state = 0;                 // on error
}
void state3 (void)
{
      assembleByte (&checkSum, 4);
      if (dataBit == 0)
            state = 0;                 // on error
}
void state4 (void)
{
      state = 0;                             // reset state

      if ((firstByte ^ secondByte ^ checkSum) == 0)
      {
            // this is where the control software is
            // placed to program the controller
            // or activate a device connected to a
            // port if the address matches
      }
}
typedef void (*ptr) (void);                 // array of function pointers

ptr states[] =                              // for machine state
{
      &state0,
```

```
                    &state1,
                    &state2,
                    &state3,
                    &state4
        };

        void Update (void)
        {
            if (count == 0)    // system reset command
            {
                state = firstByte = secondByte = checkSum = dataBit = 0;
                preambleCount = 10;            //minimum preamble bit count
                count = 1;
            }

            // measure pulse width
            //          count = 0 resets the system
            //          count = 1 or 2 counts (28 us to 84 us) data is a logic 1
            //          count is more than 2 counts data is a logic zero

            else
            {
                count = 0;                         // start fresh count

                if (PORTBbits.RB0 == 1)
                {
                    while (PORTBbits.RB0 == 1)
                    {
                        Delay10TCYx(2);     // 20 microseconds
                        count++;
                    }
                }
                else
                {
                    while (PORTBbits.RB0 == 0)
                    {
                        Delay10TCYx(2);     // 20 microseconds
                        count++;
                    }
                }
                if (PORTBbits.RB0 == 1)          // skip other half bit
                    while (PORTBbits.RB0 == 1);
                else
                    while (PORTBbits.RB0 == 0);

                if (count == 0)
                    state = 0;                        // reset system
                else if (count < 3)
                    dataBit = 1;            // bit = 1
                else
                    dataBit = 0;            // bit = 0
            }

            states[state]();                           // function call to a state function
        }

        // the only main program

        void main (void)
        {
            ADCON1 = 0x7F;        // configure PORTS A and B as digital
                                  //    this might need to be changed depending
                                  //    on the microcontroller version

            TRISB = 1;            // configure Port B
            state = 0;            // starting state is 0
            count = 0;            // reset program and start looking for
                                  //    a valid preamble

            while (1)             // main loop
                Update();         // search for preamble and a good packet
        }
```

FIGURE 5-7 Machine states for Example 5-12.

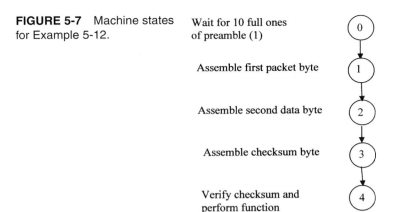

Wait for 10 full ones of preamble (1) — 0

Assemble first packet byte — 1

Assemble second data byte — 2

Assemble checksum byte — 3

Verify checksum and perform function — 4

The main program is an infinite while loop that calls the Update function with a logic zero in count. Update detects a count of zero and clears the three bytes on the packet and sets a preamble counter to 10. A preamble is at least 10 logic ones in a row, but may be many more. Update then calls a function whose address is stored in an array called states. Because the initial state is a zero, the state0 function is called which counts out at least 10 logic ones of input data. In this first initialization function call, nothing is counted and a return occurs back to the main program where the count is cleared to zero after the function call to Update.

At this point in the program the input pulse width is measured in Update. If the input bit is a logic one, the first while loop starts counting time and if the input bit is a zero, the second counts time. These loops take exactly 28 μs per iteration with a 4-MHz clock, so count increases as long as the input does not change states. If an input does not change between 28 μs and 56 μs the count is 1, likewise if the input does not change between 56 μs and 84 μs the count is two. A logic zero is between 52 μs and 64 μs. A logic one is longer than 90 μs, so if a count of 3 or greater is reached, the input has not changed for at least 84 μs, which is beyond a logic zero. This software meets the timing requirements for both the logic zero and one bits. As soon as a bit changes, the other half of the bit is checked but not measured. Only one half of a bit needs to be measured. After the bit is measured and completed, the count sets or clears the dataBit and then calls a function with states[state]() as dictated by the state of the machine. Figure 5-7 contains a state transition diagram that shows the function of each state function (state0 through state4).

Programs for controlling machines often use a number of states to accomplish the task. In this example, an array of function addresses is used to invoke different functions for each state. By doing so, the program is organized in a straightforward manner. The software should also use some method to determine if the system is hung for more than 10 milliseconds. This is accomplished with a watchdog timer, but because it has yet to be explained, this is not illustrated here. Timers and watchdogs are discussed and used later in this book in example programs. If the input pulse stuck high, as it would be if the track power is off, the system will hang and that is why a mechanism must be used to un-hang the system.

5-4   MATH LIBRARY

Although not used as much as the delays library, the math library functions have some use in programming. The math library uses only 32-bit floating point numbers, whereas the 64-bit double precision numbers are not supported. Because this is an 8-bit microcontroller, 64-bit

TABLE 5-5   Math functions in the math.h include file.

Function	Example	Comment
acos	acos (data1)	Returns the arccosine in radians of data1 (0 to π)
asin	asin (data2)	Returns the arcsine in radians of data2 ($\frac{-\pi}{2}$ to $\frac{\pi}{2}$)
atan	atan (data3)	Returns the arctangent in radians of data3 ($\frac{-\pi}{2}$ to $\frac{\pi}{2}$)
atan2	atan2 (data1, data2)	Returns the arctangent in radians of $\frac{data1}{data2}$ ($-\pi$ to π)
ceil	ceil (data1)	Returns an integer greater than or equal to float data1
cos	cos (data2)	Returns the cosine of data2 radians
cosh	cosh (data3)	Returns the hyperbolic cosine of data3
exp	exp (data4)	Returns ε^{data4}
fab	fab (data5)	Returns the absolute value of data5
floor	floor (data6)	Returns an integer less than or equal to float data6
fmod	fmod (num, mod)	Returns the remainder of num *modulo* mod
frexp	frexp (data1, intr)	Splits data1 into a fraction (returned) and integer stored at pointer location intr
ieeetomchp	ieeetomchp (data2)	Coverts data2 into microchip floating point
idexp	idexp (data3, int1)	Returns data3 * 2^{int1}
log	log (data4)	Returns the natural log of data4
log10	log10 (data5)	Returns the common log of data5
mchptoieee	mchptoieee (data6)	Converts data6 from microchip floating point
modf	modf (data7, data8)	Returns the fraction part of data7 and stores the integer part at pointer location data8
pow	pow (data1, data2)	Returns data1 ^{data2}
sin	sin (data3)	Returns the sine of radian angle data3
sinh	sinh (data4)	Returns the hyperbolic sine of data4
sqrt	sqrt (data5)	Returns the square root of data5
tan	tan (data6)	Returns the tangent of radian angle data6
tanh	tanh (data7)	Returns the hyperbolic tangent of data7

operations would require too much time for execution. Programs rarely use floating-point data in an embedded microcontroller-based system.

The math library provided with the C18 compiler performs the same functions as the standard ANSI C math library found with most compilers. Table 5-5 lists these functions, including examples and comments about their operations.

To illustrate the use of the math library, Example 5-14 solves the resonant frequency equation from electronics $\left(Fr = \dfrac{1}{2\pi\sqrt{LC}} \right)$. This program solves the resonant frequency for a capacitor value of 1.0 μF and the inductor values of 1.0 mH through 10.0 mH and stores the ten results in an array called Fr.

EXAMPLE 5-14

```
/*
 * This is Example 5-14
 */
```

```
#include <p18cxxx.h>
#include <math.h>

/* Set configuration bits
 *   - set HS oscillator
 *   - disable watchdog timer
 *   - disable low-voltage programming
 */

#pragma config OSC = HS
#pragma config WDT = OFF
#pragma config LVP = OFF

// program memory data

float Fr[10];
float L = 1.0e-3;
float C = 1.0e-6;

// main program

void main (void)
{
      int a;
      for (a = 0; a < 10; a++)
      {
            Fr[a] = 1 / (6.2831853 * sqrt( L * C ));
            L += 1.0e-6;
      }
}
```

If you write this program using the IDE and view the amount of memory used to store this task, it is 1941 bytes of program memory in length. What requires so much memory is the math function to perform the square root operation and the floating point operations. As mentioned earlier, always use characters and integers for as much of an application as possible or the size of the program may overwhelm the amount of existing memory space.

Suppose that a device needs to obtain voltage gain in decibels. The equation for this is $Vgain = 20 \log_{10} \frac{Vout}{Vin}$. A function that uses the math library to accomplish this is listed in Example 5-15. As with Example 5-14, this function requires a lot of memory because of the log10 function and floating point arithmetic operations.

EXAMPLE 5-15

```
float gainDB (float vout, float vin)
{
      return 20 * log10 (vout/vin);
}
```

5-5   SUMMARY

1. The C18 C language compiler is available in a student version as a cost-free download from the Microchip Website.
2. The IDE fully supports programming in C using the C18 compiler.
3. The #pragma config statements configure the PC18 microcontroller in C language.
4. The _asm and _endasm directives are used without { } to place assembly code into a C language program.
5. Accurate time delay functions are available in the C18 include library called delays.h.
6. The I/O ports are accessed from within C by using the port name and programming the direction register.

7. The stdlib.h header contains functions that perform conversions. Most of these are standard ANSI C conversion functions.

8. The string.h header contains functions that allow string and memory manipulation using both the program and data memory.

9. The rom near or rom far directives are used to store information in the program memory of the microcontroller.

10. The math.h header contains functions that perform the ANSI standard C mathematic operations.

11. Floating-point arithmetic is limited to 32-bit single precision floats in the C18 compiler.

5-6   QUESTIONS AND PROBLEMS

1. Compare the C18 language with ANSI C.
2. How is assembly language code placed into a C program?
3. What is the purpose of the #include <p18cxxx.h> statement?
4. What is the purpose of the #pragma config statement?
5. What is the Microchip C18 Toolsuite?
6. Detail how the Delay1KTCYx function is used to cause a 10 millisecond time delay if the system clock is 4 MHz.
7. Repeat question 6 for a 20-MHz clock.
8. Using the delays.h, write a short program that alternates bit position 0 of Port B so it's on for one second and off for one second. Assume that the microcontroller clock is 4 MHz.
9. Develop a short program that generates a 1-KHz square wave on Port B bit position 2.
10. Develop a program that performs a simple test of all eight input combinations on a 3-input NAND gate connected to the I/O ports on the microcontroller. Port B, bits 0, 1, and 2 are connected to the inputs of the NAND gate, and bit position 4 of Port A is connected to the output of the NAND gate. If the gate check is good, place a logic one on bit position 3 of Port B, and if it checks bad, place a zero on bit position 3 of Port B.
11. Using the rand function, develop a program that generates lottery numbers between 1 and 47 on Port B of the microcontroller each time a pushbutton connected to Port A bit position 4 is pressed. Assume the button generates a logic zero when pressed and a logic one when not pressed.
12. What is the difference between the directives rom near and rom far?
13. Develop a lookup table in the program memory that contains the ASCII code for 0 through 9 and A through F. Now write a function that uses the lookup table to convert a character passed to it from 0 through 15 to ASCII.
14. Explain how the entries in the program memory tables of Example 5-11 encode the Morse code.
15. Explain how the sendDitDah function operates in Example 5-11.
16. Form a single C statement that tests bit position 2 of Port B. If the bit is a one, the program must get stuck in the statement and if zero, the program must continue.
17. Explain how the Wait4Clock function in Example 5-12 waits for one complete clock cycle on Port A bit position 1.
18. What value is returned by readChar in Example 5-12 for a parity error?
19. What value is returned by readChar in Example 5-12 for a framing error?
20. Explain how the main program in Example 5-13 counts the width of the logic zero or logic one input signal on Port B, bit position 0.
21. What is meant by the C language statement &bob?
22. Explain how the states[state]() statement in Example 5-13 calls different functions depending on the value of state.

23. If the clock frequency in Example 5-13 is changed to 8 MHz from 4 MHz, what must be changed in the program to cause it to operate correctly?

24. Using the math.h header, develop a program that finds the capacitive reactance (xc) of a 1.0 μF capacitor for the frequencies from 100 Hz to 1000 Hz in steps of 100 Hz. The equation for capacitive reactance is $\frac{1}{2\pi FC}$.

25. Suppose that Port A (bits 0, 1, and 2) is a signal from a digital encoder that generates a three-bit binary code that represents 45° per binary step, that is, 000 = 0°, 001 = 45°, 010 = 90°, and so forth. Develop a function called sinPortA that reads Port A and returns the sine of the angle. Using the sin(bob) function in the math.h library, return the sine of bob assuming bob is a radian measure (360° = 2π radians). Write down the length of the function as reported by the IDE.

26. Repeat question 25 using a lookup table stored in the program memory to return the same values calculated in question 25. Compare the length of this function to the one developed in question 25.

27. From the observation in questions 25 and 26, are lookup tables an important tool in developing software for a memory-limited device such as the microcontroller?

CHAPTER 6

PIC18 Family Hardware Specifications

Before a system can be constructed around any member of the PIC18 family of microcontrollers, a detailed understanding of the pin connections and features available within the microcontroller is required. This chapter describes the DC and AC characteristics of the PIC18 family, the clocking circuitry required to operate the PIC18, and essential information on the operation of some of the many I/O devices within the PIC18. Internal devices not presented in this chapter are explained in later chapters with the appropriate applications.

Upon completion of this chapter you will be able to:

1. Power the microcontroller and describe its DC characteristics and proper handling.
2. Detail the differences in packages and pin-outs available in the PIC18 family of microcontrollers.
3. Describe the minimum and maximum voltages and currents on the I/O pins and power supply connections.
4. Use the watchdog timer.
5. Program the port pins.
6. Store and retrieve data from the date EEPROM.
7. Learn about the interrupts and develop interrupt software in C language.
8. Describe the internal peripherals available in the PIC18 family of microcontrollers.
9. Use a timer and the compare and capture unit in an application.

6-1   PIN-OUTS AND BASIC OPERATING CHARACTERISTICS

As learned in Chapter 2, the PIC18 family has many versions. This section describes details these versions, and explains the DC power requirements and driving capabilities of the microcontroller. This knowledge is necessary to design a system that will function for any length of time.

Pin-Outs

The PIC18 microcontroller has many versions, ranging from an 18-pin package to a 128-pin package. In order to choose the correct microcontroller for an application, familiarity with the various pin-outs is required. The microcontroller is manufactured in the following packages: PDIP (plastic dual inline package), SOIC (small outline integrated circuit), PLCC (plastic leadless chip carrier), QFN (quad flat no-lead), QFP (quad flat pack), TQFP (thin quad flat pack),

FIGURE 6-1 Microcontroller packages.

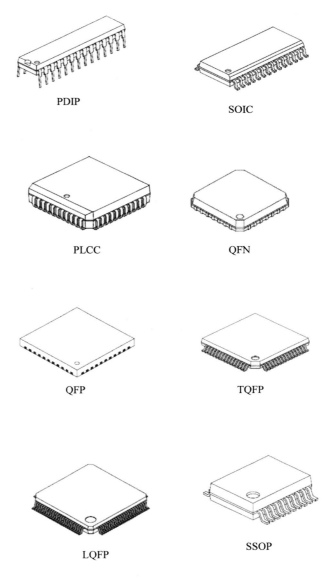

PDIP

SOIC

PLCC

QFN

QFP

TQFP

LQFP

SSOP

LQFP (low profile quad flat pack), and SSOP (small shrink outline package). All of these packages, except the PDIP and PLCC, are designed for surface mounting and are not used for prototypical work. Figure 6-1 illustrates these package designs. The PDIP is typically used for prototypical work, and surface mount packages are used for production.

Figure 6-2 illustrates some of the PIC18 microcontroller family members. The pin-outs are representative of the PIC18 family and do not include the 18F242/248/252/258/448/458, which are not recommended by Microchip for new designs because they will probably be discontinued eventually. For a complete list of the currently available 18F family members, visit the Microchip website. In general, I/O is limited to Ports A and B on the 18-pin PDIP versions; to Ports A, B, and C on the 28-pin versions; and to Port A, B, C, D, and E on the 40-pin versions. Family members in packages larger than 40 pins are not available in PDIP packages, so they are not discussed here because it would be difficult to develop prototypes. Once these smaller versions are learned, the switch to a higher pin count surface mount package is easily made because all members are

FIGURE 6-2 Several representative PIC18 PDIP, SOIC family members.

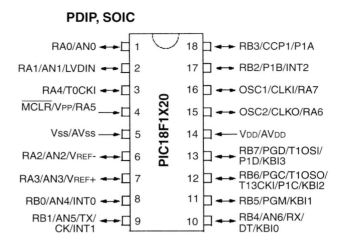

18F1230/1330

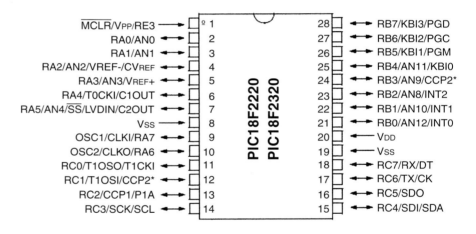

PIC18F2220/2320

otherwise identical. The only difference is the way the device is mounted (package style), the addition of more I/O pins, and additional features that appear on the lower pin count versions.

Power Supply Connections

As in any MOSFET (metal oxide semiconductor field effect transistor) integrated circuit, the power supply connections are VDD for connection to a nominal +5.0 V power supply (this can be as low as 4.2 V for maximum speed operation) and VSS for connection to ground. All VDD and VSS connections must be connected to the power supply, including versions of the microcontroller that have more than a single VDD or VSS connection. In all cases, the maximum allowable supply voltage is listed as +7.5 V. These voltages allow a wide variety of power sources, including batteries. A low-power version (PIC18L) is available that allows the VDD voltage to be as low as 2.5 V for full speed

FIGURE 6-2 (*continued*)

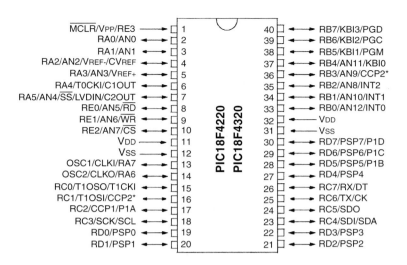

18F4220/4320

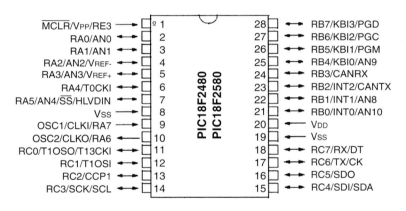

18F2480/2580

operation. The microcontroller is a CMOS device that must be handled appropriately and is subject to damage by electro-static discharge (ESD). Make sure that the microcontroller is stored in its antistatic protective package until it is connected to a circuit, and that appropriate handling methods are followed. These include proper grounding and handling only in environments that have sufficient relative humidity (above 20%). Always wear a grounding strap when handling the microcontroller and make certain the work area is properly grounded to prevent damage. If some of the input pins are unused, connect them to ground or 5.0 volts. No input connection should be left floating or open-circuited. An open connection can cause problems because it will be subject to ESD damage.

Generally, the amount of power supply current is determined by the clocking frequency and can range from 60 μA when operated with a 31-KHz clock to 12 mA or higher when operated at 40 MHz. The absolute maximum current is 300 mA and the maximum device power dissipation is 1 watt. The exact amount of current consumed varies with the frequency of operation and the loading on the I/O pins, but is included in the data sheet for a given device and given operating frequency. The loading from the I/O pins must be calculated for the application and added to the power consumed by the microcontroller to determine the total power.

FIGURE 6-2 (*continued*)

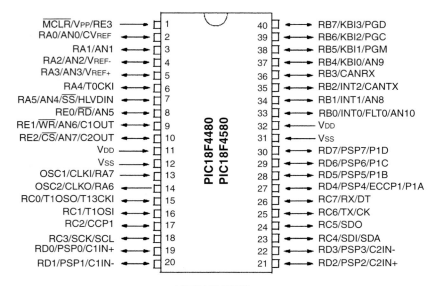

18F4480/4580

The power supply to the microcontroller should always be decoupled using a 0.1 μF capacitor connected from the VDD pin to the VSS pin as close to the integrated circuit as possible.
The PIC microcontroller produces a lot of internal switching noise that can be coupled to other
components in a system through the power supply connections. The bypass capacitor shunts the
noise to ground so it does not interfere or affect other devices in the system. This noise is produced each time an internal MOSFET switch turns on or off, causing surges or drops in current
flow from the power supply connections. All digital integrated circuits produce this type of noise
and all must be bypassed. Even a linear amplifier, such as an operation amplifier, produces noise
on its power connections and should be bypassed.

Input/Output Connections

The I/O characteristics are important for interfacing. The interface characteristics of the microcontroller are listed in Table 6-1. The maximum logic zero and one output currents listed are for
the voltage ranges listed. These output vales must be adhered to if the output is used to drive a
logic circuit. Otherwise (as when driving an LED, etc.), the true maximum output current is
25 mA for a logic zero or logic one as long as the total current for the microcontroller package is
not exceeded. If an I/O pin has a weak pull-up (Port B and Port D on some versions), as described in later sections, the maximum input current is 400 μA. If more than the recommended

TABLE 6-1 Input/Output
characteristics with
VDD = +5.0 volts.

Signal	Vmin	Vmax	Imax
0 input	0.0 V	0.8 V	±1.0 μA
1 input	2.0 V	5.0 V	±1.0 μA
0 output	0.0 V	0.6 V	−8.5 mA
1 output	4.3 V	5.0 V	3 mA

*The maximum total allowable chip current is 200 mA with up to
25 mA per pin as long as 200 mA is not exceeded for the entire
chip and the output does not connect to logic circuitry inputs.

currents are required, an amplifier is used in an I/O interface. Generally, the amount of current at an output pin is sufficient to drive four standard TTL unit loads and 10 high-speed CMOS unit loads. If other power supply voltages are used, refer to the data sheet for the microcontroller for additional details about the power supply current and pin currents.

Reset

To properly operate the microcontroller it must be reset correctly. The $\overline{\text{MCLR}}$ input pin is the **master clear** input that initializes the microcontroller so that it begins operation by fetching an instruction from memory location zero, which is called the **reset vector address**. If the power supply has a rise time of 0.05 V/ms (minimum), then a 1 KΩ to a 10 KΩ resister is connected from the $\overline{\text{MCLR}}$ input to +5.0 V and the internal circuitry automatically resets the microcontroller. If the voltage rise time cannot be guaranteed, as with all the examples in this book because the power supply is unknown, then the circuit illustrated in Figure 6-3 properly resets the microcontroller. As a bonus, the normally open pushbutton switch is used to manually reset the microcontroller (this reset button can be deleted if a manual reset is not needed). The RC time constant for the reset circuit should be between 10 ms and 20 ms. The circuit of Figure 6-3 has an RC time constant of $10K \times 1.0\ \mu F$ or 10 ms.

The microcontroller is also reset under the following conditions in addition to grounding the master clear input:

1. A watchdog timer reset (WDT)
2. A programmed brownout reset (BOR)
3. The RESET instruction (a Reset() function call in C language)
4. A stack full reset
5. A stack underflow reset

The master clear input is enabled with the #pragma config MCLRE = ON statement in C language. If it is not enabled, then the power-up reset is performed by detecting the voltage on the VDD pin, as long as the power supply has sufficient rise time to meet the requirements. The master clear input becomes a Port A bit I/O pin when disabled with #pragma config MCLRE = OFF. The actual pin number and port bit position are determined by the version of the microcontroller. In many cases the reset capacitor and resister are required to assure a proper reset.

FIGURE 6-3 PIC18 microcontroller reset circuit.

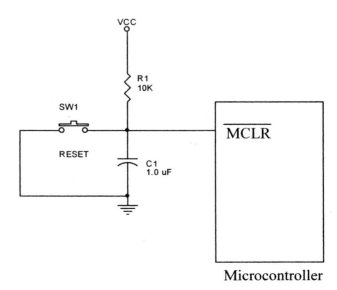

Microcontroller

The #pragma config command in C language specifies the state of the microcontroller configuration bits. These bits are located in a space of program memory that is accessible only through table read and write commands at addresses 0x300000 through 0x3FFFFF. Because of the amount of work required to change these locations, C language has implemented the #pragma config command for this task. The configuration bits can also be modified in the MPLAB IDE under the Configuration tab at the main menu, but it is much more descriptive if configuration is accomplished through the #pragma statements.

Watchdog Timer.   The **watchdog timer** (WDT) is an internal counter that resets a microcontroller after a given period of time. This timer is programmable and the WDT can be enabled or disabled using #pragma commands or with software in a program. In C language, the #pragma config statement is often used to enable or disable the WDT for an entire software system. The WDT operates at a clock time of 4 ms that is generated by an internal oscillator. The WDT is scaled from 1 to 32K as indicated in Table 6-2 and operates between 4 ms and 131 seconds.

Another configuration command is used to enable the WDT. The #pragma config WDT = ON/OFF statement is used in C and the _WDT_ON_2H or _WDT_OFF_2H (for the 18F1320) is used in assembly language. Example 6-1 illustrates both the C and assembly language versions of enabling the WDT and selecting a scaling factor of 4, which programs the WDT to reset the microcontroller after 16 ms.

EXAMPLE 6-1

```
// C language for all 18F family members

#pragma config WDT = ON
#pragma config WDTPS = 4                ; 16 ms WDT timeout


;assembly language for PIC18F1320

__CONFIG _CONFIG2H, _WDT_ON_2H & _WDTPS_4_2H
```

The WDT is started when the microcontroller is reset and begins counting out 4 ms intervals. In Example 6-1, the interval is four 4 ms periods or 16 ms. After running a program for

TABLE 6-2   Allowable watchdog scaling factors.

C Statement	Assembly Language	Scaling Factor	Time to Reset
#pragma config WDTPS = 1	_WDTPS_1_2H	1:1	4 ms
#pragma config WDTPS = 2	_WDTPS_2_2H	1:2	8 ms
#pragma config WDTPS = 4	_WDTPS_4_2H	1:4	16 ms
#pragma config WDTPS = 8	_WDTPS_8_2H	1:8	32 ms
#pragma config WDTPS = 16	_WDTPS_16_2H	1:16	64 ms
#pragma config WDTPS = 32	_WDTPS_32_2H	1:32	128 ms
#pragma config WDTPS = 64	_WDTPS_64_2H	1:64	256 ms
#pragma config WDTPS = 128	_WDTPS_128_2H	1:128	512 ms
#pragma config WDTPS = 256	_WDTPS_256_2H	1:256	1.024 sec
#pragma config WDTPS = 512	_WDTPS_512_2H	1:512	2.048 sec
#pragma config WDTPS = 1024	_WDTPS_1024_2H	1:1024	4.096 sec
#pragma config WDTPS = 2048	_WDTPS_2048_2H	1:2048	8.192 sec
#pragma config WDTPS = 4096	_WDTPS_4096_2H	1:4096	16.384 sec
#pragma config WDTPS = 8192	_WDTPS_8192_2H	1:8192	32.768 sec
#pragma config WDTPS = 16384	_WDTPS_16384_2H	1:16384	65.536 sec
#pragma config WDTPS = 32768	_WDTPS_32768_2H	1:32768	131.072 sec

Note: The complete listing for the assembly language commands appears in the PIC18F include file for the microcontroller. (The PIC18F1320 is illustrated here.)

16 ms, the WDT triggers a reset. This allows wayward software to reset the system if the WDT timer is allowed to expire. If a program functions normally, the WDT is restarted without causing a reset by placing a SLEEP (a Sleep() in C) or CLRWDT (a ClrWdt() in C) instruction in the normal flow of the program (see Example 6-2). The WDT can also be turned on (1) and off (0) using the SWDTEN bit in the WDTCON register. If the program hangs, the WDT is not reset by the ClrWdt() function and a system reset occurs, restarting the system, in this example after 16 ms. This feature is extremely important in systems that must operate continually without error. Imagine if a WDT did not exist in the antilock break system (ABS) of an automobile and the program locked up when the driver slammed on the breaks, and no ABS! With the WDT, everything would reset and restart if the system locked up. Even worse, what if the ignition control computer locked up while traveling down the freeway at 70 mph?

EXAMPLE 6-2

```
// C language for all 18F family versions

void main (void)
{
      while (1)             // system loop
      {
            ClrWdt();       // reset watchdog

            // system software goes here
      }
}

; assembly language

main:

      CLRWDT                ;reset watchdog

      ; system software goes here

      GOTO main
```

Brownout Reset. The brownout reset (BOR) restarts the system whenever the power supply voltage drops below a preprogrammed level. Once stopped by a brownout reset, the microcontroller does not restart until the power supply goes higher than the preprogrammed level. This stops the system and restarts it if the power supply input drops below a preset value for any reason. The voltage drop might be caused by an overload or, if battery powered, by a weak battery. The BOR reset prevents software errors in low power situations. The voltage that trips a brownout reset is programmed using the same configuration technique used to program the WDT. Table 6-3 lists the parameters and voltage levels for the BOR. The BOR is enabled with a BOR = ON or disabled with a BOR = OFF configuration value.

Suppose that the power supply voltage of 4.2 V is used as the BOR trip voltage in a given system. Example 6-3 illustrates the C language configuration for setting the BOR.

TABLE 6-3 Brownout reset voltage levels.

C Language	Assembly Language	Brownout Voltage
#pragma config BORV = 45	_BORV_45_2L	4.5 V
#pragma config BORV = 42	_BORV_42_2L	4.2 V
#pragma config BORV = 27	_BORV_27_2L	2.7 V
#pragma config BORV = 20	_BORV_20_2L	2.0 V

Note: These voltage levels may be different in different microcontrollers, so refer to the data sheet for the given microcontroller or the configuration bit appendix for the C18 compiler. (The PIC18F1320 is illustrated here.)

EXAMPLE 6-3

```
#pragma config BORV = 42    // set brownout to 4.2 volts
#pragma BOR = ON
```

Stack Resets.   The stack never overflows or underflows during normal program operation. If a system program becomes erratic for whatever reason, the stack can underflow or overflow. The underflows are caused by too many returns or pops, and the overflows are caused by too many function calls or pushes. If one of these errors occurs, the system automatically resets if the stack reset is enabled. The stack reset is enabled/disabled with a #pragma config STVR = ON/OFF statement. An overflow occurs if more than 31 return addresses are placed on the stack, and an underflow occurs if too many return addresses are removed from the stack.

Sleep.   A microcontroller is placed into a sleep state or mode by using either the Sleep() function in C language or the SLEEP instruction in assembly language. In the sleep mode, the microcontroller's oscillator is turned off and the program stops operating to minimize the power consumption. This is useful in battery-operated systems. During the sleep mode, the watchdog timer can be programmed to continue running so when the watchdog times out, the microcontroller wakes from the sleep state. Events that generate an interrupt will also wake the microcontroller from the sleep mode because an interrupt also exits the sleep mode. Many applications benefit from the sleep mode to conserve power, but if a device is not operated from a battery, the sleep mode is rarely used. A good example of the sleep mode is the remote control for your television. When a button is pressed to change a channel or other function, an interrupt is generated that causes the microcontroller to exit the sleep mode and perform the function. When the function is completed, the microcontroller again enters the sleep mode. This reduces power consumption to a low level, allowing the battery potentially to last for years.

Clock

The microcontroller is a synchronous system that requires a timing source or clock to operate. Most PIC18 family members have the feature of selecting the clock input, which is provided from up to 10 different sources. (Some versions, indicated with an asterisk in the following list, do not have an internal oscillator and some versions may have additional modes.) The oscillator determines the execution speed of the microcontroller and is used in determining time delays and other events inside of the microcontroller. The clock sources are:

1. Low-power crystal (LP)
2. Crystal or ceramic resonator (XT)
3. High-speed crystal or ceramic resonator (HS)
4. High-speed crystal or ceramic resonator with PLL (HSPLL)
5. External resister/capacitor with Fosc/4 output on OSC2 (RC)
6. External resister/capacitor with I/O on OSC2 (RCIO)
7. Internal oscillator with Fosc/4 on RA6 and I/O on RA7 (INTIO1)
8. Internal oscillator with I/O on RA6 and RA7 (INTIO2)
9. External clock with Fosc/4 (EC)
10. External clock with I/O on RA6 (ECIO)

The first four modes use an external crystal or ceramic resonator connected to the OCS1 and OSC2 pins on the microcontroller. These are the most common ways to provide a clock to the microcontroller. If a crystal is used, a parallel resonant crystal not a series resonant, must be selected. Series resonant crystals do not oscillate when the system if first powered, so they are rarely used in a system design. Figure 6-4 illustrates the connection of a crystal to the OSC1 and OSC2 pins. Here 27 pF capacitors are used with the crystal. The value of the capacitors may need

FIGURE 6-4 Using a crystal or ceramic resonator as the clock input.

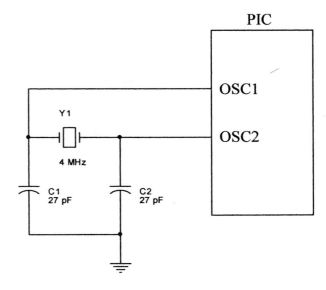

to be slightly adjusted (15 pF – 33 pF) for reliable oscillator operation, but 27 pF works reliably with most crystal or resonator frequencies.

If the PLL (phased-locked loop) mode is selected, an internal PLL multiplies the crystal frequency by a factor of 4. Therefore, if a 4-MHz crystal is connected to OSC1 and OSC2 and the PLL mode is selected, the microcontroller internally operates at a 16-MHz clock rate. This type of clock operation may be required in a system that needs to reduce **electromagnetic interference** (EMI) by keeping the external clock frequency as low as possible.

The clock also operates with an external RC circuit used to time the frequency. Figure 6-5 depicts such a scheme. A close approximation of the operating frequency is $\frac{1}{4.2\,RC}$. This will vary

FIGURE 6-5 RC circuit used for timing.

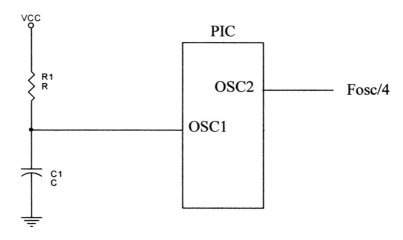

2-MHz operation is attained with R = 3.9K and C = 30 pF; Fosc/4 is 500 KHz with these values

R = 3K to 100K
C > 20 pF

FIGURE 6-6   Driving the PIC with an external clock input.

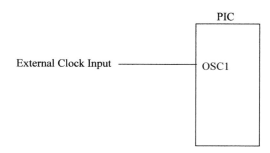

somewhat with the supply voltage and the tolerances on the components. Because there will be variations due to component tolerances, an RC timing circuit should be used only in systems that do require precise instruction timings. In the RC mode, the OSC2 pin is programmed as an I/O pin or as an oscillator output divided by 4 (Fosc/4).

Two other modes of operation use an external clocking source wherein the external clock is connected to OSC1. The OSC2 is used as either an I/O connection or as the Fosc/4 signal, which is one-fourth the clock frequency. Figure 6-6 illustrates the external clock input signal.

To select the operation of the clock, the configuration bits are programmed. Example 6-4 illustrates the configuration code for a C language system that uses a crystal oscillator as a timing source of 4 MHz. The OSC value is the same as listed at the beginning of this section, that is, HS for mode 3, RC for mode 5, and so forth. If a program is developed using only assembly language, refer to the include file for the microcontroller selected for the system. The include file contains the information needed in assembly language for programming the configuration register for the oscillator mode.

EXAMPLE 6-4

```
#pragma config OSC = HS          // select a high-speed crystal oscillator
//     OR
#pragma config OSC = RC          // select an RC oscillator
```

If internal oscillator operation is selected (modes 7 and 8), the OSC1 and OSC2 pins are available for I/O (Port A bits 6 and 7) or as Fosc/4 and PA7. (Not all versions of the PIC have an internal oscillator.) Two internal clock frequencies are available, one operates at 8 MHz and the other operates at 31 KHz. The 8-MHz signal can be scaled down from 4 MHz to 125 KHz by an internal scaler. Figure 6-7 depicts the oscillator control register and the oscillator tuning register, which are used to control the internal oscillators. Example 6-5 shows how to use C language to program the PIC to use an internal oscillator at a frequency of 4 MHz. The center frequency is chosen in this example. In addition to the code in Example 6-5, the #pragma config OSC = INIO2 or #pragma config OSC = INTIO1 statement must appear in the program to select the internal oscillator operation. Mode INTIO2 causes the OSC1 and OSC2 to function as I/O pins, and INTIO1 causes the OSC2 pin to provide Fosc/4 and OSC1 to function as an I/O pin. To select the 31-KHz internal oscillator, the INTRC pragma is used in a program.

EXAMPLE 6-5

```
OSCCON = 0x62;    // select internal oscillator frequency
OSCTUNE = 0;      // tune oscillator at center
```

Now that the clock and reset connections are understood, an example is needed to tie things together. Figure 6-8 illustrates the microcontroller using a 4-MHz crystal and a reset pushbutton and circuit. Example 6-6 lists the C language program used to program the micro-

FIGURE 6-7 Internal oscillator control registers.

(a) OSCCON register

IDLEN	IRCF2	IRCF1	IRCF0	OSTS	IOFS	SCS1	SCS0

IDLEN = 0   Run mode enabled
IDLEN = 1   Idle mode enabled

IECF2 IRCF1 IRCF0
  0   0   0 = 31 KHz
  0   0   1 = 125 KHz
  0   1   0 = 250 KHz
  0   1   1 = 500 KHz
  1   0   0 = 1 MHz
  1   0   1 = 2 MHz
  1   1   0 = 4 MHz
  1   1   1 = 8 MHz

OSTS = 0   Oscillator start-up timer running
OSTS = 1   Oscillator start-up timer expired

IOFS = 0   Internal oscillator unstable
IOFS = 1   Internal oscillator stable

SCS1 SCS0
  0   0 = Primary oscillator
  0   1 = Timer 1 oscillator
  1   0 = Internal oscillator
  1   1 = Internal oscillator

(b) OSCTUNE register

		T5	T4	T3	T2	T1	T0

T bits:

    011111 =   maximum frequency

    000001
    000000 =   center frequency
    111111

    100000 =   minimum frequency

controller so that the WDT is enabled to cause a reset once per second and the brownout voltage is set to 4.2 V.

This circuit seems to have an extra capacitor (C4). This capacitor is needed to decouple the microcontroller from the power supply as described early in this chapter. Because the microcontroller is a synchronous circuit, it generates high-frequency noise (**switching noise**) on its power supply connections. To filter the noise, C4 provides a path to ground for any high-frequency noise generated by the microcontroller. Standard practice requires a 0.1 μF capacitor for bypassing the noise on each analog or digital integrated circuit in a system. If a capacitor is not used to bypass the power supply connection, problems and mysterious program crashes will often occur.

EXAMPLE 6-6

```
#pragma config MCLRE = ON        // enable master clear input
#pragma config OSC = HS          // select crystal oscillator
#pragma config WDT = ON          // set watchdog
#pragma config WDTPS = 256       // watchdog time is 1 second
```

FIGURE 6-8 Typical reset and oscillator circuit.

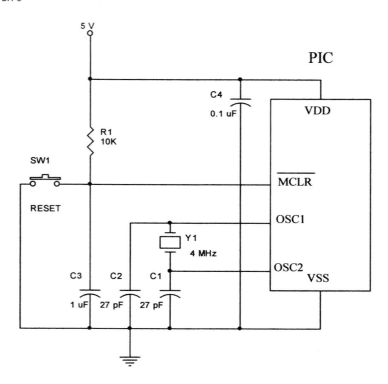

```
#pragma config BORV = 42          // set brownout reset voltage
#pragma BOR = ON                  // brownout is on

void main (void)

     // initialize system here

     while (1)                    // main program loop
     {
          ClrWdt();       // reset watchdog

          // system software goes here

     }
```

The main loop in this program contains the ClrWdt function. Because a properly designed system functions in the main loop, the WDT is cleared and never fires to reset the microcontroller. In C language, the main program loop for any operating system is an infinite while loop, even for such complex operating systems as Windows and Linux. The WDT can also be enabled using the WTCONbits.SWDTEN = 1; instruction in a C language program instead of the #pragma statement.

6-2   I/O PORT PINS

The number of I/O ports and port pins vary by PIC18 family member, but all versions have at least Port A and Port B. The pins for the I/O ports are labeled as RA0, for Port A bit position zero. The R in RA0 indicates register so the pin is for register A, bit zero.

Port A

Port A pins are used in Figure 6-8. The $\overline{\text{MCLR}}$ input is RA5 (on the PIC18F1220) and is typically used for a manual reset input so it is often not available as an I/O pin. Likewise, the RA7 and RA6 bits (on the PIC18F1220) are used in Figure 6-8 for the OCS1 (RA7) and OSC2 (RA6) pins. Port A has three registers that define its operation and are used for its operation: TRISA, PORTA, and LATA. The TRISA register defines the direction of the Port A pins, where a logic one in a bit position defines the pin as an input pin and a zero in a bit position defines it as an output pin. Some versions use a DDRA (data direction) register, which is an alias to the TRISA register to control the TRISA bits. For a particular version of the microcontroller, refer to the .lkr file for the definitions). The PORTA register is used to read the pins on Port A or write the pins. The LATA register is used internally to read, modify, and then write to the Port A pins and is not typically directly addressed in software. The latch register allows the port to change without glitches. An example of a read, modify, and write operation is a C-language instruction such as PORTA++, where Port A is read, incremented, and then written back to Port A. To better understand the placement and operation of a port, refer to Figure 6-9, which illustrates the internal structure of a port. To write data to the port, write to either the LATA register

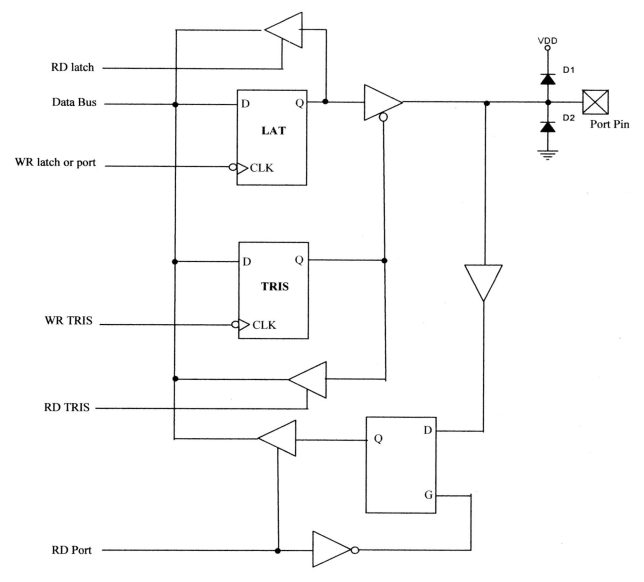

FIGURE 6-9   Internal structure of a port pin.

or the PORTA register. To read the port pins, read PORTA. To read the contents of the port latch, read LATA. Typically only PORTA is used as a register in a program for accessing Port A data.

In C language, the entire port is read or written using PORTA, or just a single bit is read or written using PORTAbits.RA0 to read/write bit position 0 of Port A. Example 6-7 shows how to write a 0x45 to Port A, a zero to Port B bit 3, and a one to Port B bit 5. This assumes that the ports are defined as digital port pins and programmed for output operation. The example also illustrates how to address individual TRIS register bits to program a port pin as an input (1) or as an output (0).

EXAMPLE 6-7

```
TRISBbits.TRISB3 = 0;    // program RB3 as output
TRISBbits.TRISB5 = 0;    // program RB5 as output

PORTA = 0x45;            // Port A is 0x45
PORTBbits.RB3 = 0;       // RB3 = 0
PORTBbits.RB5 = 1;       // RB5 = 1
```

Table 6-4 lists all of the Port A pins and the function of each for the PIC18F1220 microcontroller. Other versions may have different port pin functions, so always refer to the data sheet for the microcontroller chosen for a project. Also, open the .lkr file to see which register names have been defined for the TRIS register (either TRIS or DDR). Many I/O pins have two or three functions as selected by programming as will be discussed later in this chapter. The RA5 (master clear) pin cannot be programmed as an output pin. The analog inputs are used with an internal analog-to-digital converter that is also discussed later in this chapter.

On a microcontroller reset, in some versions of the PIC18, bits 0 through 3 are programmed as analog inputs. To program these bits for digital input, the ADCON1 register is addressed and programmed with ones for each Port A pin that is a digital input, and a zero bit for each analog input. By default, the port pins are programmed as analog inputs because the ADCON1 register is cleared to zero on a reset. The rightmost seven bits of ADCON1 program are AN0–AN6. Some of these analog inputs (AN0–AN3) are on Port A and some (AN4–AN6) are on Port B. To configure Port A as a digital input Port where RA0–RA2 are inputs and RA3–RA4 are outputs, refer to Example 6-8. Here it is assumed that MCLR is enabled and the oscillator mode is HS, so both OSC1 and OSC2 are connected to a crystal.

EXAMPLE 6-8

```
ADCON1 = 0x0f;          // Ports A and B pins are digital

TRISA = 0x07;           // Port A bits 0--2 are inputs
                        // Port A bits 3--4 are outputs
```

TABLE 6-4 Port A pin functions.

Pin Name	Bit Position	Functions
RA0/AN0	0	I/O pin	analog input
RA1/AN1/LVDIN	1	I/O pin	analog input	low voltage detect input
RA2/AN2/Vref-	2	I/O pin	analog input	Vref-
RA3/AN3/Vref+	3	I/O pin	analog input	Vref+
RA4/T0CKI	4	I/O pin	external clock input to timer 0
RA5/MCLR//Vpp	5	I/O pin	master clear	programming voltage input
RA6/OSC2/CLKO	6	I/O pin	OSC2	clock output pin
RA7/OSC1/CLKI	7	I/O pin	OSC1	clock input pin

TABLE 6-5 Port B pin functions.

Pin Name	Bit Position	Function
RB0/AN4/INT0	0	I/O pin	analog input	interrupt input 0
RB1/AN5//INT1	1	I/O pin	analog input	interrupt input 1
RB2/P1B/INT2	2	I/O pin	P1B data	interrupt input 2
RB3/CCP1	3	I/O pin	capture/compare or PWM output
RB4/AN6	4	I/O pin	analog input
RB5/PGM	5	I/O pin	single supply programming
RB6/PGC	6	I/O pin	serial programming clock
RB7/PGD	7	I/O pin	serial programming data

*These pin definitions vary on different family members.

Port B

Port B is programmed in the same manner as port A. The TRISB register programs the direction of the Port B bits, the PORTB register is used to read or write to the port, and the LATB register is used to read or write to the port latch. As mentioned with Port A, the ADCON1 register programs the function of the analog/digital pins on both Port A and Port B.

The functions of the Port B pins (for the PIC18F1220) are listed in Table 6-5. As with Port A, Port B pins have multiple functions. Not all versions have all of the functions listed here and some have additional functions. As each feature is explained, the multiple functions will become clear. The port pins are used primarily as simple digital input/output pins.

The weak pull-ups are selected by programming a bit in the INTCON2 register called the RBPU bit. If a logic zero is placed in the RBPU bit of INTCON2, then the pull-ups are enabled for the Port B pins. These pull-ups are disabled after a reset or when a port pin is programmed as an output pin. If the weak pull-ups are enabled, the microprocessor connects the pin to 5 V through an internal pull-up resistance. This allows input devices, such as switches, to be connected to the Port B pins without the use of an external pull-up resister. This saves money because of the reduced component count.

Example 6-9 shows how to enable the weak pull-up resisters in C language. As with other initialization software, this is placed early in a program before the main program loop.

EXAMPLE 6-9

```
INTCON2bits.RBPU = 1;          // enable weak pull-ups
```

Ports C, D, E, and Beyond

Many PIC18 family members have more than two I/O ports and may contains Ports C, D, and E. These additional ports are programmed in the same manner as Ports A and B. Each has a TRIS, PORT, and LAT register for general digital I/O programming. As with Ports A and B, many of the pins on these additional ports have more than one function. The data sheet for a particular family member lists the multiple functions of each I/O pin. Generally, the more pins on the microcontroller, the more I/O ports available for interfacing.

I/O Example

This section demonstrates a complete example, illustrating the hardware and software, in order to connect things explained in this and prior chapters. This first example system, which can be constructed, shows how to design and program a simple dice toss system from a pushbutton, and 14 LED diodes, for displaying a pair of dice. This is a simple system, with a short program written in both C and assembly language to highlight the differences. This example, uses the PIC18F1220 because of its cost ($2.78) and small integrated circuit size (18-pin PDIP). Figure 6-10 illustrates

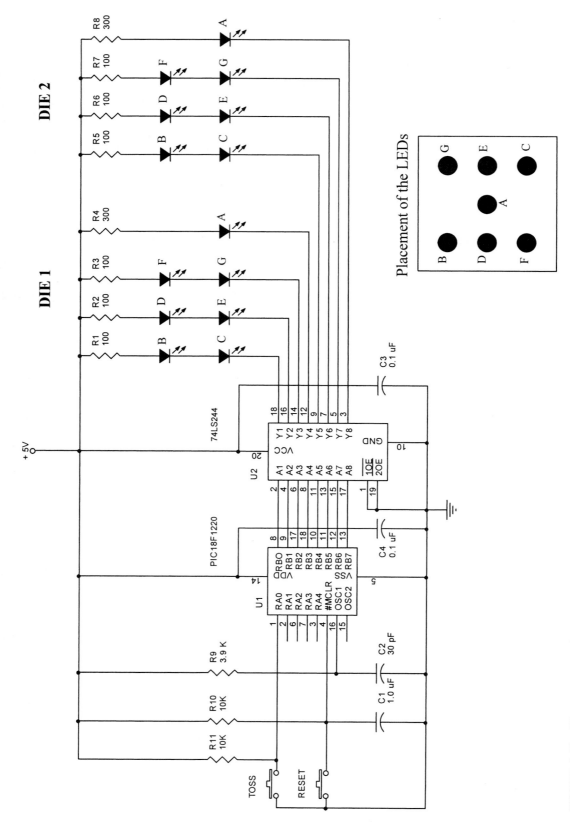

FIGURE 6-10 Dice toss example.

the complete schematic of the dice toss system. The power supply is listed as 5 V, but three AA batteries connected in series can be used with this system.

This circuit uses an RC-controlled oscillator that runs at about 2 MHz, which is fast enough for this application. The application has no critical timing functions, so an RC oscillator is an inexpensive method of obtaining a clock signal. Two pushbuttons are used for the system, but the RESET pushbutton can be eliminated to reduce the cost. To toss the dice, press the TOSS pushbutton; when it's released, a random pair of dice faces is displayed. This dice toss system can be used with any game or even to play craps. The only hardware design required is to select the resister values for biasing the LEDs. Most low-cost LEDs require a nominal current of 10 mA and a forward voltage of 1.65 V to light. Because the PIC provides only 8.5 mA maximum per output pin (see the details in the first few pages of this chapter), a buffer is added between the PIC and the LEDs. The buffer could be eliminated if the forward current is reduced to 8.5 mA instead of 10 mA and the LEDs will still light, although they may be slightly dimmer. If more than 8.5 mA of current is used from an output pin, the signal from the output pin is not guaranteed to be a valid TTL voltage level. (Even though the outputs in this example are not used for TTL signals, one example limiting the outputs to these current levels is a good practical experience in interfacing).

The single LED (LEDA) requires a 300 Ω resister for bias. The forward voltage drop across an LED varies from 1.5 V to 2 V even though the data sheet claims that it is a nominal 1.65V. For the calculation, 3 V is dropped across the resister and 2 V is dropped across the LED when the output of the buffer becomes a logic zero. The calculated value is $\left(\text{using Ohm's law } R = \frac{E}{I}\right) \frac{3V}{10mA}$, or 300 Ω. A 100 Ω is used in series with the pairs of LEDs. The voltage drop across the resister is 1 V, with 2 V dropped across each LED yielding a value of 100 Ω $\left(\frac{1.0V}{10mA}\right)$.

The inputs to the LEDs are active low. That is, to light an LED a logic zero is required from the PIC. The buffers, in the circuit are 74LS244 octal buffers, which are noninverting buffers. The correct bit pattern to light an LED is stored in a lookup table in the program memory in this application. To store data in the program memory, use the tokens "rom" and "near" before the type of data, in this case char. The rom token causes the compiler to store the table in the program memory (which is static), and the near token causes the compiler to use a 16-bit near address, which should be used in systems with fewer than 64K bytes of program memory because if is more efficient. If the near token is not used, the address is a 24-bit address. The C language version of the program appears in Example 6-10, which shows the 36 combinations of a pair of dice as stored in the lookup table. For a comparison of assembly and C language, Example 6-11 illustrates the assembly language version of the same software. If these versions are compared, it is evident which took less time to write. Both versions use the WDT, although it isn't needed in an application such as this one.

EXAMPLE 6-10

```
/*
 * Dice Toss example written for a PIC18F1220
 */

#include <p18cxxx.h>

/* Set configuration bits
 *   - set RC oscillator
 *   - disable watchdog timer
 *   - disable low-voltage programming
 *   - disable brownout reset
 *   - enable master clear
 */

#pragma config OSC = RC
#pragma config WDT = ON
#pragma config WDTPS = 256
#pragma config LVP = OFF
#pragma config BOR = OFF
#pragma config MCLRE = ON
```

```
// ****************** program memory data ************************

// Lookup table for dice

// 7 <0111> = 1, E <1110> = 2, 6 <0110> = 3,
// A <1010> = 4, 2 <0010> = 5, 8 <1000> = 6

rom near char lookup[] =    // all 36 dice combinations
{
      0x77, 0x7E, 0x76,          // 1,1 1,2 1,3
      0x7A, 0x72, 0x78,          // 1,4 1,5 1,6
      0xE7, 0xEE, 0xE6,          // 2,1 2,2 2,3
      0xEA, 0xE2, 0xE8,          // 2,4 2,5 2,6
      0x67, 0x6E, 0x66,          // 3,1 3,2 3,3
      0x6A, 0x62, 0x68,          // 3,4 3,5 3,6
      0xA7, 0xAE, 0xA6,          // 4,1 4,2 4,3
      0xAA, 0xA2, 0xA8,          // 4,4 4,5 4,6
      0x27, 0x2E, 0x26,          // 5,1 5,2 5,3
      0x2A, 0x22, 0x28,          // 5,4 5,5 5,6
      0x87, 0x8E, 0x86,          // 6,1 6,2 6,3
      0x8A, 0x82, 0x88          // 6,4 6,5 6,6
};

// ****************** data memory data ***************************

int count;                  //random number

// ******************* main program *****************************


void main (void)
{
//
// initialize system
//
      ADCON1 = 0x0F;        // all digital port pins
      TRISA = 0x01;         // Port A, bit 0, is input
      TRISB = 0;            // Port B is output
      PORTB = 0xFF;         // all LEDs are off
      count = 0;            // start count at zero

// ************** operating system loop **************************

      while (1)             // the one and only program loop
      {
            ClrWdt();       // reset watchdog

            if (PORTAbits.RA0 == 0)   // if TOSS = 0
            {
                  count++;              // generate random number

                  if (count == 36)    // keep count between 0 and 35
                        count = 0;

                  PORTB = lookup[count];   // display dice code
            }
      }
}
```

EXAMPLE 6-11

```
;*************************************************************************
;Dice Toss example written in assembly language

      LIST P=18F1220, F=INHX32 ;directive to define processor and file format
      #include <P18F1220.INC>    ;processor-specific variable definitions

;*************************************************************************
;Configuration bits

      __CONFIG _CONFIG1H, _IESO_OFF_1H & _FSCM_OFF_1H & _RC_OSC_1H
      __CONFIG _CONFIG2L, _BOR_OFF_2L & _PWRT_OFF_2L
```

```
            __CONFIG _CONFIG2H, _WDT_ON_2H & _WDTPS_4_2H
            __CONFIG _CONFIG3H, _MCLRE_ON_3H
            __CONFIG _CONFIG4L, _DEBUG_OFF_4L & _LVP_OFF_4L & _STVR_OFF_4L
            __CONFIG _CONFIG5L, _CP0_OFF_5L & _CP1_OFF_5L
            __CONFIG _CONFIG5H, _CPB_OFF_5H & _CPD_OFF_5H
            __CONFIG _CONFIG6L, _WRT0_OFF_6L & _WRT1_OFF_6L
            __CONFIG _CONFIG6H, _WRTC_OFF_6H & _WRTB_OFF_6H & _WRTD_OFF_6H
            __CONFIG _CONFIG7L, _EBTR0_OFF_7L & _EBTR1_OFF_7L
            __CONFIG _CONFIG7H, _EBTRB_OFF_7H

;********************************************************************************
;Data memory definitions

            UDATA_ACS

COUNT       RES   1    ;count for random number

;Program memory definitions
;
;lookup table

Look        CODE_PACK
Lookup:
            DB    0x77, 0x7E, 0x76         ; 1,1 1,2 1,3
            DB    0x7A, 0x72, 0x78         ; 1,4 1,5 1,6
            DB    0xE7, 0xEE, 0xE6         ; 2,1 2,2 2,3
            DB    0xEA, 0xE2, 0xE8         ; 2,4 2,5 2,6
            DB    0x67, 0x6E, 0x66         ; 3,1 3,2 3,3
            DB    0x6A, 0x62, 0x68         ; 3,4 3,5 3,6
            DB    0xA7, 0xAE, 0xA6         ; 4,1 4,2 4,3
            DB    0xAA, 0xA2, 0xA8         ; 4,4 4,5 4,6
            DB    0x27, 0x2E, 0x26         ; 5,1 5,2 5,3
            DB    0x2A, 0x22, 0x28         ; 5,4 5,5 5,6
            DB    0x87, 0x8E, 0x86         ; 6,1 6,2 6,3
            DB    0x8A, 0x82, 0x88         ; 6,4 6,5 6,6

;********************************************************************************
;Reset vector
; This code will start executing when a reset occurs.

RESET_VECTOR CODE   0x0000

            goto   Main    ;go to start of main code

            CODE

;********************************************************************************
;Start of main program

Main:
      MOVLW  0x0F
      MOVWF  ADCON1             ;all digital ports

      MOVLW  0x01
      MOVWF  TRISA              ;Port A, bit zero, is input

      MOVLW  0
      MOVWF  TRISB              ;Port B is output

      CLRF   PORTB              ;all LEDs off
      CLRF   COUNT              ;clear count

MainLoop:

      CLRWDT                    ;reset watchdog

      BTFSC  PORTA,0            ;test bit 0 of Port A
      BRA    MainLoop           ;if bit 0 is a one

      INCF   COUNT              ;increment count

      MOVLW  .36
      SUBWF  COUNT, 0
      BNZ    DoNotClear         ;if count is not 36
      CLRF   COUNT
```

```
DoNotClear:

        MOVLW   UPPER(Lookup)    ;get lookup table address
        MOVWF   TBLPTRU
        MOVLW   HIGH(Lookup)
        MOVWF   TBLPTRH
        MOVLW   LOW(Lookup)
        MOVWF   TBLPTRL
        MOVF    COUNT,0          ;add count to table address
        ADDWF   TBLPTRL
        MOVLW   0
        ADDWFC  TBLPTRH
        ADDWFC  TBLPTRU
        TBLRD*                   ;lookup code
        MOVF    TABLAT,0         ;get code to WREG
        MOVWF   PORTB            ;display it

        GOTO    MainLoop


;*****************************************************************************
;End of program

        END
```

Suppose that Figure 6-10 is redesigned using all the features available in the PIC18F1220 to reduce the cost as much as possible. One change to the circuitry is to reduce the current for the LEDs to 8.5 mA and remove the buffer. Another is to move the TOSS switch to a Port B pin so the pull-up resister is not needed. The port B pins have internal pull-up resisters designed for this purpose. Instead of the RC oscillator, the internal oscillator can be used to eliminate the cost of the resister and capacitor. Finally, the RESET pushbutton can be removed. The RC circuit on reset is still needed because the power supply is not known and may not rise in a short enough time to properly reset the microcontroller. Figure 6-11 shows this new circuitry and Example 6-12 illustrates the C language version of the software. Only a few additional lines of code are needed to make the changes.

EXAMPLE 6-12

```
/*
 * Dice Toss example written for a PIC18F1220
 */

#include <p18cxxx.h>

/* Set configuration bits
 *   - set internal clock
 *   - enable watchdog timer
 *   - disable low-voltage programming
 *   - disable brownout reset
 *   - enable master clear
 */

#pragma config OSC = INTIO2
#pragma config WDT = ON
#pragma config WDTPS = 256
#pragma config LVP = OFF
#pragma config BOR = OFF
#pragma config MCLRE = ON

// program memory data

// Lookup table for dice
//
// 7 <0111> = 1, E <1110> = 2, 6 <0110> = 3,
// A <1010> = 4, 2 <0010> = 5, 8 <1000> = 6
//
```

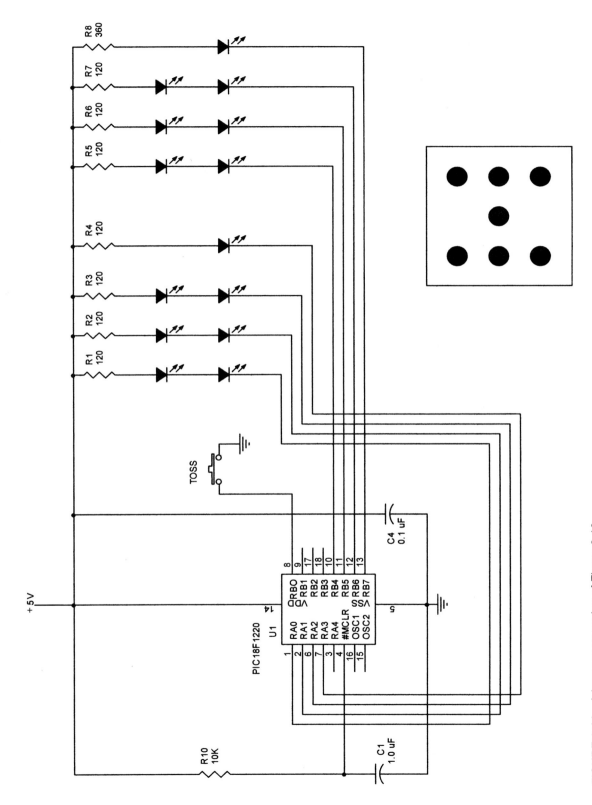

FIGURE 6-11 A lower-cost version of Figure 6-10.

```
rom near char lookup[] =    // all 36 dice combinations
{
        0x77, 0x7E, 0x76,           // 1,1 1,2 1,3
        0x7A, 0x72, 0x78,           // 1,4 1,5 1,6
        0xE7, 0xEE, 0xE6,           // 2,1 2,2 2,3
        0xEA, 0xE2, 0xE8,           // 2,4 2,5 2,6
        0x67, 0x6E, 0x66,           // 3,1 3,2 3,3
        0x6A, 0x62, 0x68,           // 3,4 3,5 3,6
        0xA7, 0xAE, 0xA6,           // 4,1 4,2 4,3
        0xAA, 0xA2, 0xA8,           // 4,4 4,5 4,6
        0x27, 0x2E, 0x26,           // 5,1 5,2 5,3
        0x2A, 0x22, 0x28,           // 5,4 5,5 5,6
        0x87, 0x8E, 0x86,           // 6,1 6,2 6,3
        0x8A, 0x82, 0x88            // 6,4 6,5 6,6
};

// data memory data

int count;                  //random number

// main program

void main (void)
{
//
// initialize system
//
        OSCCON = 0x64;      // frequency set to 4 MHz
        ADCON1 = 0x0F;      // all digital port pins
        TRISB = 0x01;       // Port B, bit 0, is input
        TRISA = 0;          // Port A is output
        PORTB = 0xF0;       // all LEDs are off
        PORTA = 0x0F;

        INTCON2bits.RBPU = 1;       // pull-ups on

        count = 0;          // start count at zero

//
// operating system loop
//

        while (1)           // the one and only program loop
        {
            ClrWdt();       // reset watchdog

            if (PORTBbits.RB0 == 0)      // if TOSS = 0
            {
                count++;                 // generate random number

                if (count == 36)
                    count = 0;

                PORTA = lookup[count];   // get die code
                PORTB = lookup[count];   // get die code
            }
        }
}
```

6-3   INTRODUCTION TO INTERRUPTS

Interrupts are used in many applications for just about all but the simplest I/O. The PIC18 family has two basic interrupts levels: the low-priority interrupt, which vectors to program memory location 0x000018, and the high-priority interrupt, which vectors to program memory location 0x000008. Interrupt priority takes precedence only when both interrupts occur simultaneously and, in such a case, the high-priority interrupt is serviced or handled first. A high-priority interrupt will always interrupt a low-priority interrupt.

An **interrupt** is either a hardware or software event that is initiated when a function call interrupts the currently executing program. Because an interrupt calls a function, the point of interruption is placed on the program stack so a return will return to the point of interruption at the end of the interrupt function. The function called by the interrupt is often referred to as an **interrupt service function (ISF)** or an **interrupt service procedure (ISP)**. In the PIC18 family, only two such directly accessed functions can actually exist in a system because only two interrupt vectors are available. An **interrupt vector** is the address of an interrupt service procedure. In either the low- or high-priority interrupt service procedure, the cause or source of the interrupt is determined by testing the **interrupt flag bits**, which are located in various interrupt control registers. This testing of the interrupt flags for a particular interrupt is often called **polling** and is used to prioritize multiple interrupts to either the high or low interrupt vector.

When an interrupt is detected by the microcontroller, the corresponding global interrupt enable bit is cleared, blocking an interrupt of equal or lower level. If a high-priority interrupt occurs, no other interrupt is possible until its global interrupt enable bit is set, which may at times be forced from within the interrupt. If a low-priority interrupt occurs, it can be interrupted by a high-priority interrupt. The interrupt control bits are located in the INTCON register along with other interrupt control and interrupt flag bits and the RCON register, as depicted in Figure 6-12. All versions of the PIC18 microcontroller have an INTCON and RCON (reset control) register. The reset control register indicates the state of the various types of resets in a system and is also used to enable only high-priority interrupts (IPEN = 0) or high- and low-priority interrupts

FIGURE 6-12 Reset and interrupt control register.

INTCON

GIE GIEH	PEIE GIEL	TMR0IE	INT0IE	RBIE	TMR0IF	INT0IF	RBIF

GIE/GIEH = 0	Disables all interrupts when IPEN = 0; disables high-priority interrupts when IPEN = 1
GIE/GIEH = 1	Enables all high-priority interrupts when IPEN = 1; enables unmasked interrupts when IPEN = 0
PEIE/GIEL = 0	Disables all interrupts when IPEN = 0; disables low-priority interrupts when IPEN = 1
PEIE/GIEL = 1	Enables all low-priority interrupts when IPEN = 1; enables unmasked interrupts when IPEN = 0
TMR0IE = 0	Disables Timer 0 overflow interrupt
TMR0IE = 1	Enables Timer 0 overflow interrupt
INT0IE = 0	Disables INT0 pin interrupt
INT0IE = 1	Enables INT0 pin interrupt
RBIE = 0	Disables Port B change interrupt
RBIE = 1	Enables Port B change interrupt
TMR0IF = 0	Timer 0 did not overflow
TMR0IF = 1	Timer 0 did overflow (must be cleared with software)
INT0IF = 0	INT0 pin interrupt did not occur
INT0IF = 1	INT0 pin interrupt did occur (must be cleared with software)
RBIF = 0	Port B pin change interrupt did not occur
RBIF = 1	Port B pin change interrupt did occur (must be cleared with software) (At least one of RB4—RB7 pins changed)

FIGURE 6-12   (*continued*)   **RCON**

IPEN	LWRT		$\overline{\text{RI}}$	$\overline{\text{TO}}$	$\overline{\text{PD}}$	$\overline{\text{POR}}$	$\overline{\text{BOR}}$

IPEN = 0   Disables priority interrupts
IPEN = 1   Enables priority interrupts

LWRT = 0   Disables table write to internal program memory
LWRT = 1   Enables table write to internal program memory

$\overline{\text{RI}}$ = 0   RESET instruction was executed (software must set $\overline{\text{RI}}$ after RESET)
$\overline{\text{RI}}$ = 1   No RESET instruction executed

$\overline{\text{TO}}$ = 0   WDT timeout occurred
$\overline{\text{TO}}$ = 1   Cleared by power up, CLRWDT, or SLEEP

$\overline{\text{PD}}$ = 0   SLEEP executed
$\overline{\text{PD}}$ = 1   Cleared by power up or CLRWDT

$\overline{\text{POR}}$ = 0   Cleared by power on reset (must be set by software after a power-on reset)
$\overline{\text{POR}}$ = 1   No power on reset occurred

$\overline{\text{BOR}}$ = 0   A brownout reset occurred (must be set by software afterwards)
$\overline{\text{BOR}}$ = 1   No brownout reset occurred

Note: The function of the LWRT bit may be different in various PIC18 versions.

(IPEN = 1). This register is also tested to determine what type of reset occurred if a program needs to determine the type of reset. A sample use of the reset information is to determine whether the reset is a cold boot or a warm boot. A cold boot is from the initial system power and a warm boot might occur after a brownout reset or watchdog reset.

When an interrupt is accepted by the microcontroller, it occurs at any point in the software. Because of the seemingly random nature of interrupts, all registers used by the interrupt service procedure must be saved, or anything changed in the interrupt service procedure will change in the software that is interrupted. In assembly language this requires quite a bit of work, but in C language most of the saving is accomplished by the compiler. (The exact state save for an interrupt is defined at the end of the .lkr file). There are cases where a program must still save registers in C language. The registers listed in Table 6-6 are saved by the interrupt structure of the C language compiler. Missing from this table are the product registers (PROD), because these

TABLE 6-6   Registers automatically saved by an interrupt in C language.

Resource	Description
PC	Addresses instructions
WREG	Intermediate variables
STATUS	Condition bits
BSR	Memory bank selection
FSR0	Pointer to the data memory
FSR1	Pointer to the data stack
FSR2	Frame pointer

registers are not automatically saved. If the interrupt service procedure uses the product or table registers, they must be implicitly saved or the interrupt service procedure will change things in other software. An example is when the interrupt occurs after the multiply instruction, but before the PRODL and PRODH registers are saved. If the interrupt service procedure uses the PRODL or PRODH register, the values will change in the program that was interrupted.

Example 6-13 shows how to save the product or table registers in an interrupt service procedure. MyHighInt saves the PROD registers and MyLowInt saves the registers used by the Math library. If a function that returns a 16-bit variable is called from the interrupt service procedure, PROD must be saved and if a function returns a 24-bit or 32-bit variable, then the "MATH_DATA" section must be saved. If 8-bit data or no data are returned from a function, no special effort is needed to save the context. Every attempt should be made to keep the interrupt service procedure as short and uncomplicated as possible. Note that the standard implementation of the .lkr file already saves the PROD register for interrupts.

EXAMPLE 6-13

```
#pragma interrupt MyHighInt save=PROD

#pragma interruptlow MyLowInt save=section("MATH_DATA")

#pragma interrupt HighInt save=PROD, save=section("MATH_DATA")
```

Suppose that a counter (called count) must be incremented in the memory every time a signal of a very short duration occurs on RB0 (the INT0 input). The signal is long enough to cause an interrupt, but in a few microseconds returns to a logic one so software would not always be able to detect the signal. INT0 is an edge-triggered input, and as long as the rise time is sufficient to be detected by the microcontroller as an edge, an interrupt will occur. The minimum pulse width should be no less than 25 ns. An interrupt is used to accomplish this task because the system must also be able to perform other tasks besides looking at the signal attached to RB0. Example 6-13 shows the C language software required to initialize the RB0 pin as an INT0 input that causes a high priority interrupt. Note that the INT0 input can be only a high-priority interrupt. Figure 6-13 shows the contents of the INTCON2 register that selects the edge-triggering level of the INT0 pin.

FIGURE 6-13 INTCON2 register.

INTCON2

| $\overline{RBPU}$ | INTEDG0 | INTEDG1 | INTEDG2 | | TMR0IP | | RBIP |

$\overline{RBPU} = 0$	Port B pullups disabled
$\overline{RBPU} = 1$	Port B pullups enabled
INTEDG0 = 0	INT0 is negative edge-triggered
INTEDG0 = 1	INT0 is positive edge-triggered
INTEDG1 = 0	INT1 is negative edge-triggered
INTEDG1 = 1	INT1 is positive edge-triggered
INTEDG2 = 0	INT2 is negative edge-triggered
INTEDG2 = 1	INT2 is positive edge-triggered
TMR0IP = 0	Timer 0 interrupt priority = low
TMR0IP = 1	Timer 0 interrupt priority = high
RBIP = 0	Port B change interrupt priority = low
RBIP = 1	Port B change interrupt priority = high

EXAMPLE 6-14

```
ADCON1 = 0x0F;                  // make port pins digital

TRISB = 1;                      // make RB0 input

RCONbits.IPEN = 1;              // IPEN = 1

INTCON2bits.INTEDG0 = 0;        // make INT0 negative edge-triggered
INTCONbits.INT0IE = 1;          // enable INT0
INTCONbits.GIEH = 1;            // enable high-priority interrupts

// INT0 is now armed and active
```

Example 6-14 programs Port B, RB0 as an input port pin and enables the INT0 pin as a negative interrupt input. The last instruction enables the interrupts so that when a pulse appears on INT0, the high-priority interrupt occurs. As mentioned, the vector address for the high-priority interrupt is at program memory location 0x0008. Example 6-15 shows how to set up a program that uses the high-priority interrupt. It also shows the definition of a low-priority interrupt. The first thing that appears is the prototype for the interrupt service procedure, followed by the two #pragma statements to define the contents of the high-priority interrupt vector as an interrupt. Both the #pragma interrupt MyHighInt (identifies MyHighInt as an interrupt service procedure) and #pragma code high_vector = 0x08 (identifies the location of the interrupt vector) statements are required. The GOTO must be in assembly language within the vector. The rest of the software is fairly straightforward except the #prama code statement just before the code for the interrupts in the program. The #pragma code statement must appear before the code in a program that uses interrupts or it will not compile correctly. Also, only one single bit can be changed at a time in the INTCON register or the results are unpredictable.

Every time that the INT0 input pin receives a negative edge, the high-priority interrupt function is executed. This function has only two instructions. One increments the count, which counts the negative edges, and the other clears the interrupt request located in the INT0IF bit (INT0 interrupt flag) in the INTCON (interrupt control) register. If you fail to clear the INT0IF bit, no additional INT0 interrupts will occur.

EXAMPLE 6-15

```
/*
 * A high-priority interrupt example
 */

#include <p18cxxx.h>

/* Set configuration bits
 *   - set RC oscillator
 *   - disable watchdog timer
 *   - disable low-voltage programming
 *   - disable brownout reset
 *   - enable master clear
 */

#pragma config OSC = RC
#pragma config WDT = OFF
#pragma config LVP = OFF
#pragma config BOR = OFF
#pragma config MCLRE = ON

void MyHighInt (void);              // prototypes for interrupts
void MyLowInt (void);               // service procedures

#pragma interrupt MyHighInt         // MyHighInt is an interrupt
#pragma code high_vector=0x08       // high_vector is the vector at 0x08

void high_vector (void)
{
     _asm GOTO MyHighInt _endasm
}
```

```
#pragma interruptlow MyLowInt      // MyLowInt is an interrupt
#pragma code low_vector = 0x18     // low_vector is the vector at 0x18

void low_vector (void)
{
     _asm GOTO MyLowInt _endasm
}

// data memory data

int count;

// high-prioity interrupt

#pragma code                       // must be here or the program will not
                                   // compile correctly!

void MyHighInt (void)
{
     count++;
     INTCONbits.INT0IF = 0;        // clear INT0IF flag
}

void MyLowInt (void)
{
     // low interrupt goes here
}

// main program

void main (void)
{
     ADCON1 = 0x0F;               // make port pins digital

     TRISB = 1;                   // make RB0 input

     count = 0;                   // start with count of 0

     RCONbits.IPEN = 0;           // IPEN = 0 (only high-priority on)

     INTCON2bits.INTEDG0 = 0;     // make INT0 negative edge-triggered
     INTCONbits.INT0IE = 1;       // enable INT0
     INTCONbits.GIE = 1;          // enable interrupts

     // do other initialization here

     while (1)
     {

          // do main program loop

     }
}
```

Other registers exist besides RCON, INTCON, and INTCON2 to control the interrupts to the microcontroller. These other register select priority levels and indicate the status of other interrupts in the system. Figure 6-14 illustrates control register three (INTCON3). As with INTCON2, the INTCON3 register contains bits for setting interrupt priority, enabling interrupts, and determining the status of an interrupt input pin.

If only one interrupt priority is used in a system, it must be the high-priority interrupt. In this case the GIEH bit is often labeled GIE (global interrupt enable) and the GIEL bit is ignored. To select a single priority level for interrupts, place a zero in the IPEN bit. There are cases where it might also be desirable to enable interrupts from within an interrupt service procedure. To do so, set either the GIEH or GIEL bit in the procedure.

Suppose that three interrupt inputs are connected to INT0 (not programmable), INT1, and INT2. All three are set as high-priority interrupts. All three are negative edge-triggered inputs. Software to install them and enable them is listed in Example 6-16. Notice how the IF bits are checked to send the program to the correct interrupt procedure for the three different input pins. In this software INT0 has the highest priority and INT2 has the lowest.

INTCON3

INT2IP	INT1IP		INT2IE	INT1IE		INT2IF	INT1IF

INT2IP = 0	INT2 pin priority is low
INT2IP = 1	INT2 pin priority is high
INT1IP = 0	INT1 pin priority is low
INT1IP = 1	INT1 pin priority is high
INT2IE = 0	disables the INT2 interrupt
INT2IE = 1	enables the INT2 interrupt
INT1IE = 0	disables the INT1 interrupt
INT1IE = 1	enables the INT1 interrupt
INT2IF = 0	INT2 has not occurred
INT2IF = 1	INT2 has occurred (must be cleared with software)
INT1IF = 0	INT1 has not occurred
INT1IF = 1	INT1 has occurred (must be cleared with software)

FIGURE 6-14   Interrupt control register three.

EXAMPLE 6-16

```
/*
 * Priority of multiple interrupts
 */

#include <p18cxxx.h>

/* Set configuration bits
 *   - set RC oscillator
 *   - disable watchdog timer
 *   - disable low-voltage programming
 *   - disable brownout reset
 *   - enable master clear
 */

#pragma config OSC = RC
#pragma config WDT = OFF
#pragma config LVP = OFF
#pragma config BOR = OFF
#pragma config MCLRE = ON

void MyHighInt (void);

#pragma interrupt MyHighInt
#pragma code high_vector = 0x08

void high_vector (void)
{
      _asm GOTO MyHighInt _endasm
}

#pragma code

void Int0 (void)
{
      INTCONbits.INT0IF = 0;        // clear INT0IF flag
      // do INT0 software here
}
```

```
void Int1 (void)
{
      INTCON3bits.INT1IF = 0;        // clear INT1IF flag
      // do INT1 software here
}

void Int2 (void)
{
      INTCON3bits.INT2IF = 0;        // clear INT2IF flag
      // do INT2 software here
}

void MyHighInt (void)                // polling interrupt inputs
{
      if (INTCONbits.INT0IF == 1)
          Int0();                    // highest priority

      else if (INTCON3bits.INT1IF == 1)
          Int1();                    // next to highest priority

      else if (INTCON3bits.INT2IF == 1)
          Int2();                    // lowest priority
}

// main program

void main (void)
{
      ADCON1 = 0x0F;                 // make ports pins digital
      TRISB = 7;                     // make RB0, RB1, and RB2 inputs

      RCONbits.IPEN = 1;             // IPEN = 1

      INTCON2bits.INTEDG0 = 0;       // make INT0 negative edge-triggered
      INTCON2bits.INTEDG1 = 0;       // make INT1 negative edge-triggered
      INTCON2bits.INTEDG2 = 0;       // make INT2 negative edge-triggered

      INTCON3bits.INT1IP = 1;        // make INT1 high priority
      INTCON3bits.INT2IP = 1;        // make INT2 high priority

      INTCONbits.INT0IE = 1;         // enable INT0
      INTCON3bits.INT1IE = 1;        // enable INT1
      INTCON3bits.INT2IE = 1;        // enable INT2

      INTCONbits.GIEH = 1;           // enable high-priority interrupts

      // do other initialization

      while (1)
      {

          // main program loop
      }
}
```

Each internal peripheral device has three bits in three different registers to control the operation of each peripheral in the interrupt structure. There are IPR registers to set the priority of the interrupt, PIE registers to set the interrupt enable bit, and PIR registers that contain the flag bits that are tested to determine which input caused the interrupt. Figures 6-15, 6-16, and 6-17 show all of these registers and depict the operation of each control/status bit.

As can be seen from these three illustrations, many internal peripherals are available to use in a system that include timers, an analog-to-digital converter (ADC), an embedded universal asynchronous/synchronous receiver/transmitter (EUSART or USART), and many other types of interrupts. Chapter 8 is devoted to the discussion of the interrupts and their use in an embedded system.

The simplest example of an interrupt system is a burglar alarm. This type of interrupt is a low-priority interrupt because it does not happen often and when it does it can wait to be serviced.

FIGURE 6-15 IPR (interrupt priority) registers.

IPR1

PSPIP	ADIP	RCIP	TXIP	SSPIP	CCP1IP	TMR2IP	TMR1IP

PSPIP	=	Parallel slave port read/write interrupt (0 = low priority, 1 = high priority)
ADIP	=	ADC interrupt (0 = low priority, 1 = high priority)
RCIP	=	EUSART receive interrupt (0 = low priority, 1 = high priority)
TXIP	=	EUSART transmit interrupt (0 = low priority, 1 = high priority)
SSPIP	=	Master synchronous serial port interrupt (0 = low priority, 1 = high priority)
CCP1IP	=	CCP1 interrupt (0 = low priority, 1 = high priority)
TMR2IP	=	Timer 2 to PR2 match interrupt (0 = low priority, 1 = high priority)
TMR1IP	=	Timer 1 overflow interrupt (0 = low priority, 1 = high priority)

IPR2

OSCFIP	CMIP		EEIP	BCLIP	HLVDIP	TMR3IP	ECCP1IP

OSCFIP	=	Oscillator failure interrupt (0 = low priority, 1 = high priority)
CMIP	=	Comparator interrupt (0 = low priority, 1 = high priority)
EEIP	=	EEPROM flash write interrupt (0 = low priority, 1 = high priority)
BCLIP	=	Bus collision interrupt (0 = low priority, 1 = high priority)
HLVDIP	=	High/low voltage detect interrupt (0 = low priority, 1 = high priority)
TMR3IP	=	Timer 3 overflow interrupt (0 = low priority, 1 = high priority)
ECCP1IP	=	CCP1 interrupt (0 = low priority, 1 = high priority)

IPR3

IRXIP	WAKIP	ERRIP	TXB2IP TXBnIP	TXB1IP	TXB0IP	RXB1IP RXBnIP	RXB0IP FIFOWMIP

IRXIP	=	CAN invalid receive interrupt (0 = low priority, 1 = high priority)
WAKIP	=	CAN bus activity wakeup interrupt (0 = low priority, 1 = high priority)
ERRIP	=	CAN bus error interrupt (0 = low priority, 1 = high priority)
TXB2IP	=	CAN transmit buffer 2 interrupt, mode 0 (0 = low priority, 1 = high priority)
TXBnIP	=	CAN transmit buffer interrupt, modes 1 or 1 (0 = low priority, 1 = high priority)
TXB1IP	=	CAN transmit buffer 1 interrupt (0 = low priority, 1 = high priority)
TXB0IP	=	CAN transmit buffer 0 interrupt (0 = low priority, 1 = high priority)
RXB1IP	=	CAN receive buffer 1 interrupt, mode 0 (0 = low priority, 1 = high priority)
RXBnIP	=	CAN receive buffer interrupt, mode 1 and 2 (0 = low priority, 1 = high priority)
RXB0IP	=	CAN receive buffer 0 interrupt, mode 0 (0 = low priority, 1 = high priority)
FIFOWMIP	=	FIFO watermark interrupt, mode 1 and 2 (0 = low priority, 1 = high priority)

The circuitry for this type of interrupt is composed of switches placed by each door and window. The switches are SPST (single-pole, single-throw) proximity switches typically constructed using magnetic reed switches. When the magnet is close to the switch it closes, and when the magnet is moved away from the switch it opens. The magnet is mounted on the door or window and the switch is mounted on the frame that surrounds the door or window. This causes the switch to open when the door or window is opened.

The circuitry for this alarm is illustrated in Figure 6-18, along with the alarm and the switches. When any switch is opened, it causes an interrupt. The interrupt service procedure causes the alarm to sound. The weak pull-up feature for Port B is used here to pull up the Port B pin RB2 input to 5 V, so when a switch is open the RB2 pin goes to 5 V. The RB2 input is programmed as

FIGURE 6-16 Peripheral interrupt enable registers.

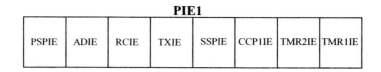

PIE1

PSPIE	ADIE	RCIE	TXIE	SSPIE	CCP1IE	TMR2IE	TMR1IE

PSIE	=	Parallel slave port read/write interrupt (1 = enable, 0 = disable)
ADIE	=	ADC interrupt (1 = enable, 0 = disable)
RCIE	=	EUSART receive interrupt (1 = enable, 0 = disable)
TXIE	=	EUSART transit interrupt (1 = enable, 0 = disable)
SSPIE	=	Master synchronous serial port interrupt (1 = enable, 0 = disable)
CCP1IE =		CCP1 interrupt (1 = enable, 0 = disable)
TMR2IE =		Timer 2 to PR2 match interrupt (1 = enable, 0 = disable)
TMR1IE =		Timer 1 overflow interrupt (1 = enable, 0 = disable)

PIE2

OSCFIE	CMIE		EEIE	BCLIE	HLVDIE	TMR3IE	ECCP1IE

OSCFIE	=	Oscillator failure interrupt (1 = enable, 0 = disable)
CMIE	=	Comparator interrupt (1 = enable, 0 = disable)
EEIE	=	Data EEPROM flash write interrupt (1 = enable, 0 = disable)
BCLIE	=	Bus collision interrupt (1 = enable, 0 = disable)
HLVDIE	=	High/low voltage detect interrupt (1 = enable, 0 = disable)
TMR3IE	=	Timer 3 overflow interrupt (1 = enable, 0 = disable)
ECCP1IE =		CCP1 interrupt (1 = enable, 0 = disable)

PIE3

IRXIE	WAKIE	ERRIE	TXB2IE TXBnIE	TXB1IE	TXB0IE	RXB1IE RXBnIE	RXB0IE FIFOWMIE

IRXIE	=	CAN invalid received message interrupt (1 = enable, 0 = disable)
WAKIE	=	CAN bus activity wakeup interrupt (1 = enable, 0 = disable)
ERRIE	=	CAN bus error interrupt (1 = enable, 0 = disable)
TXB2IE	=	CAN transmit buffer 2 interrupt, mode 0 (1 = enable, 0 = disable)
TXBnIE	=	CAN transmit buffer interrupt, modes 1 and 2 (1 = enable, 0 = disable)
TXB1IE	=	CAN transmit buffer 1 interrupt (1 = enable, 0 = disable)
TXB0IE	=	CAN transmit buffer 0 interrupt (1 = enable, 0 = disable)
RXB1IE	=	CAN receive buffer 1 interrupt, mode 0 (1 = enable, 0 = disable)
RXBnIE	=	CAN receive buffer interrupt, modes 1 and 2 (1 = enable, 0 = disable)
RXB0IE	=	CAN receive buffer 0 interrupt, mode 0 (1 = enable, 0 = disable)
FIFOWMIE =		FIFO watermark interrupt, modes1 and 2 (1 = enable, 0 = disable)

the INT2 input for the PIC18F1220 microcontroller as shown in this illustration. The RB4 pin is programmed as an output to turn the alarm on. And the normally open pushbutton connected to RB5 is pressed to turn the alarm bell off. The only way to turn the alarm off is to interrupt its power, or to push the reset alarm button, which could be hidden somewhere or it could be a key switch. The entire system software is listed in Example 6-17. The alarm could be added to any other application for the microcontroller with a minimum amount of effort.

In this system, once the interrupts are enabled, the program remains in the infinite while loop. The only thing that occurs in the loop is that the pushbutton connected to RB5 is tested to turn the alarm off. How then does the alarm turn on? The alarm is turned on when an interrupt occurs and breaks out of the main program loop. This interrupt turns the alarm on. At the end of

FIGURE 6-17 Peripheral interrupt request flag registers.

PIR1

PSPIF	ADIF	RCIF	TXIF	SSPIF	CCP1IF	TMR2IF	TMR1IF

PSPIF	=	Parallel slave port read/write interrupt (0 = inactive, 1 = active)
ADIF	=	ADC interrupt (0 = inactive, 1 = active)
RCIF	=	EUSART receive interrupt (0 = inactive, 1 = active)
TXIF	=	EUSART transmit interrupt (0 = inactive, 1 = active)
SSPIF	=	Master synchronous serial port interrupt (0 = inactive, 1 = active)
CCP1IF	=	CCP1 interrupt (0 = inactive, 1 = active)
TMR2IF	=	Timer 2 to PR2 match interrupt (0 = inactive, 1 = active)
TMR1IF	=	Timer 1 overflow interrupt (0 = inactive, 1 = active)

PIR2

OSCFIF	CMIF		EEIF	BCLIF	HLVDIF	TMR3IF	ECCP1IF

OSCFIF	=	Oscillator failure interrupt (0 = inactive, 1 = active)
CMIF	=	Comparator interrupt (0 = inactive, 1 = active)
EEIF	=	Data EEPROM flash write interrupt (0 = inactive, 1 = active)
BCLIF	=	Bus collision interrupt (0 = inactive, 1 = active)
HLVDIF	=	High/low voltage detect interrupt (0 = inactive, 1 = active)
TMR3IF	=	Timer 3 overflow interrupt (0 = inactive, 1 = active)
ECCP1IF	=	CCP1 interrupt (0 = inactive, 1 = active)

PIR3

IRXIF	WAKIF	ERRIF	TXB2IF TXBnIF	TXB1IF	TXB0IF	RXB1IF RXBnIF	RXB0IF FIFOWMIF

IRXIF	=	CAN invalid received message interrupt (0 = inactive, 1 = active)
WAKIF	=	CAN bus activity wakeup interrupt (0 = inactive, 1 = active)
ERRIF	=	CAN bus error interrupt (0 = inactive, 1 = active)
TXB2IF	=	CAN transmit buffer 2 interrupt, mode 0 (0 = inactive, 1 = active)
TXBnIF	=	CAN transmit buffer interrupt, modes 1 and 2 (0 = inactive, 1 = active)
TXB1IF	=	CAN transmit buffer 1 interrupt (0 = inactive, 1 = active)
TXB0IF	=	CAN transmit buffer 0 interrupt (0 = inactive, 1 = active)
RXB1IF	=	CAN receive buffer 1 interrupt, mode 0 (0 = inactive, 1 = active)
RXBnIF	=	CAN receive buffer interrupt, modes 1 and 2 (0 = inactive, 1 = active)
RXB0IF	=	CAN receive buffer 0 interrupt (0 = inactive, 1 = active)
FIFOWMIF	=	FIFO watermark interrupt (0 = inactive, 1 = active)

the interrupt service procedure, control is returned back to the main program loop where the alarm can be turned off if the pushbutton is pressed. A watchdog timer is used because it would not be a good idea to have an alarm program hang by a random error.

EXAMPLE 6-17

```
/*
 * Burglar alarm interrupt
 */

#include <p18cxxx.h>

/* Set configuration bits
 *  - set RC oscillator
 *  - enable watchdog timer
```

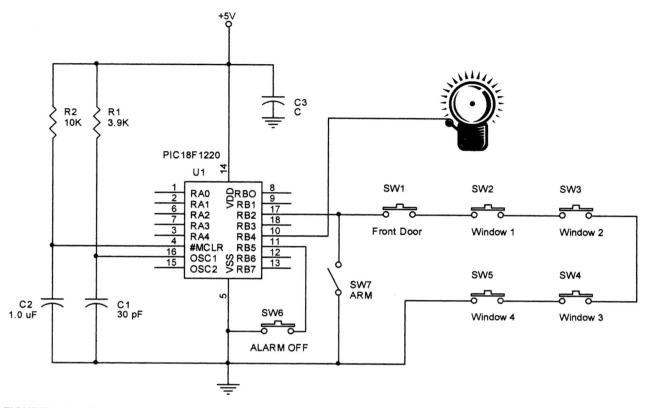

FIGURE 6-18   Simple interrupt-processed burglar alarm in the armed position.

```
 *   - disable low-voltage programming
 *   - disable brownout reset
 *   - enable master clear
 */

#pragma config OSC = RC
#pragma config WDT = ON
#pragma config WDTPS = 4
#pragma config LVP = OFF
#pragma config BOR = OFF
#pragma config MCLRE = ON

void MyHighInt (void);
void MyLowInt (void);

#pragma interrupt MyHighInt
#pragma code high_vector=0x08

void high_vector (void)
{
     _asm GOTO MyHighInt _endasm
}

#pragma interruptlow MyLowInt      // MyLowInt is an interrupt
#pragma code low_vector=0x18       // low_vector is the vector at 0x18

void low_vector (void)
{
   _asm GOTO MyLowInt _endasm
}

#pragma code
```

```
void MyHighInt (void)
{

}
void MyLowInt (void)
{
        if (INTCON3bits.INT2IF == 1)     // is it INT2 pin?
        {
                INTCON3bits.INT2IF = 0;   // clear INT2IF flag
                PORTBbits.RB4 = 1;        // turn alarm on
        }
}

// main program
void main (void)
{
        ADCON1 = 0x0F;                    // make port pins digital
        TRISB = 0x24;                     // make RB2 and RB5 inputs
                                          // make RB4 output

        PORTB = 0x00;                     // alarm off

        INTCON2bits.RBPU = 1;             // Port B pullups on

        RCONbits.IPEN = 1;                // IPEN = 1

        INTCON2bits.INTEDG2 = 1;          // make INT2 positive edge-triggered
        INTCON3bits.INT2IP = 0;           // make INT2 low priority
        INTCON3bits.INT2IE = 1;           // enable INT2

        INTCONbits.GIEH = 1;              // enable high-priority interrupts
        INTCONbits.GIEL = 1;              // enable low-priority interrupts

        while (1)                         // main program loop
        {
                ClrWdt();                 // pet spot (woof/pant)
                if (PORTBbits.RB5 == 0)   // pushbutton pressed
                    PORTBbits.RB4 = 0;    // alarm off
        }
}
```

6-4   OTHER INTERNAL PERIPHERAL DEVICES

The PIC18 family has many internal peripherals, from timers for timing events to a multiple channel analog-to-digital converter for sampling analog events. Because of the shear number of internal peripherals, some are introduced in this section and then described and used in applications in later chapters. Throughout the text all of the internal peripherals are eventually used in applications.

Timers

The microcontroller contains four or five timers that are used for many tasks. Some of these tasks include prescaling clock signals, generating signals of multiple frequencies, causing interrupts at a specific time, and so forth. This section introduces Timer 1 along with an interrupt to illustrate a real-time clock (RTC). This application illustrates the utility of a timer. This is only one of the internal timers—the other three or four timers are available for additional tasks.

Timer 1 is a 16-bit timer that operates in three modes. Figure 6-19 shows the internal structure of Timer 1. Timer 1 operates as a timer, a synchronous counter, or as an asynchronous counter. A **timer** is a counter, which operates from the internal instruction cycle clock or external

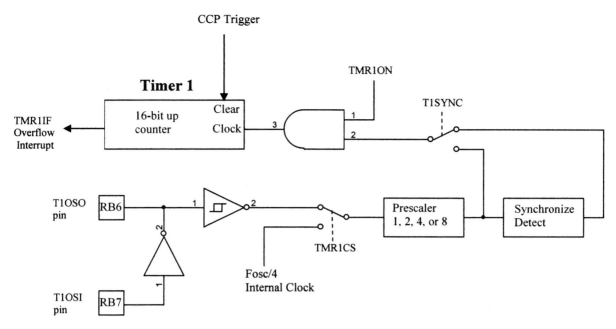

FIGURE 6-19 Timer I internal structure.

clock input. For each instruction cycle or external clock, the timer counts and, if it hits its the maximum count and rolls over to zero (called an overflow), an interrupt cycle is initiated. For example, suppose the clock input to the microcontroller is 4 MHz, which causes the timer to count once every microsecond because the clock is almost always divided by four (an instruction clock) for most internal units. An 8-bit timer counts from 0x00 to 0xFF and then rolls over to zero and starts counting again from zero. With the 4-MHz clock input to the microcontroller, this generates an interrupt once every 256 μs. A **synchronous counter** is a counter that counts the rising edges of an external clock input applied to the T1OSI pin (RB7). An asynchronous counter is a counter that operates with an external input, but the internal counter is not synchronized with the microcontroller clock. In the asynchronous mode, the timer cannot be used with compare or capture operations.

Suppose that a real-time clock (RTC) is needed. A real-time clock counts time in seconds, minutes, and hours and holds the real time of day, hence the name real-time clock. An RTC is also used for time delays. The RTC can be operated with an external crystal connected to the clock input of Timer 1 or from the crystal used to time the microcontroller's execution.

Before this application is written, the control register for Timer 1 must be known. Figure 6-20 illustrates the contents of the Timer 1 control register. In order to use Timer 1, the Timer 1 control register (T1CON) and the interrupt control registers are used in this sample program for a real-time clock. The RD16 bit selects whether the entire 16 bits of the counter are read or written at once, or whether each 8-bit half is read or written separately. When read or written at once, (RD = 1), the high part of the count is output first and held within the microcontroller until the low part is written when the timer is updated with both parts of the count.

This application uses a 32.768-KHz crystal, which is commonly available for use with time sources such as clocks and wristwatches. This crystal is connected to the Timer 1 clock input pins (T1OSO and T1OSI) as illustrated in Figure 6-21. Notice that this system contains two clocks. The main clock, which times the execution of instructions, is connected as an RC oscillator that runs at 2 MHz on the OSC1 input pin. The other is a crystal oscillator connected to the RB6 and RB7 pins. The RB6 pin is the T1OSO (T1 oscillator output) pin and the RB7 pin is the T1OSI (T1

FIGURE 6-20 Timer I control register.

T1CON

RD16		T1CKPS1	T1CKPS0	T1OSCEN	$\overline{\text{T1SYNC}}$	TMR1CS	TMR1ON

RD16   16-bit read/write mode
                1 = r/w is 16 bits
                0 = r/w is two 8 bits

T1CKPS1 ⎫
T1CKPS2 ⎭   00 = ÷1; 01 = ÷2; 10 = ÷4 ; and 11 = ÷8

T1OSCEN   Oscillator enable bit
                1 = enabled
                0 = disabled

$\overline{\text{T1SYNC}}$ Timer 1 synchronization
                1 = no synchronization
                0 = synchronize external clock input

TMR1CS   Timer 1 clock source
                1 = external clock
                0 = internal clock

TMR1ON   Timer 1 on bit
                1 = Timer 1 on
                0 = Timer 1 off (stopped)

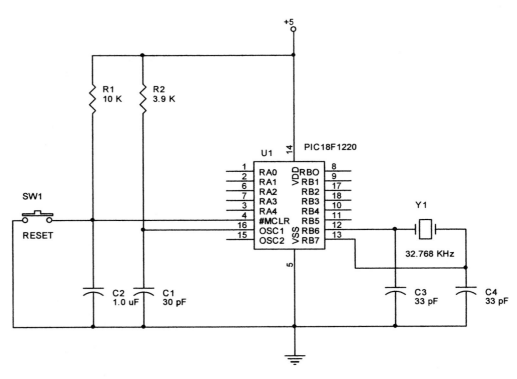

FIGURE 6-21   Circuit for a real-time clock.

oscillator input) pin. A count of 32,768 or 0x8000 causes an interrupt in one second if the timer is programmed to start at this value. The Timer 1 counter is 16 bits, so in order for it to cause an interrupt once per second, it is preloaded with 0x8000. Programmed in this way, 32,768 input clocks are required to cause an overflow interrupt once per second. The software that maintains a 24-hour clock in memory at access locations seconds, minutes, and hours in C language appears in Example 6-18.

EXAMPLE 6-18

```c
/*
 * Real-time clock (RTC) for a PIC18F1220
 */

#include <p18cxxx.h>

/* Set configuration bits
 *   - set RC oscillator
 *   - disable watchdog timer
 *   - disable low-voltage programming
 *   - disable brownout reset
 *   - enable master clear
 */

#pragma config OSC = RC
#pragma config WDT = OFF
#pragma config LVP = OFF
#pragma config BOR = OFF
#pragma config MCLRE = ON

// ************ INTERRUPT PROTOYPE ********************

void MyHighInt (void);

#pragma interrupt MyHighInt
#pragma code high_vector = 0x08

// ************ HIGH-PRIORIY INTERRUPT VECTOR **********

void high_vector (void)
{
    _asm GOTO MyHighInt _endasm
}

// ********** DATA MEMORY ***************************

// time of day counters

#pragma udata access IntMyData     // place in the access bank

near char seconds;                 // seconds
near char minutes;                 // minutes
near char hours;                   // hours

// ************* FUNCTIONS ************************

#pragma code

// function to increment the clock

void DoTime (void)                 // do the clock once per second
{
    TMR1H = 0x80;                  // preload Timer 1 with 0x8000
                                   // TIMR1L is already 0x00
    PIR1bits.TMR1IF = 0;           // clear Timer 1 interrupt

    seconds++;                     // increment seconds

    if (seconds == 60)             // if seconds hits 60
    {
        seconds = 0;               // clear seconds back to zero
        minutes++;                 // increment minutes
```

```
                    if (minutes == 60)        // if minutes hits 60
                    {
                            minutes = 0;       // clear minutes
                            hours++;           // increment hours
                            if (hours == 24)   // if hours hits 24
                                hours = 0;     // clear hours
                    }
            }
    }

// interrupt service procedure
void MyHighInt (void)             // context saved by the compiler
{
        if (PIR1bits.TMR1IF == 1)
            DoTime();

        // test other interrupt flags here
}

// main program
void main (void)
{
        ADCON1 = 0x0f;             // program for digital pins

        TRISA = 0xFF;              // Port A is input
        RCONbits.IPEN = 1;         // IPEN = 1

        TMR1L = 0;                 // preload Timer 1 with 0x8000
        TMR1H = 0x80;

        T1CON = 0x0F;              // select external clock and enable Timer 1

        seconds = minutes = hours = 0;   // start clock at midnight

        IPR1bits.TMR1IP = 1;       // make Timer 1 high priority
        PIE1bits.TMR1IE = 1;       // enable Timer 1 interrupt

        INTCONbits.GIEH = 1;       // enable high-priority interrupts

        while (1)
        {
            // do other stuff here
        }
}
```

Each time a second elapses, the Timer 1 interrupt fires, calling the DoTime function in the high-priority interrupt service procedure. In DoTime, Timer 1 is reloaded with 0x8000, the Timer 1 interrupt flag is cleared, and the RTC is incremented. On each interrupt, the second counter increments, once per minute the minute counter increments, and once an hour the hour counter increments. The software is fairly straightforward and keeps time as time would be kept using discrete counters. The accuracy is determined by the frequency of the crystal, which is usually ±0.005% or a few seconds a month.

The only slight problem with this clock is that a function is needed to read the contents of the clock and convert it into an ASCII character string for display. Example 6-19 illustrates a function that obtains the time in either the 24-hour format or the 12-hour AM/PM format and stores the string in memory at timeString as a null-terminated C-style string. This function is called whenever the time is needed in this format for display. The time is stored as 22:12:01 for the 24-hour format or as 10:12:01 PM in the 12-hour format. The mode parameter selects the 24/12 format when the get-TimeString function is invoked. If the mode is a one, then the 12-hour format is returned.

Notice how ASCII code is created by adding a 0x30 to the values obtained from the hours, minutes, and seconds variables after a divide by 10 or a modulo division by 10. If the number of hours is a 12, and it is divided by 10, the result is 1, which is not in ASCII code. After adding 0x30 to the 1, the result is 0x31, which is an ASCII-coded "1". If the 12 is divided using modulo division (%), the remainder of 2 is returned. The 2 is also converted to ASCII code by adding a 0x30. Likewise, an inclusive-OR operation (|) can be used to convert to ASCII.

EXAMPLE 6-19

```
#pragma udata              // select data memory

char timeString[12];       // time string

#pragma code
void getTimeString (char mode)
{
    char ptr = 0;              // point to array element 0
    char tempHours = hours;
    char amPM = 'A';

    if (mode == 1)             // 12-hour mode
    {
        if (hours >= 12)       // convert to 12-hour format
        {
            tempHours -= 12;
            amPM = 'P';
        }
        if (tempHours == 0)
            tempHours = 12;
    }
    if ((tempHours / 10) == 0)      // blank leading zero
        timeString[ptr++] = ' ';
    else
        timeString[ptr++] = tempHours / 10 + 0x30;

    timeString[ptr++] = tempHours % 10 + 0x30;

    timeString[ptr++] = ':';
    timeString[ptr++] = minutes / 10 + 0x30;
    timeString[ptr++] = minutes % 10 + 0x30;

    timeString[ptr++] = ':';
    timeString[ptr++] = seconds / 10 + 0x30;
    timeString[ptr++] = seconds % 10 + 0x30;

    if (mode == 1)             // 12-hour mode
    {
        timeString[ptr++] = ' ';
        timeString[ptr++] = amPM;
        timeString[ptr++] = 'M';
    }

    timeString[ptr] = 0;       // terminate string with NULL
}
```

An alternate method for handling the time of day is to use a single 24-bit counter in the memory to hold the hours, minutes, and seconds. There are 86,400 (60 * 60 * 24) seconds in a day. This number easily fits into a 24-bit **short long** integer. This reduces the amount of time spent in the interrupt service procedure, although the complexity of extracting the time is slightly more difficult. Example 6-20 illustrates the interrupt service procedure with this type of clock. (The Windows operating system uses this technique for its RTC, but the variable is a 64-bit variable that counts milliseconds from midnight of 100 AD to midnight of 9999 AD). Example 6-21 shows the getTimeString function for this clock. Here hours are obtained by dividing the time by 3600. The minutes are obtained by taking the remainder of the time divided by 3600 and dividing the remainder by 60. Seconds are obtained by taking the remainder of the remainder of the time divided by 3600 divided by 60.

The method used for a RTC depends on how the RTC is used in other functions in a program. If the RTC is used for precision time delays, then the first method described is best. If it is merely used for the time of day, then the second method is best. A good example of precision time delays appears in Appendix D for the USB-enabled system, which supports accurate time delays in millisecond increments for a real-time system.

EXAMPLE 6-20

```
#pragma udata access IntMyData      // place in the access bank

near short long time;               // 24-bit counter

#pragma udata

char timeString[12];                // time string

#pragma code

// new interrupt service procedure for the RTC

void DoTime (void)
{
    TMR1H = 0x80;                   // preload Timer 1 with 0x8000
    PIR1bits.TMR1IF = 0;           // clear Timer 1 interrupt
    time++;
    if (time == 86400)             // if a new day
        time = 0;
}
```

EXAMPLE 6-21

```
void getTimeString (char mode)
{
    char ptr = 0;                          // point to array element
    char hours = time / 3600;              // get hours
    char amPM = 'A';
    char minutes = (time % 3600) / 60;     // get minutes
    char seconds = (time % 3600) % 60;     // get seconds

    if (mode == 1)                         // 12-hour mode
    {
        if (hours >= 12)                   // convert to 12 hour
        {
            hours -= 12;
            amPM = 'P';
        }

        if (hours == 0)
            hours = 12;
    }

    if ((hours / 10) == 0)
        timeString[ptr++] = ' ';
    else
        timeString[ptr++] = hours / 10 | 0x30;

    timeString[ptr++] = hours % 10 | 0x30;

    timeString[ptr++] = ':';
    timeString[ptr++] = minutes / 10 | 0x30;
    timeString[ptr++] = minutes % 10 | 0x30;

    timeString[ptr++] = ':';
    timeString[ptr++] = seconds / 10 | 0x30;
    timeString[ptr++] = seconds % 10 | 0x30;

    if (mode == 1)                         // 12-hour mode
    {
        timeString[ptr++] = ' ';
        timeString[ptr++] = amPM;
        timeString[ptr++] = 'M';
    }

    timeString[ptr] = 0;
}
```

The remaining timers, along with many other applications, are illustrated in later sections. Some of these other applications use interrupts and some do not. The timers.h header can be used to define the timer. Examples of the timers.h library appear in later sections of this chapter, in other chapters of this textbook, and in Appendix B.

ADC

Another internal peripheral is the **analog-to-digital converter** (ADC). The ADC converts an analog input voltage into a digital number. Most PIC18 family members contain a 10-bit converter that converts the analog input to a digital value using a step resolution of $\frac{|Vref+| - |Vref-|}{1023}$. If Vref+ is 5 V and Vref− is ground, the voltage step value is $\left(\frac{5}{1023}\right)$ 0.00489 V per step. Therefore, if the input voltage is 1.0 V, the converter generates a digital output of 1.0 V / 0.00489 V or 205 decimal. The converter also has up to 16 inputs, depending on the family member. These inputs are used to monitor voltages from different sources. There is also a pair of control bits that select the source of the Vref inputs to the converter. The internal structure of the ADC appears in Figure 6-22. Note that a few family members contain a 12-bit converter, which

FIGURE 6-22 16-channel, 10-bit analog-to-digital converter.

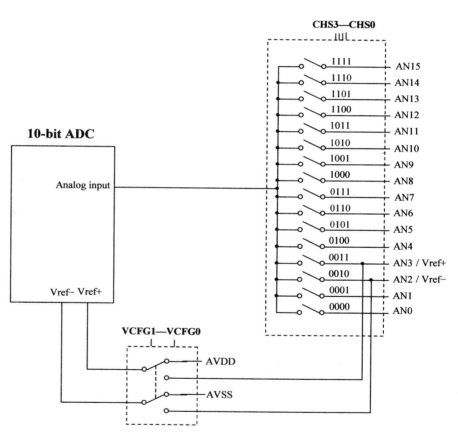

CHS3—CHS0   Select the input pin ANn

VCFG1—VCFG0 Select the voltage reference source

$$00 = \text{AVDD, AVSS}$$
$$01 = \text{Vref+, AVSS}$$
$$10 = \text{Vref−, AVDD}$$
$$11 = \text{Vref+, Vref−}$$

uses the same technique provided here, except the step voltage is determined by dividing by 4095 $(2^{12} - 1)$ in place of 1023 $(2^{10} - 1)$.

Three control registers (ADCON0, ADCON1, and ADCON2) define the operation of the ADC. The ADCON1 register determines whether the Port A and B input pins function as either digital or analog inputs. The ADC control registers appear in Figure 6-23.

ADCON0

		CHS3	CHS2	CHS1	CHS0	GO/$\overline{\text{DONE}}$	ADON

CHS3—CHS0　　　　　Select the input channel as depicted in Figure 6-19

GO/$\overline{\text{DONE}}$　　　　　ADC status bit (0 = busy converting, 1 = done)

ADON　　　　　ADC on bit (0 = disabled, 1 = enabled)

ADCON1

		VCFG1	VCFG0	PCFG3	PCFG2	PCFG1	PCFG0

VCFG1—VCFG0　　　　　Select the voltage reference as depicted in Figure 6-19

PCFG3—PCFG0　　　　　Select the digital/analog input pins (see table)

PCFGn	AN15	AN14	AN13	AN12	AN11	AN10	AN9	AN8	AN7	AN6	AN5	AN4	AN3	AN2	AN1	AN0
0000	A	A	A	A	A	A	A	A	A	A	A	A	A	A	A	A
0001	D	D	A	A	A	A	A	A	A	A	A	A	A	A	A	A
0010	D	D	D	A	A	A	A	A	A	A	A	A	A	A	A	A
0011	D	D	D	D	A	A	A	A	A	A	A	A	A	A	A	A
0100	D	D	D	D	D	A	A	A	A	A	A	A	A	A	A	A
0101	D	D	D	D	D	D	A	A	A	A	A	A	A	A	A	A
0110	D	D	D	D	D	D	D	A	A	A	A	A	A	A	A	A
0111	D	D	D	D	D	D	D	D	A	A	A	A	A	A	A	A
1000	D	D	D	D	D	D	D	D	D	A	A	A	A	A	A	A
1001	D	D	D	D	D	D	D	D	D	D	A	A	A	A	A	A
1010	D	D	D	D	D	D	D	D	D	D	D	A	A	A	A	A
1011	D	D	D	D	D	D	D	D	D	D	D	D	A	A	A	A
1100	D	D	D	D	D	D	D	D	D	D	D	D	D	A	A	A
1101	D	D	D	D	D	D	D	D	D	D	D	D	D	D	A	A
1110	D	D	D	D	D	D	D	D	D	D	D	D	D	D	D	A
1111	D	D	D	D	D	D	D	D	D	D	D	D	D	D	D	D

ADCON2

ADFM					ADCS2	ADCS1	ADCS0

ADFM　　　　　ADC result format (0 = left justified, 1 = right justified)

ADCS2—ADCS0　　Clock select bits

　　　　　000 = Fosc/2
　　　　　001 = Fosc/8
　　　　　010 = Fosc/32
　　　　　011 = FRC (internal RC clock; 1 MHz max)
　　　　　100 = Fosc/4
　　　　　101 = Fosc/16
　　　　　110 = Fosc/64
　　　　　111 = FRC (Clock from RC oscillator, 1 MHz max)

FIGURE 6-23　ADC control registers.

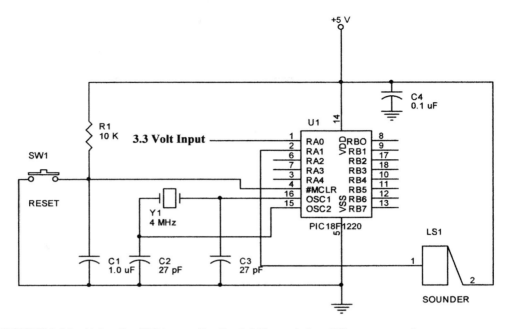

FIGURE 6-24   Using the PIC to monitor the 3.3 V supply in a PC power supply.

To illustrate a simple application for the ADC, suppose that the 3.3 V source in a PC power supply needs to be monitored. The PIC can manage this task with one of the input pins to the ADC. Suppose that the circuit depicted in Figure 6-24 is used to monitor the voltage from the power supply. The circuit uses a 4-MHz external crystal oscillator and only the AN0 input (RA0) to monitor the voltage.

The software required to measure the voltage is stored in the function getVoltage() that returns the floating-point value of the voltage on AN0. Also included in Example 6-22 is the software required to program the AN0 pin because an analog input is located at the start of the main function.

EXAMPLE 6-22

```
/*
 * Sampling the 3.3 V supply for a PIC18F1220
 */

#include <p18cxxx.h>

/* Set configuration bits
 *   - set OSC input external oscillator
 *   - disable watchdog timer
 *   - disable low-voltage programming
 *   - disable brownout reset
 *   - enable master clear
 */

#pragma config OSC = HS
#pragma config WDT = OFF
#pragma config LVP = OFF
#pragma config BOR = OFF
#pragma config MCLRE = ON

#pragma code

float getVoltage (void)
{
        ADCON0bits.GO = 1;    //start a conversion
```

```
        while (ADCON0bits.GO == 1);   // wait for completion
        return (ADRESL | (ADRESH << 8)) * 0.00489;
}

// main program
void main (void)
{
        ADCON0 = 0x01;              // select input AN0, enable ADC
        ADCON1 = 0x0e;              // AN0 is analog, VDD and VSS are references
        ADCON2 = 0x84;              // convert using 1 MHz

        TRISA = 1;                  // Port A bit 0 = input for ADC
        PORTA = 0;                  // alarm off

        while (1)
        {
            if (getVoltage() > 3.465 || getVoltage() < 3.135)
                PORTAbits.RA1 = 1;   // alarm on
            else
                PORTAbits.RA1 = 0;   // alarm off
        }
}
```

When the getVoltage function is called, the GO bit in ADCON0 is set to start a conversion. It is then tested to see if the conversion is complete (GO = 0) using a while loop. The program can be expanded so that once per second the voltage is checked in the interrupt service procedure of an RTC. If the voltage rises 5% or drops by 5%, an alarm is sounded by a piezoelectric buzzer. In Example 6-22, an alarm is attached to RA1. Additional examples of the ADC are presented in later chapters.

Data EEPROM

The data EEPROM is available on many versions of the microcontroller and is used to store persistent data. Passwords, setup information, or any other type of data that is stored long term is stored in the data EEPROM. Two sizes of on-chip data EEPROMs are currently available: 256 or 1024 bytes. The 256-byte device is accessed though a single 8-bit address placed in the EEADR register and the 1024-byte device uses two 8-bit address registers for access: EEADR and EEADRH. EEADR holds the low order 8-bit of the 10-bit address and EEADRH holds the upper 2 bits of the address. Because many devices have only 256 only bytes of EEPROM, only the EEADR register is used for selecting a memory location on the EEPROM in this section.

The data EEPROM stores data for 40 years without any power and can be rewritten up to 1 million times (the data sheet lists a minimum of 100,000 so that is really the only guaranteed number). Why is it explained in this section of the text instead of with memory? The EEPROM is treated as an I/O device and cannot be directly written or read as normal memory. To read or write the EEPROM requires additional effort and an understanding of I/O instructions. Example 6-23 lists a function that reads any location from a 256-byte EEPROM. The function returns the contents of the location on the EEPROM passed to it as a parameter.

EXAMPLE 6-23

```
char eeRead (char address)
{
        EECON1bits.EEPGD = 0;       // select data EEPROM
        EEADR = address;            // set up the EEPROM address
        EECON1bits.RD = 1;          // select a read operation
        return EEDATA;              // read the EEPROM data
}
```

This short function places a zero in the EEPGD bit. The EEPGD bit selects either the data EEPROM or the program FLASH memory (EEPGD = 1). The program FLASH memory is discussed in Chapter 10 with a boot loader program.

To write the EEPROM requires a little more effort than reading, so the function to write a location in the data EEPROM is longer. Example 6-24 lists the function to write a byte to the data EEPROM.

EXAMPLE 6-24

```
void eeWrite (char address, char data)
{
    INTCONbits.GIEH = 0;            // disable all interrupts
    EECON1bits.EEPGD = 0;          // select data EEPROM
    EECON1bits.WREN = 1;           // unprotect writing
    EEADR = address;               // set up the EEPROM address
    EEDATA = data;                 // set up the EEPROM data
    EECON2 = 0x55;                 // erase the current byte
    EECON2 = 0xAA;
    EECON1bits.WR = 1;             // select a write operation
    while (PIR2bits.EEIF == 0);    // wait until finished
    PIR2bits.EEIF = 0;             // clear the EE interrupt flag
    EECON1bits.WREN = 0;           // write-protect data EEPROM
    INTCONbits.GIEH = 1;           // re-enable all interrupts
}
```

As stated, more work is required to write the data EEPROM. Because the EEPROM logic causes an interrupt that is not normally used, the first and last steps disable and re-enable the interrupt system. Here both the low- and high-priority interrupts are disabled with the GIEH bit of the interrupt control register. The WREN bit protects the EEPROM from erroneous writes that could occur. To be able to write the EEPROM, this bit is set and then cleared to protect the EEPROM near the end of the function.

The EEPROM location selected by EEADR is erased by writing a 0x55 and then a 0xAA to the EECON2 register. This stores a 0xFF into the selected memory location, which is considered the erase state. The actual write to the EEPROM stores only the zero bits of a number into the memory. Once the write is initiated, the EEIF bit (EEIF = 0) in the PIR2 register indicates that the EEPROM is busy writing data. The EEPROM write process requires approximately 4 ms to accomplish this. An interrupt flag bit is used to indicate that the EEPROM is busy performing a write if a program decides to interrupt this operation. In this example, the software waits for the EEIF flag to become a logic one before writing a zero to it and then continuing. Chapter 10 discusses programming the internal flash program memory in much the same way as the data EEPROM.

Because variable names cannot be assigned to the EEPROM in a program, one method of handling the task is to use the #define statement to assign numeric addresses to labels. Example 6-25 illustrates how to accomplish this task and then uses the eeRead function to read the variable into the data register file.

EXAMPLE 6-25

```
// *********** EEPROM DATA ********************

#define timeMode 0          // timeMode assigned address 0
#define unitAddressL 1       // unitAddressL assigned address 1
#define unitAddressH 2       // unitAddressH assigned address 2

// ********* DATA MEMORY VARIABLE ************

#pragma udata

char TimeMode;
char UnitAddressL;
char UnitAddressH;
```

```
// later in the program

        TimeMode = eeRead (timeMode);      // read EEPROM address 0x00
        UnitAddressL = eeRead (unitAddressL)
        UnitAddressH = eeRead (unitAddressH)

// still later in the program

        eeWrite (timeMode, TimeMode)        // burn TimeMode into EEPROM
```

Compare and Capture Unit (CCP)

The compare and capture unit can be used with the pulse width modulator (PWM) or to capture the count in Timer 1 or Timer 3 to measure durations. For example, suppose that Timer 3 is operating at a millisecond rate and an input changes its rate of change and the input needs to be measured. The CCP unit can measure this change by capturing the contents of a timer when the input changes. This allows the rate of change to be measured in milliseconds because the timer is counting in milliseconds.

Figure 6-25 shows a PIC18F1220 that measures the AC power line frequency for a power monitoring system. In this circuit, the AC power line voltage is isolated from the microcontroller using an optical coupler which converts the 120 VAC signal into a TTL-compatible signal applied to CCP1 input (RB3) to the microcontroller. The CCP unit is programmed to capture the contents of Timer 3 whenever a positive edge appears on the CCP1 pin. In this example, Timer 3 is operated at an 8 μs rate. The count captured from the CCP1 pin is used to determine the frequency of the input and stored at the memory location frequency. The content of Timer 3 is divided into 125,000 to calculate the frequency. Example 6-26 illustrates the software used to accomplish this measurement using the CCP unit in the microcontroller. This program uses the timers.h and capture.h files for the Timer 3 and CCP functions. The timer header has functions to initialize, read, and write any timer. The capture header file has functions that open the capture unit and allow it to be read.

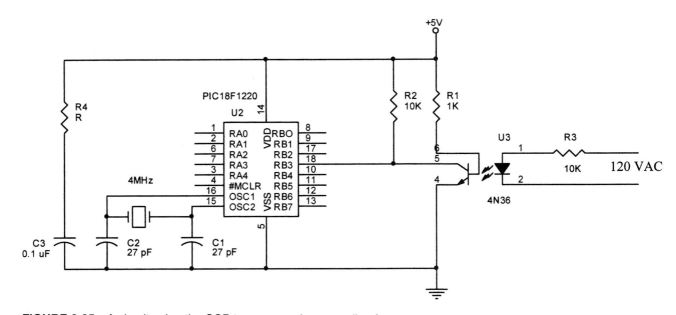

FIGURE 6-25   A circuit using the CCP to measure the power line frequency.

EXAMPLE 6-26

```
/*
 * Measure the AC line frequency
 */

// In order for this code to fit into the boot block
// the linker file was changed to use the c018.o initialization
// file instead of the C018i.0 file
//
// The effect is that no initialized data memory can exist in the
// program.

#include <p18cxxx.h>
#include <timers.h>
#include <capture.h>

/* Set configuration bits
 * - set external oscillator
 * - disable watchdog timer
 * - disable low-voltage programming
 * - master clear enabled
 */
#pragma config OSC = HS              // external 4-MHz oscillator
#pragma config WDT = OFF
#pragma config LVP = OFF
#pragma config MCLRE = ON
#pragma interrupt MyHighInt          save=PROD
#pragma code high_vector = 0x08      // high_vector is at 0x0008

void high_vector (void)              // high-prioity vector
{
        _asm GOTO MyHighInt _endasm  // goto high software
}

float frequency;                     // power line frequency

#pragma code

void MyHighInt (void)
{
     int count;
     if (PIR1bits.CCP1IF == 1)
     {
          PIR1bits.CCP1IF = 0;
          count = ReadCapture1();
          WriteTimer3 (0);
          frequency = 125000.0 / (count);   // in hertz
     }
}

void main (void)
{
     TRISBbits.TRISB3 = 1;                   // RB3 is input

     OpenCapture1 (C1_EVERY_RISE_EDGE &      // interrupt on rising edge
                   CAPTURE_INT_ON);

     OpenTimer3 (TIMER_INT_OFF &             // 8 us
                 T3_PS_1_8 &
                 T3_SOURCE_INT);

     WriteTimer3 (0);                        // clear Timer 3

     INTCONbits.GIEH = 1;            // enable high-priority interrupts

     while (1)
     {

     }
}
```

Additional Internal Devices

This chapter introduced the microcontroller and some of the internal devices; a discussion of other internal devices are covered in later chapters that highlight their operations. These other devices include the USART (universal synchronous/asynchronous receiver/transmitter) and various other devices that may or may not be unique to a particular version of the microcontroller.

6-5   **SUMMARY**

1. The PIC18 is manufactured in various sizes from an 18-pin PDIP to a 128-pin TQFP package.
2. The power supply to the PIC18 can be as low as 4.2 V for full-speed operation, with a current of up to 60 μA for a low clock speed to 12 mA for a 40-MHz operation.
3. The output drive current is sufficient for four standard TTL unit loads to ten MOS unit loads.
4. The microcontroller is reset without any components if the power supply rises to 5.0 V quickly (0.05/ms), otherwise it is often reset using an external resister/capacitor combination and a reset pushbutton.
5. The microcontroller is reset by the watchdog timer reset (WDT), a brownout reset (BOR), the RESET instruction (a Reset() in C), a stack full reset, a stack underflow reset, or by the rest circuit.
6. The watchdog timer is a programmable counter that resets the microprocessor if it is allowed to overflow. The amount of overflow time is adjustable and can range as high as 131 seconds.
7. The brownout reset is programmed to reset the microcontroller if the power supply voltage drops below a predefined point.
8. If the stack is overflowed or underflowed, the microcontroller can be programmed to reset.
9. The clock for the PIC18 can be from an external source, a crystal oscillator or resonator, an RC timing circuit, or an internal oscillator.
10. The #pragma config statement specifies the configuration of the microcontroller in a C program.
11. The I/O pins on the PIC18 are grouped into 8-bit ports that vary in number from one family member to another. All versions contain Ports A and B, and other versions contains Ports C to J.
12. An I/O port contains a TRIS register for programming the direction of the port pins, a PORT register for reading or writing to a port, and a port LAT register for writing to an internal port latch. Many versions also contain an ADCON1 register that programs port pins for either analog or digital operation.
13. An interrupt is a function call caused by either a hardware signal or some internal event. When an interrupt occurs, the microcontroller uses either the high- or low-priority interrupt vector for calling an interrupt service procedure.
14. The interrupt structure is controlled by several registers that program the priority level, indicate the status, and also enable or disable an interrupt.
15. The timers are used to cause an interrupt when they overflow, or count synchronous or asynchronous events.
16. The microcontroller contains a 10-bit analog-to-digital converter that has up to 16 channels of input data.
17. The data EEPROM is not memory as are the SRAM data register file or the program memory. The data EEPROM is accessed as if it were an I/O device through an address register and a data register.

6-6   **QUESTIONS AND PROBLEMS**

1. Go to the Microchip Web page and determine what packages are available for a PIC18F1320.
2. What is a PDIP package?
3. What is a SOIC package?
4. What is a SSOP package?
5. Again visit the Microchip Web site and locate the lowest and highest price versions of the PIC18 family members. The lowest priced PIC18 = _____ and the highest priced PIC18 = _____.
6. What are the signal names of the power supply connections to the PIC18?
7. What is the range of voltages allowed for high-speed operation of the microcontroller?
8. What is the current required to operate the PIC18 at its maximum frequency?
9. How much current is required for a digital input signal to the PIC microcontroller?
10. The microcontroller can drive a load to a logic zero with a maximum current of _____.
11. The microcontroller can drive a load to a logic one with a maximum current of _____.
12. If the reset circuit (see Figure 6-3) uses a 15K resister and a 1.0 μF capacitor, the time constant is _____.
13. What is the purpose of the $\overline{\text{MCLR}}$ pin on the microcontroller?
14. Where is the vector address located for a reset?
15. What instruction is normally placed at the reset vector address and why?
16. What is a watchdog timer (WDT) and what occurs when the WDT overflows?
17. The WDT is programmed for a minimum count of _____ and a maximum count of _____.
18. At what clock time is the WDT operated?
19. What is a BOR and what voltages can be programmed into the BOR?
20. What are the stack resets and is there any reason for them to occur?
21. What is the purpose of a #pragma config MCLRE = ON statement in C language?
22. What is the maximum clock frequency for the PIC18?
23. Is there a minimum clock frequency for the PIC18?
24. One instruction cycle is composed of _____ clock times.
25. What is the XT mode for the clock on a PIC18?
26. What is the RCIO mode for the clock on a PIC18?
27. Select a resister and a capacitor to operate the PIC18 at a frequency of approximately 750 KHz.
28. How is the HSPLL clock mode selected in a C language program for the PIC18?
29. Which pins on the microcontroller are connected to a crystal or ceramic resonator?
30. Search the Internet and locate a source for the ceramic resonator. What frequency values are available for the source (list the Web address of the source)?
31. Explain how the HSPLL mode for the clock functions.
32. Explain how the dice toss example selects a random number between zero and 35.
33. If an LED requires a voltage of 2.5 V and a current of 20 mA, what value bias resister is required if the power supply voltage is 5.0 V?
34. If an LED requires a voltage of 2.5 V and a current of 20 mA, what value bias resister is required if the power supply voltage is 12 V?
35. What are the purposes of the TRISA, LATA, and PORTA devices for I/O Port A?
36. Develop a short sequence of instructions that program Port B as an output port where all of the pins are for digital output information.
37. Develop a short sequence of instructions that program bit positions 2 and 3 of Port B as input information and the remaining Port B bits as output information.
38. Describe an interrupt.
39. What occurs when a high-priority interrupt is activated?

40. Explain the purpose of #pragma interrupt MyHighInt and #pragma code high_vector = 0x08.

41. If a function that returns a 16-bit number is used in an interrupt service procedure, what must be done in the #pragma statement for the interrupt?

42. What is accomplished in C language by the INTCONbits.GIEH = 1 instruction?

43. What are the purposes of the IPR, PIE, and PIR registers?

44. Describe how Timer 1 is used to cause a 1-second interrupt in the real time clock example in Section 6-4.

45. Rewrite the real time clock procedure so that the counts in seconds, minutes, and hours are in binary-coded decimal instead of binary. (Refer to the description of DAW elsewhere in the text).

46. What happens if the interrupt request flag is not cleared to zero from within an interrupt service procedure?

47. Develop a high-priority interrupt that calls the function tick(void) one time per millisecond. Assume that the same 32,768-Hz crystal is used as a time base for Timer 1.

48. If the Vref+ input to the PIC is 3.3 V and the Vref− is ground, what is the step voltage for the ADC?

49. Describe how a program could sample the ADC input on RA0 2000 times a second.

50. How is a conversion started with the ADC?

51. How is an input channel selected for the ADC?

52. Explain how data are read from the data EEPROM.

53. How long does the data EEPROM retain information without power?

54. How much time is required to write to the data EEPROM?

55. Why are interrupts disabled in the function listed in Example 6-24?

56. Explain how the CCP measures the AC power line frequency in Example 6-26.

CHAPTER 7

Basic Input and Output

This chapter describes how to interface and use many different basic I/O devices in a PIC18 system. It provides the software and hardware information needed to interface the switch-base and display devices to the microcontroller and the software foundation for controlling these devices. This chapter also provides many sensors to sense the world around us and control various aspects of the environment.

Upon completion of this chapter you will be able to:

1. Debounce and develop software for a switch.
2. Interface and program any size keypad.
3. Write software for single LEDs and segmented LED displays to the PIC18.
4. Control an LCD connected to the microcontroller.
5. Develop software to control stepper motors and DC motors.
6. Interface and control relays and solenoids.
7. Use sensors to sense various events.
8. Use the RC5 code for remote controls in a sample application.
9. Measure various gaseous entities in the environment.

7-1   SWITCH-BASED INPUTS

Switches, were used in the programming examples in previous chapters, but no information was provided about interfacing the switch to the I/O port. Switches must be interfaced correctly or problems, such as sensitivity to noise and mechanical switch bouncing, will occur.

Switch Interface

A switch is often a single-pole, single-throw (SPST) device that connects the pole pieces (**contacts**) together when activated. The pole pieces open or disconnect when the switch is deactivated. Figure 7-1 illustrates a few variations of mechanical switches often used with microcontrollers. Each switch or pushbutton is connected so that either a logic zero or logic one is present at the output connection or connections. Illustrated are SPST, SPDT (single-pole, double-throw), and DPDT (double-pole, double-throw) switches and pushbuttons. In all cases, a 10K resister is used to pull up the switch output to a logic one when the switch contacts are open. The value of

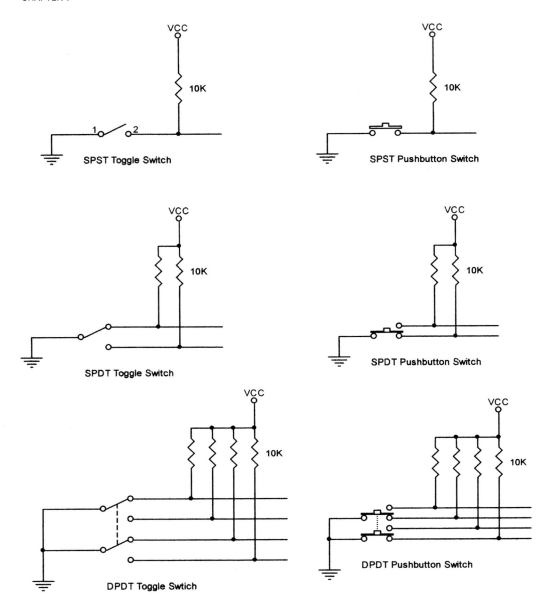

FIGURE 7-1   Various switches.

the resister can range from a low of about 1K (lower values will cause excessive current flows) and a maximum of 47K (larger values will cause a problem with noise). You may recall from Chapter 6 that Port B pins have a weak pull-up resister that is used as the pull-up resisters for a switch or anything else that requires a pull-up. Some versions of the PIC also have weak pull-ups available on Port D pins.

For simple inputs to a microcontroller, a SPST toggle switch or pushbutton is often used in a system. The SPST switch is the least expensive switch available. Switches present problems when being used to enter certain types of data because they mechanically bounce. A switch internally contains metal pole pieces that, when connected, physically bounce apart a few times. These bounces must be eliminated in systems where the number of connections or the timing of a switch is important.

Switch Debouncing

There are various schemes for removing a bounce from a switch. Figure 7-2 shows what a bounce appears as electrically on a switch, and it also shows noise. Noise occurs on switch signal lines because it is usually impossible to shield a switch and its connections to a system from environmental noise. In most cases noise, as well as bounces, must be removed from a switch connection for reliable system operation.

As can be seen in the illustration, switches can be extremely noisy devices. The amount of noise depends on the environment and Figure 7-2 represents the extreme case. Contact bounces also vary. When a switch is new, the number of bounces are minimal, but as a switch ages, the bouncing becomes more severe. The contact break noise, which is virtually nonexistent in a new switch, is produced as a switch ages and becomes worse with time.

How are these problems removed to produce the clean signal shown in the bottom of the illustration? A software time delay is often used to skip over these problem areas. Circuitry can also

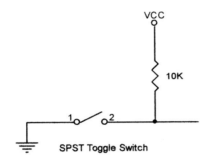

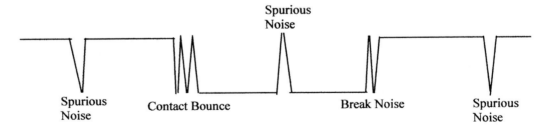

Actual Switch Signal

Debounced Switch Signal

FIGURE 7-2   Switch bouncing and noise.

FIGURE 7-3 Electronic contact bounce eliminators.

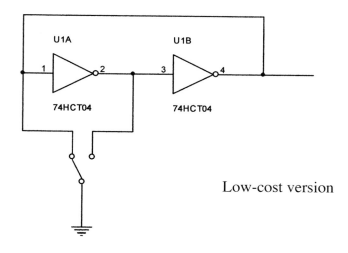

Low-cost version

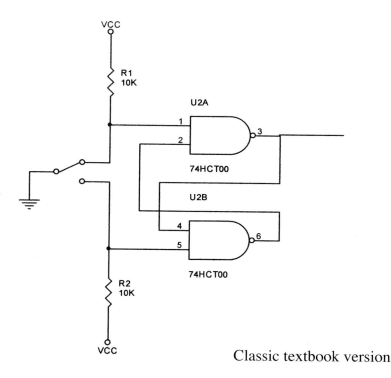

Classic textbook version

be used, but it costs money and therefore is rare today. Figure 7-3 shows some circuitry that is used to eliminate bounces on mechanical switches. The circuits are effective, but they cost money.

A better, and less costly, way to remove the problem of bouncing and noise is through software. A function that removes the bounces and noise from a switch connected to Port A at any bit position is illustrated in Example 7-1. This example uses a time delay to remove noise and bounce problems from the switch. When the switch is a good logic zero, a return from the function occurs. The clock is assumed to be 4 MHz and the time delay is 15 ms. Switches have a maximum bounce time of 8 ms, so this example uses a 15-ms delay to ensure that the bounces are skipped. Generally 10 to 20 ms are acceptable delay times for bounce removal. If the delay time is much longer than 20 ms, the switch action functions awkwardly, because it must be held down too long. This software uses a couple of #define statements to specify the port and also the count for the time delay function. The define statements allow the port and the time delay to be easily changed.

EXAMPLE 7-1

```
// >>> Don't forget the #include <delays.h> statement <<<

// *********************** Switch *****************************

// to use this function, make sure that it is invoked as follows
//
//      Switch (0x04)       ← switch on bit 2
//
//      or
//
//      Switch (0x40)       ← switch on bit 6
//
//      or
//
//      Switch (0x03)       ← switches on bits 0 and 1
//

// ********************* CONSTANTS *****************************

#define KEYPORT PORTA      // change to match the actual port
#define DELAY 15           // change as needed for time delay

void Switch (char bit)
{
      do                          // wait for release
      {
            while ((KEYPORT & bit) != bit);
            // while ((KEYPORT & bit) != bit) ClrWdt(); if needed
            Delay1KTCYx (DELAY);
      }
      while ((KEYPORT & bit) != bit);
      do                          // wait for press
      {
            while ((KEYPORT & bit) == bit);
            // while ((KEYPORT & bit) == bit) ClrWdt(); if needed
            Delay1KTCYx (DELAY);
      }
      while ((KEYPORT & bit) == bit);
}
```

The first do–while loop ensures that the switch is in the up or logic one position, and if it is not, the software stalls in this loop. In some cases the while loop will have a ClrWdt() function to clear the watchdog timer. (If you need this, comment out the while statement before the // while statement). The first do–while loop removes any noise pulses that drop the logic one level to a zero as shown in Figure 7-2. For a good release to occur, the switch must be open both before and after the 15-ms time delay. This is how bouncing and noise are removed. Once the switch is open, the second do–while loop tests the switch for a logic zero or closed condition. (This can also have a ClrWdt() function.) If a good logic zero is detected (a zero of longer than 15 ms), a return from the function occurs. This effectively removes any bounces or noise from an input bit connected to a switch or pushbutton. The only time this typically fails is if the noise happens exactly at a 15-ms interval which is extremely rare, but possible. This software assumes that the switch is active low; if it is active high, exchange the positions of the two do–while loops in the function. This software will also remove bounces and noise from multiple switches if the correct value for a bit variable is chosen. For example, to remove bouncing and noise from switches connected to Port A bits 1 and 5, use the function call Switch (0x22). The 0x22 (0010 0010) selects bits 1 and 5 for testing. The return occurs when either switch is activated.

Keypads

Numeric keypads are sometimes used in applications and they, too, must be de-bounced. In addition to de-bouncing, a numeric keypad must return the code for a key. Figure 7-4 illustrates a telephone-style numeric keypad interfaced to Ports A and B. This is a 4×3 key matrix. Keypads are available in a variety of styles and sizes from 2×2 to almost any size. A keypad matrix is constructed with SPST pushbutton switches that connect a vertical column with a horizontal row when a key is pressed. The keypad matrix is the most common keypad connection.

In this interface, Port A is used as an input port to read the rows and Port B is used as an output port to select the columns. The pull-up resistors must be placed on the input port for proper operation. (If Port B is used as the input port with the weak pull-up enabled, the pull-up resisters are eliminated). Any size matrix is connected in a likewise fashion, except the choice of I/O ports may differ. The software used to detect a key closure also de-bounces the keypad. To de-bounce the keypad, a logic zero is placed on all three Port B bits. The software to de-bounce any key is identical to that found in Example 7-1, except the bits tested are RA0 through RA3 instead of a single bit. Once the key closure is debounced, the position is calculated with software by looking at each column separately and testing Port A to determine if any of the keys in a

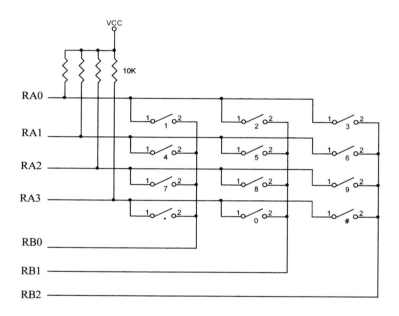

1	2	3
4	5	6
7	8	9
10	0	11

0	4	8
1	5	9
2	6	10
3	7	11

Actual Keyboard

Keyboard as seen by software
before the lookup table

FIGURE 7-4   Telephone-style 4×3 keypad matrix.

column are closed. Example 7-2 lists the function that returns the key code for a given key. This function uses the Switch function listed in Example 7-1 for key debouncing, but the bit pattern is changed to 0x0F in the bit variable of Switch, which tests the rightmost 4 bits of Ports A.

EXAMPLE 7-2

```
//
// key codes for a telephone-style keypad
// stored as static constants in the program memory
//

rom near char lookupKey[] =
{
        1, 4, 7, 10,        // left column
        2, 5, 8, 0,         // middle column
        3, 6, 9, 11         // right column
};

//
// uses function Switch from Example 7-1
//

unsigned char Key (void)
{
        #define MASK 0x0f               // set mask
        #define ROWS 4                  // set number of rows

        char a;
        char keyCode;

        PORTB = keyCode = 0;            // clear Port B and keyCode

        Switch (MASK);                  // debounce and wait for any key

        PORTB = 0xFE;                   // select a left-most column

        while ((PORTA & MASK) == MASK)  // while no key is found
        {
                PORTB = (PORTB << 1) | 1;  // get next column
                keyCode += ROWS;           // add rows to keycode
        }

        for (a = 1; a != 0; a <<= 1)
        {                               // find row
                if ((PORTA & a) == 0)
                    break;
                keyCode++;
        }

        return lookupKey[keyCode];      // look up correct key code
}
```

The keyboard software scans the keyboard from left to right and from the top down. This is shown in Figure 7-4 next to the actual key codes for this keyboard. Once a column is found, the number 0, 4, or 8 is placed in keyCode to indicate the starting key code (top key code) for the column. After finding the column, the row is searched for a logic zero at Port A to locate the key and its row. If a key is not found, the keyCode is incremented to produce a number between 0 and 11 that corresponds to a key on the keypad. If a key is found, the lookup table is accessed to convert from the scanned key codes to the actual key codes that match the key cap legends. The software generates the key codes 0, 1, 2, and 3 for the leftmost column; 4, 5, 6, and 7 for the middle column; and 8, 9, 10, and 11 for the rightmost column. Because the key caps do not match this code, a lookup table is used to convert to the correct codes for the keypad.

The Key function is written using a few #define statements that make changes in the keypad configuration easier to accomplish. For example, if a 5×6 keypad is used, the 5 rows are connected to RA0 through RA4 and the 6 columns are connected to RB0 through RB5. The #define statements are changed so that MASK is 0x1F and ROWS is 5 and the software functions for the new keypad size. The lookup table also must be changed to contain 30 entries instead of 12 entries, because a 5 × 6 keypad has 30 keys instead of 12.

Suppose that a number is read from the keypad in a program. The # key is used for enter and the * key is used for a backspace for any typing errors. A function that uses Key to read a number for the keypad appears in Example 7-3. This function returns a 16-bit unsigned integer when the # key is typed. This is difficult to use because there is no display device, which are explained in the next section. It does, however, show how to use the keypad to accept a number.

EXAMPLE 7-3

```
unsigned int GetNumber (void)
{
      char number[5];
      int retval = 0;
      char count = 0;
      char temp = Key();
      while (temp != 11)                        // while not enter
      {
            if (temp == 10 && count != 0)
                  count--;                        // do backspace
            else
            {
                  number[count] = temp;           // accept digit
                  if (count != 5)
                        count++;
            }
            temp = Key();
      }
      for (temp = 0; temp < count; temp++)       // convert to binary
            retval = retval * 10 + number[temp];
      return retval;
}
```

7-2   DISPLAY DEVICES

Many display devices available today, including light emitting diodes (LEDs), liquid crystal displays (LCDs), fluorescent displays, and so forth. This section describes these display devices along with the circuitry and software required to interface them to the microcontroller.

LEDs

Light emitting diodes first appeared in the mid-1960s. Today, LEDs are available in many different colors and soon will be in the form of a flexible organic panel called an OLED (organic LED). Because of their low cost and long life, in a relatively short time LEDs may replace incandescent light bulbs and fluorescent tubes for general purpose lighting. The LED is often interfaced to an embedded system such as an indicator or segmented numeric display. The main advantage is high visibility and longevity. The LED normally functions for about 20 years. Other advantages are relatively low power requirements and the low level of heat generated compared to an incandescent lamp.

A standard red, green, or yellow LED diode requires 10 mA of current and 1.65 V to fully illuminate. The LED requires (10 mA × 1.65 V) 16.5 mW of power to illuminate. An equivalent incandescent lamp requires 6.3 V at 100 mA or 630 mW. The LED requires 40 times less power to operate. Figure 7-5 illustrates a single LED diode interfaced to a PIC18 port pin. One interface uses 10 mA of current and an inverter that provides enough current, which is slightly more than allowed by a port pin, and the other uses 8.5 mA, which is allowed at a port pin. In both cases, the series current limiting resistor is chosen using Ohm's law $\left(R = \dfrac{E}{I} \right)$. Because the LED drops

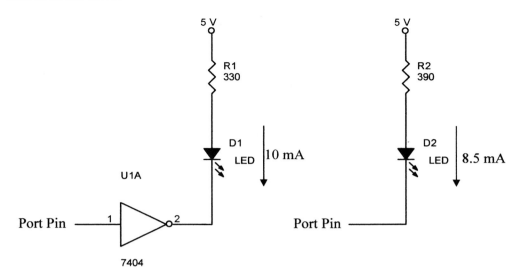

FIGURE 7-5 Driving a single LED using a +5.0 V power supply.

1.65 V, the voltage across the resister is 3.35 V so the current of 10 mA determines that the resister size is 335 Ω $\left(\dfrac{3.35 \text{ V}}{10 \text{ mA}}\right)$. A 335 Ω resister is not available, so a 330 Ω standard value resister is chosen. (Standard resister charts can be found on the Internet.) The resister value for 8.5 mA is chosen in the same manner. Using 8.5 mA causes the LED to generate nearly the same light as 10 mA and is acceptable in most applications. As explained in Chapter 6, depending on the total power for the PIC and the use of the output pin, the current may be up to 25 mA.

Suppose that a 12-V, instead of 5-V, power supply is used to drive the LED. The maximum voltage on any pin of the PIC18 is 7.5 V, which means that 12 V will destroy the microcontroller. In order to interface to a 12-V power supply for the LED, a driver is needed. The most cost-effective driver in many cases is a simple transistor inverter as illustrated in Figure 7-6. Here, a 2N2222 transistor isolates the microcontroller from the 12-V power supply and provides the proper current for the LED. This example uses a blue LED, which requires 3.5 V and 20 mA for proper operation. A transistor is a current amplifier that multiplies the input current through its base to the emitter junction by the gain of the device to produce a collector current. In this example, a collector current of 20 mA is required and the minimum current gain for this transistor (2N2222) is 100. The base current needed to generate a 20-mA collector current is therefore $\left(\dfrac{20 \text{ mA}}{100}\right)$ or 0.2 mA, which is easily provided by the microcontroller. The base resistor is calculated by using 0.2 mA and 3.6 V across the base resister. The base emitter voltage drop on a transistor is 0.7 V. Recall from Chapter 6 that the minimum logic one output voltage from the microcontroller is 4.3 V and the maximum current is 3.0 mA.

Suppose that a turn signal bar is needed for the back window of an automobile or to attach above the license plate on a motorcycle. To construct this turn signal bar, 20 bright red LEDs are selected and PIC18F1220 is used for the microcontroller. Figure 7-7 shows a schematic diagram of this system. Ten port pins control the LEDs and two input pins interface to the electrical system. The software must detect the left, right, and brake signals from the vehicle. The voltage from the left and right tail lamps is dropped to 5 V by using 5.1 V zener diodes. This is followed by a pair of transistor inverters. The 12 V appears when a taillight is on and an open appears when it is off. The inverters are needed using the weak-pulls in Port B for generating the inputs to the microcontroller. For the left or right indication, the LEDs must flash to the left or right, lighting only the left or right LEDs. For a brake light indication, all the LEDs must light. The bright red LEDs require

FIGURE 7-6   Driving an LED
with a transistor amplifier.

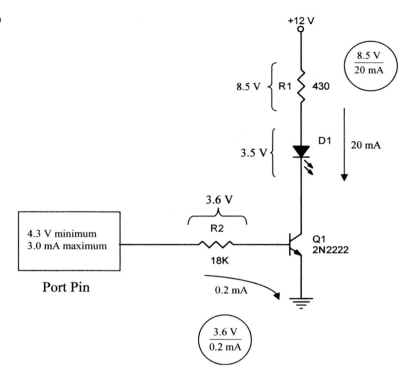

2 V with a current of 30 mA. The 74LS244 noninverting buffer provides up to 32 mA of drive current at an output (Y) pin. In the schematic, a 7805 is used to supply +5.0 V to the microcontroller and buffers. Alternately, a MAX603 from Maxim Technology (http://www.maxim-ic.com) could be used as the regulator. Maxim provides free sample parts for students.

The software for this application uses time delays written for the RC oscillator clock frequency of 2 MHz. Example 7-4 lists the C language program that controls the LED turn signals and stop light. In this example, DELAY is defined as 20, which causes a time delay of 40 ms. This time can be adjusted for the preferred flash rate, but 20 should be acceptable. Also notice that the WDT is used in case the program hangs so the system automatically reboots and recovers. Tail lamps are important and any lockup must be prevented. In this example, the WDT is reset in three different places.

EXAMPLE 7-4

```
/*
 * Turn signal example written for a PIC18F1220
 * RC clock is 2 MHz
 */

#include <p18cxxx.h>
#include <delays.h>

/* Set configuration bits
 * - set RC oscillator
 * - enable watchdog timer
 * - disable low-voltage programming
 * - disable brownout reset
 * - enable master clear
 */

#pragma config OSC = RC
#pragma config WDT = ON
```

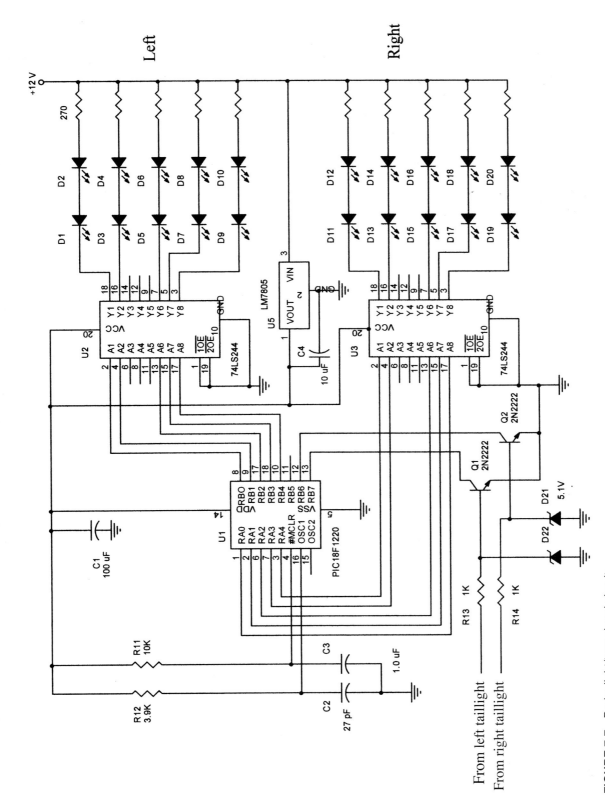

FIGURE 7-7 Brake light/turn signal circuit.

```c
#pragma config WDTPS = 64        // 256 ms WDT
#pragma config LVP = OFF
#pragma config BOR = OFF
#pragma config MCLRE = ON

#pragma code

// main program

void main (void)
{
    #define DELAY 20    // define delay count of 20 (40 ms)

    ADCON1 = 0x0F;       // Ports A and B are digital

    TRISA = 0;           // Port A is output
    TRISB = 0xC0;        // Port B is output except for
                         // bits 6 and 7

    INTCON2bits.RBPU = 1;        // pull-ups on

    while (1)            // main loop
    {
        ClrWdt();        // reset WDT

        if ((PORTB & 0xC0) == 0)                         // if STOP
        {
            PORTA = 0;       // all LEDs on
            PORTB = 0;
        }
        else if ((PORTB & 0x40) == 0x40)        // if LEFT
        {
            PORTB = 0x10;            // ....X
            Delay1KTCYx(DELAY);
            PORTB = 0x18;            // ...XX
            Delay1KTCYx(DELAY);
            PORTB = 0x1C;            // ..XXX
            Delay1KTCYx(DELAY);
            PORTB = 0x1E;            // .XXXX
            Delay1KTCYx(DELAY);
            PORTB = 0x1F;            // XXXXX

            while ((PORTB & 0xC0) != 0xC0)
            {
                ClrWdt();           // reset WDT
                if ((PORTB & 0xC0) == 0)                 // if STOP
                    break;
            }
        }
        else if ((PORTB & 0x80) == 0x80)        // if RIGHT
        {
            PORTA = 0x10;           // X....
            Delay1KTCYx(DELAY);
            PORTA = 0x18;           // XX...
            Delay1KTCYx(DELAY);
            PORTA = 0x1C;           // XXX..
            Delay1KTCYx(DELAY);
            PORTA = 0x1E;           // XXXX.
            Delay1KTCYx(DELAY);
            PORTA = 0x1F;           // XXXXX

            while ((PORTB & 0xC0) != 0xC0)
            {
                ClrWdt();           // reset WDT
                if ((PORTB & 0xC0) == 0)                 // if STOP
                    break;
            }
        }
```

```
            else // if nothing
            {
                    PORTA = 0xFF;           // all LEDs off
                    PORTB = 0xFF;
            }

        }
}
```

7-segment LED Displays

Another common display type is the **7-segment LED** display. These devices are used when the display must be fairly large and have good visibility. An example is physical therapy equipment for the elderly, who often have poor eyesight. The 7-segment LED display is available in sizes ranging from 0.25″ to 5.0″ (see the lite-on manufacturer Web site, http://www.liteon.com). Figure 7-8 illustrates a single LED display. Common anode and common cathode versions are available. The common anode version has all its anodes of the LEDs connected to 5.0 V with a logic zero lighting a segment. The common cathode version connects all its cathodes to ground and a logic one is applied to the segments to light them. Both types are available and are commonly found in applications. The LED display contains seven segments lettered from a to g. Some displays also contain a decimal point labeled dp.

Figure 7-9 illustrates a single 7-segment display connected to a PIC18F1220. Also connected to the microcontroller is a single pushbutton switch. The switch is connected to RA0 and the 7-segment display is connected to Port B. This display is driven with a 74LS244 buffer to provide the 10 mA per segment required by the display. A program that controls the display is listed in Example 7-5. This program de-bounces the pushbutton so that each time it is pressed, the number on the 7-segment display increments one time. Notice how a lookup table in the program memory is used to determine the 7-segment code for the display.

EXAMPLE 7-5

```
/*
 * 7-segment display demo example written for a PIC18F1220
 */

#include <p18cxxx.h>
#include <delays.h>
```

FIGURE 7-8 7-segment LED display.

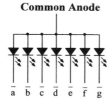

Common Anode

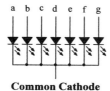

Common Cathode

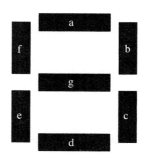

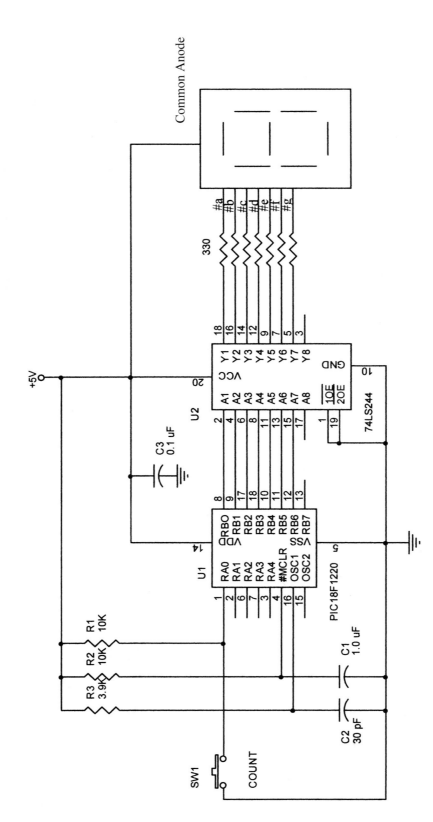

FIGURE 7-9 Simple single 7-segment LED system.

```c
/* Set configuration bits
 * - set RC oscillator
 * - disable watchdog timer
 * - disable low-voltage programming
 * - disable brownout reset
 * - enable master clear
 */

#pragma config OSC = RC
#pragma config WDT = OFF
#pragma config LVP = OFF
#pragma config BOR = OFF
#pragma config MCLRE = ON

// **************** PROGRAM MEMORY DATA ************************

rom near char look7[] = // 7-segment lookup table
{
      0x40,             // 0    active low signals
      0x79,             // 1    x g f e d c b a
      0x24,             // 2
      0x30,             // 3
      0x19,             // 4
      0x12,             // 5
      0x02,             // 6
      0x78,             // 7
      0x00,             // 8
      0x10,             // 9
};

// **************** DATA MEMORY DATA ***************************

int count;

#pragma code

// **************** FUNCTIONS **********************************

void Switch (char bit)
{
      do                // wait for release
      {
          while ((PORTA & bit) != bit);
          Delay1KTCYx(30);      // 15-ms delay

      }while          ((PORTA & bit) != bit);

      do                // wait for press
      {
          while ((PORTA & bit) == bit);

          Delay1KTCYx(30);

      }while ((PORTA & bit) == bit);
}

// **************** MAIN PROGRAM ******************************

void main (void)
{
      ADCON1 = 0x7F;   // Ports A and B are digital
      TRISA = 1;       // Port A, bit 0 is input
      TRISB = 0;       // Port B is output
      count = 0;       // start count at zero

      while (1)        // main loop
      {
          PORTB = look7[count]; // display number
          Switch (1);           // wait for pushbutton
          count++;
          if  (count >= 10)
              count = 0;
      }
}
```

Sometimes more than one display is needed. Suppose a system is required that displays a 4-digit number on a 4-digit LED display. If the same technique is used as in Figure 7-9, many I/O pins, buffers, and resisters are needed. In order to reduce the number of components and I/O pins required, the displays are usually multiplexed. A multiplexed display shares the segment driver with all of the display digits and displays the information on only one display at a time. Because of the persistence of human vision, the eye and retina remember a flash of light for approximately 20 ms. If a light flashes faster than 50 times a second, it appears to be constantly lit. If each digit of the display is flashed at least 50 times per second, all four digits appear to be constantly lit.

This same effect occurs when a motion picture is viewed at a movie theater. In film, 24 still pictures (frames) are projected each second. Each frame is flashed at the screen twice by a bowtie-shaped rotating shutter. The flash rate of a motion picture is 48 Hz and no flickering is noticed—only smooth motion. Older silent movies used 18 frames that were flashed twice per second and were known as flickers because the flash rate was 36 Hz and they flickered. American analog television has a flash rate of 60 Hz with 30 frames sent per second.

Figure 7-10 illustrates a multiplexed display connected to a PIC18F1220. This circuit functions as a simple volt meter and displays a voltage that is applied to the RA0 pin on the 4-digit display. Because no input scaling of the voltage is used, the range of acceptable input voltage is 0–5 V. The left decimal point on the second display from the left is hardwired to ground. Refer to Chapter 6 for a description of a simple analog input connected to the ADC.

This circuit uses a DS2003 Darlington driver to provide current to the segments. (National semiconductor at www.national.com provides free samples with a shipping charge to students.) Because the Darlington amplifiers invert the input signal, the 7-segment lookup table must be adjusted to active high 7-segment codes. Each segment requires 40 mA of current. The nominal segment current per display digit is 10 mA, but because each digit is on for only 1/4 of the time, the current is increased by a factor of four, for a 4-digit display, to maintain an average or nominal current of 10 mA per segment. (The human eye averages the brightness levels.) The series segment current-limiting resister is calculated by using 3 V across the resister and 40 mA through it for a value of 75 Ω. The LED display drops about 2 V. The anode driver is a 2N2907 PNP transistor with a minimum current gain of 100. Because at times an 8 is displayed lighting all seven segments, the amount of current that must be carried by the anode driver is 7 times 40 mA, or 280 mA. The base current is therefore 2.8 mA, with 4.3 V dropped across the base resister for a value of 1.5 KΩ.

The software for the system appears in Example 7-6. The C language package does not contain a function that converts a floating point number into a series of digits. The for loop, in the example, does this by getting the integer part of the floating point voltage as a digit, after which the digit is subtracted from the voltage. Then before the next iteration, the voltage is multiplied by 10 so the digit immediately to the right of the decimal point becomes available as an integer. This process continues to retrieve all four digits for display.

EXAMPLE 7-6

```
/*
 * Voltmeter written for a PIC18F1220
 * using a 4-MHz clock
 */

#include <p18cxxx.h>
#include <delays.h>

/* Set configuration bits
 * - set HS oscillator
 * - disable watchdog timer
 * - disable low-voltage programming
 * - disable brownout reset
 * - enable master clear
 */
```

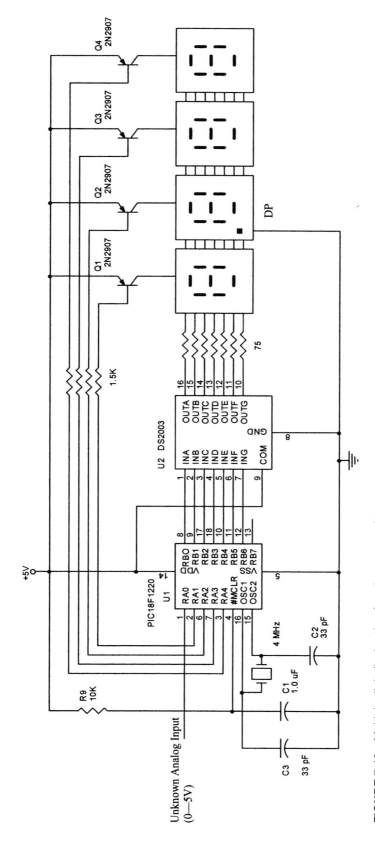

FIGURE 7-10 Multiple digit display that functions as a volt meter.

```c
#pragma config OSC = HS
#pragma config WDT = OFF
#pragma config LVP = OFF
#pragma config BOR = OFF
#pragma config MCLRE = ON

// program memory data

rom near char look7[] = // 7-segment lookup table
{
    0x3F,       // 0   active high signals
    0x06,       // 1   x g f e d c b a
    0x5B,       // 2
    0x4F,       // 3
    0x66,       // 4
    0x6D,       // 5
    0x7D,       // 6
    0x07,       // 7
    0x7F,       // 8
    0x6F        // 9
};

// data memory data

float volts;

#pragma code

// main program

float getVoltage (void)
{
    ADCON0bits.GO = 1;    //start a conversion

    while (ADCON0bits.GO == 1);   // wait for completion

    return (ADRESL + (ADRESH << 8)) * 0.00489;
}

void main (void)
{
    char a;
    char selectPattern;

    ADCON0 = 1;        // select input AN0, enable ADC
    ADCON1 = 0x0e;     // AN0 is analog, VDD and VSS are references
                       //     all other Port A and B pins are digital
    ADCON2 = 0x8C;     // convert using 1 MHz
    TRISA = 0;         // Port A is output
    TRISB = 0;         // Port B is output

    while (1)          // main loop
    {
        volts = getVoltage();           // get voltage

        selectPattern = 0xFC;

        for (a = 0; a < 4; a++)         // convert to digits
        {                               //     and display them
            PORTA = selectPattern;      // select a digit

            PORTB = look7[ (int) volts ];

            volts -= (int) volts;
            volts *= 10;

            selectPattern = (selectPattern << 1) | 2;

            Delay1KTCYx(3);             // 3 ms delay
        }
    }
}
```

A time delay is used to display each digit once every 3 ms. This is a flash rate of 83.3 Hz for the four digits of the display, which is sufficient to prevent flickering. The flash rate must not be 50 or 60 Hz or any multiple. In the United States, power line frequencies are 60 Hz, causing fluorescent lamps to flash at 120 Hz. (A fluorescent lamp lights only when the voltage across it is 90 V or higher, so it turns on and off 120 times per second, once for each alternation of the voltage.) If the flash rate of the display is any multiple of 120 Hz, problems will occur when viewing the display under fluorescent lighting. The same is true for 100 Hz in other countries because of the 50-Hz AC voltage. The flash rate can be any value up to about 30 KHz. If a higher flash rate is used, the signal starts to interfere with radio communications and the FCC (Federal Communications Commission), which regulates radio frequencies, will not approve the system for commercial sale. Digital data contains significant harmonics out to about the 15th harmonic. A flash rate of 30 KHz times 15 is 450 KHz. The FCC regulates and licenses (sells) radio frequency spectrum from 450 KHz and upwards. If there is interference from any digital device, the FCC cannot sell the spectrum space, so most digital devices must be type accepted and approved by the FCC.

If a voltage greater than 5 V needs to be measured, a voltage divider is used to drop the voltage down to the 0–5 V range. For example, to scale the input by a factor of 10 (0–50 V), connect a 9K and 1K resister in series from the input to ground and take the voltage off of the 1K resister. To divide by a factor of 100, use a 91K, a 9K, and a 1K in series and take the voltage across the 1K. The 1K, 9K, and 91K resisters are available as ±1% parts for this purpose. Figure 7-11 illustrates this connection to the microcontroller shown with the switch set to the 0–500 V range. A bilateral switch can be used to select the input and replace the mechanical switch. Unfortunately, the analog inputs to the microcontroller are clamped at TTL logic levels so the inputs cannot be connected to read all three voltages directly.

LCD Display

The LCD (liquid crystal display) has replaced the LED display in many applications. However, the LCD display is difficult to see in low-light situations and the size of the display tends to be fairly small, so the LED is still used in limited applications. Many LCD manufactures have begun selling LCD displays with bright backlights to overcome this problem. If the price of the OLED (organic LED) becomes low enough, LCD and conventional LED displays will disappear.

Figure 7-12 illustrates the connection of the Optrex DMC-20481, LCD display interfaced to a PIC18F1220. The DMC-20481 is a 4-line by 20-characters-per-line display that accepts ASCII code as input data. It also accepts commands that initialize it and control its operation. As seen in Figure 7-12, the LCD display has only a few connections. The data and control connections, attached to Port B, are used to input display data and read information from the display. The LCD operates with either an 8-bit or 4-bit data connection. In this example, a 4-bit data connection, is chosen to reduce the number of pins needed from the microcontroller and is the common way to connect an LCD display to a microcontroller. When using a 4-bit data connection, data bits D4 through D7 on the LCD are used for data and connection; data bits D0 through D3 have no function and are not connected.

The display has four control pins. The VEE connection adjusts the contrast of the LCD display and is normally connected to a 10-KΩ potentiometer, as illustrated. The RS (register select) input selects data (RS = 1) or instructions (RS = 0). The E (enable) input must be a logic one for the DMC-20481 to read or write information, and functions as a clock that is pulsed high. Finally, the R/$\overline{\text{W}}$ pin selects a read or a write operation. Normally, the RS pin is placed at a 1 or 0, the R/$\overline{\text{W}}$ pin is set or cleared, data are placed on the data input pins, and then the E pin is pulsed to either clock data into the DMC-20481 or read data from it. This display also has two inputs (LEDA [anode] and LEDK [cathode]) for backlighting LED diodes. The backlight requires 240 mA of current, which requires a 12 Ω series current-limiting resister. Some displays have an electro-luminance (EL) backlight that requires 120 VAC for proper operation. The EL backlight

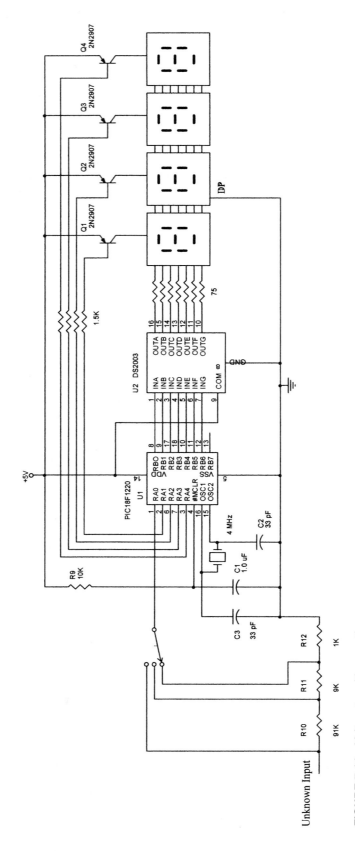

FIGURE 7-11 Volt meter with more than one range.

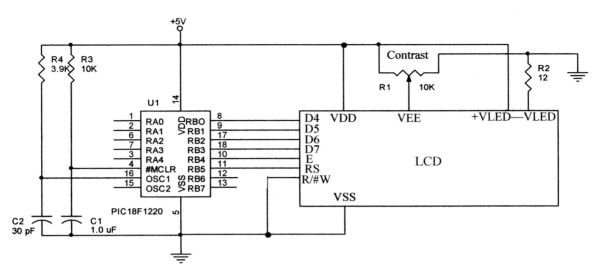

FIGURE 7-12 LCD interfaced to the microcontroller.

is a fluorescent tube. Some displays that use LED backlights have the current-limiting resister built into the display.

In order to program and use the DMC-20481 in a system, it must be initialized. This applies to any display that uses the HD44780 (Hitachi) display driver integrated circuit or its equivalent. The entire line of small display panels from Optrex and most other manufacturers is programmed in the same manner. When the LCD is first powered, the HD44780 executes a **self-check** sequence that displays black blocks (ASCII code 0x7F) on the first line of the display. These blocks must be removed and the display must be initialized as indicated in the following steps before it is used in a system:

1. Wait at least 15 ms after VCC rises to 5 V.
2. Output the function set command (0x30 [8-bit] or 0x20 [4-bit]), and wait at least 4.1 ms.
3. Output the function set command (0x30 or 0x20) a second time, and wait at least 100 µs.
4. Output the function set command (0x30 or 0x20) a third time, and wait at least 40 µs.
5. Output the function set command (0x38 for an 8-bit interface or 0x28 for a 4-bit interface) a fourth time, and wait at least 40 µs.
6. Output a 0x01 to home the cursor and clear the display, and wait at least 1.64 ms.
7. Output the enable display with the cursor off (0x0C), and wait at least 40 µs.
8. Output a 0x06 to select auto-increment, which shifts the cursor toward the right each time a character is sent to the display, and wait at least 40 µs.

The software to accomplish the initialization of the LCD display is listed in Example 7-7. It is long, but the display controller requires this long initialization dialog. Various instructions to the LCD have different time delay requirements. Table 7-1 describes the required delays for the instructions. This software sends and receives data through Port B. To send the 0x20 reset command, the 2 is sent first and then the 0 is sent through the four data connections.

EXAMPLE 7-7

```
// Assumes the clock frequency is 2 MHz
//     for an instruction clock of 2 us

#define LCD PORTB                 // define data port
#define RS PORTBbits.RB5          // define RS
#define E PORTBbits.RB4           // define E
```

```
void SendLCDdata (char data, char rs)
{
      LCD = data >> 4;            // send left nibble
      RS = rs;                    // control RS
      E = 1;                      // pulse E
      E = 0;
      Delay10TCYx (2);            // delay 40 us
      LCD = data & 0x0F;          // send right nibble
      RS = rs;                    // control RS
      E = 1;                      // pulse E
      E = 0;
      Delay10TCYx (2);            // delay 40 us

}

void InitLCD (void)
{
      char a;

      Delay1KTCYx (10);           // wait 20 ms (see text)

      for (a = 0; a < 3; a++)
      {
            SendLCDdata (0x20, 0);                 // send 0x20
            Delay1KTCYx (3);                       // wait 6 ms
      }

      SendLCDdata (0x28, 0);      // send 0x28
      SendLCDdata (0x01, 0);      // send 0x01
      Delay1KTCYx (1);            // wait 2 ms
      SendLCDdata (0x0C, 0);      // send 0x0C
      SendLCDdata (0x06, 0);      // send 0x06
}
```

The delays are slightly longer than required because of the delay in the SendLCD function to ensure that the time provided is long enough. The clock in this example is an RC clock and it can vary in frequency. As with other examples, the time delays must be adjusted if the clock frequency (2 MHz) is not the same as used in the circuit provided. Note that a small time delay may be needed before and after each change on the E signal to reduce EMI for equipment that must be FCC-type accepted.

TABLE 7-1   Instructions for most LCD displays.

Instruction	Code	Description	Time
Clear display	0000 0001	Clears the display and homes the cursor	1.64 ms
Cursor home	0000 0010	Homes the cursor	1.64 ms
Entry mode set	0000 01AS	Sets cursor movement direction (A = 1, increment) and shift (S = 1, display shift)	40 μs
Display on/off	0000 1DCB	Sets display on/off (D = 1, on) (C = 1, cursor on) (B = 1, cursor blink)	40 μs
Cursor/display shift	0001 SR00	Sets cursor movement and display shift (S = 1, shift display) (R = 1, right)	40 μs
Function set	001L NFXX	Programs LCD circuit (L = 1, 8-bit interface) (N = 1, 2 lines) (F = 1, 5 × 10 characters) (F = 0, 5 × 7 characters)	40 μs
Set CGRAM address	01XX XXXX	Sets character generator RAM address	40 μs
Set DRAM address	10XX XXXX	Sets display RAM address	40 μs
Read busy flag	B000 0000	Reads busy flag (B = 1, busy)	0
Write data	Data	Writes data to the display RAM or the character-generator RAM	40 μs
Read data	Data	Reads data from the display RAM or character-generator RAM	40 μs

Once the LCD display is initialized, a few procedures are needed to display information and control the display. After initialization, a time delay of 40 μs occurs in the SendLCDData function, which is used when sending data or many commands to the display. The clear display command still needs a time delay because the busy flag is not used with the command in the functions in Example 7-7. Instead of time delays, the busy flag can be tested to see whether the display has completed an operation. If the busy flag is used, the R/W pin is connected to the Port B, bit 6 pin in this example, and controlled to read the LCD. A procedure to test the busy flag appears in Example 7-8. The BUSY procedure tests the LCD display and returns only when the display has completed a prior instruction.

EXAMPLE 7-8

```
#define LCD PORTB                   // define data port
#define RS PORTBbits.RB5            // define RS
#define E PORTBbits.RB4             // define E
#define RW PORTBbits.RB6            // define RW
#define LCD_TRIS TRISB              // define LCD_TRIS

void SendLCDdataWbusy (char data, char rs)
{
        RW = 0;                     // select write mode
        LCD = data >> 4;            // send left nibble
        RS = rs;                    // set RS
        E = 1;                      // pulse E
        E = 0;
        Delay10TCYx (2);            // delay 40 us
        LCD = data & 0x0F;          // send right nibble
        RS = rs;                    // set RS
        E = 1;                      // pulse E
        E = 0;
        Delay10TCYx (2);            // delay 40 us
        RW = 1;                     // select read mode
        RS = 1;                     // set R/W
        LCD_TRIS = 0x0F;            // set RB0-RB4 to input
        RS = 0;                     // read busy command
        E = 1;                      // read high nibble
        E = 0;

        data = LCD;                 // read busy bit

        while ((data & 8) == 8)
        {
                E = 1;              // read low nibble
                E = 0;
                E = 1;              // pulse E
                E = 0;              // read high nibble
                data = LCD;         // read LCD

        }
        LCD_TRIS = 0;               // program PORT B as output
}
```

Once this new procedure is available, data can be sent to the display by writing without any time delays. The initialization dialog has sent the cursor for auto-increment, so if WRITE is called more than once, the characters written to the display will appear one next to the other, as they would on a video display.

The only other procedure is needed for a basic display is the clear and home cursor procedure called CLS, shown in Example 7-9. This procedure uses the SEND macro from the initialization software to send the clear command to the display. With CLS and the procedures presented thus far, you can display any message on the display, clear it, display another message, and basically operate the display. As mentioned earlier, the clear command requires a time delay (at least 1.64 ms) for proper operation.

EXAMPLE 7-9

```
void CLS (void)
{
      SendLCDdata (0x01, 0);      // send 0x01
      Delay1KTCYx (1);            // wait 2 ms
}
```

Additional procedures that can be developed might select a display RAM position. The display RAM address starts at 0 and progresses across the display until the last character address on the first line is location 19. Once the cursor or display address is changed, individual characters on the display can be changed or read, and data can be read from the display.

The bytes are not used for display locations can be used to store data. Because the microcontroller has a limited number of DATA RAM locations, the unused locations in the data RAM can supplement it. The LCD contains 128 bytes of memory, addressed from 0x00 to 0x7F. Not all of this memory is always used. For example, the one-line × 20-character display uses only the first 20 bytes of memory (0x00–0x13.) The first line of any of these displays always starts at address 0x00. The second line of any display powered by the HD44780 always begins at address 0x40. For example, a two-line × 40-character display uses addresses 0x00–0x27 to store ASCII-coded data from the first line. The second line is stored at addresses 0x40–0x67 for this display. In the four-line displays, the first line is at 0x00, the second is at 0x40, the third is at 0x14, and the last line is at 0x54. The largest display device that uses the HD44780 is a two-line × 40-character display. The four-line by 40-character display uses an M50530 or a pair of HD44780s. Information on these devices can be found on the Internet, so they are not covered in this textbook.

The program listed in Example 7-10 displays the message "Hello" on line 1 of the 4 × 20 display and "I'm the PIC18" on line 2. This example includes one additional function to display a C-style null string. When a display location is addressed, the leftmost bit of the address must be a logic one (see Table 7-1 for the set DRAM address). For example, to address the first character in line 1 use a 0x80. Likewise, to address the first character in line 2 use address 0xC0 (0x40 + 0x80).

EXAMPLE 7-10

```
/*
 * LCD example written for a PIC18F1220
 *      for a 2-MHz internal RC clock
 */

#include <p18cxxx.h>
#include <delays.h>

/* Set configuration bits
 * - set RC oscillator
 * - disable watchdog timer
 * - disable low-voltage programming
 * - disable brownout reset
 * - enable master clear
 */

#pragma config OSC = RC
#pragma config WDT = OFF
#pragma config LVP = OFF
#pragma config BOR = OFF
#pragma config MCLRE = ON

#pragma code

char str1[] = "Hello";
char str2[] = "I'm the PIC18";

// main program
```

```
#define LCD PORTB                              // define data port
#define RS PORTBbits.RB5                       // define RS
#define E PORTBbits.RB4                        // define E

void SendLCDdata (char data, char rs)
{
      LCD = data >> 4;                         // send left nibble
      RS = rs;                                 // control RS
      E = 1;                                   // pulse E
      E = 0;
      Delay10TCYx (2);                         // delay 40 us
      LCD = data & 0x0F;                       // send right nibble
      RS = rs;                                 // control RS
      E = 1;                                   // pulse E
      E = 0;
      Delay10TCYx (2);                         // delay 40 us

}

void InitLCD (void)
{
      char a;

      Delay1KTCYx (10);                        // wait 20 ms (see text)

      for (a = 0; a < 3; a++)
      {
            SendLCDdata (0x20, 0);             // send 0x20
            Delay1KTCYx (3);                   // wait 6 ms
      }

      SendLCDdata (0x28, 0);                   // send 0x28
      SendLCDdata (0x01, 0);                   // send 0x01
      Delay1KTCYx (1);                         // wait 2 ms
      SendLCDdata (0x0C, 0);                   // send 0x0C
      SendLCDdata (0x06, 0);                   // send 0x06
}

void CLS (void)
{
      SendLCDdata (0x01, 0);                   // send 0x01
      Delay1KTCYx (1);                         // wait 2 ms
}

void String (char *str, char position)
{
      int ptr = 0;
      SendLCDdata (position, 0);               // send position
      while (str[ptr] != 0)
            SendLCDdata (str[ptr++], 1);       // send character
}

void main (void)
{
      ADCON1 = 0x0F;                           // select all digital signals
      TRISB = 0;                               // Port B is output
      PORTB = 0;                               // all Port B pins = 0
      InitLCD();                               // initialize LCD

      while (1)
      {
            String (str1, 0x80);
            String (str2, 0xC0);
      }
}
```

Suppose that data must scroll into the display from left to the right like on a tickertape display. A function that sends a string to line 1 of the LCD in this fashion appears in Example 7-11. This is accomplished by sending four characters per second to the display. This function assumes that the line has been cleared before being called. The loop counter in the for loop is char. If a char is used instead of an integer, the amount of memory required for the loop is reduced significantly. This is an important consideration when developing software for the PIC.

EXAMPLE 7-11

```
void scroll (char *str)
{
      char a;
      char b;

      for (a = 0; a < strlen (str); a++)
      {
            for (b = 0; b <= a; b++)
            {
                  SendLCDdata (0x80 | a-b, 0);   // position line 1
                  SendLCDdata (str[ strlen (str) − b], 1);
            }
            Delay1KTCYx (125);       // wait 1/4 second
      }
}
```

If the number of pins available on the PIC are limited to fewer than six, either a serial LCD can be obtained or the circuit in Figure 7-13 is used. This circuit uses only two pins on the microcontroller to operate an LCD, but requires one extra component. This circuit uses a shift register to receive data from the microcontroller. One port pin is a data input to the shift register and the other is a clock signal. By controlling the data and clock signals, a number is sent to the shift register and assembled as parallel signals to the LCD.

This circuit is designed so that when the data signal from the microcontroller is zero or the value of Q5 is zero, the E signal is zero. In order to obtain the E clock signal to the LCD, both Q5 and RB0 must both be a logic one. The two diodes and the 1K resister form an AND gate. Diode logic gates are still used occasionally in modern digital circuitry to save the cost of adding another integrated circuit.

Example 7-12 lists the software required to send data to the LCD. Because the data are serial, most of the time delays required to operate the LCD are already included in the function to send data to the LCD. The only part of the software that is significantly different is the function to clear the 74HCT164 shift register.

EXAMPLE 7-12

```
/*
 * Serial LCD example written for a PIC18F1220
 */

#include <p18cxxx.h>
#include <delays.h>

/* Set configuration bits
 * - set RC oscillator
 * - disable watchdog timer
 * - disable low-voltage programming
 * - disable brownout reset
 * - enable master clear
 */

#pragma config OSC = RC
#pragma config WDT = OFF
#pragma config LVP = OFF
#pragma config BOR = OFF
#pragma config MCLRE = ON

#pragma code

// main program
```

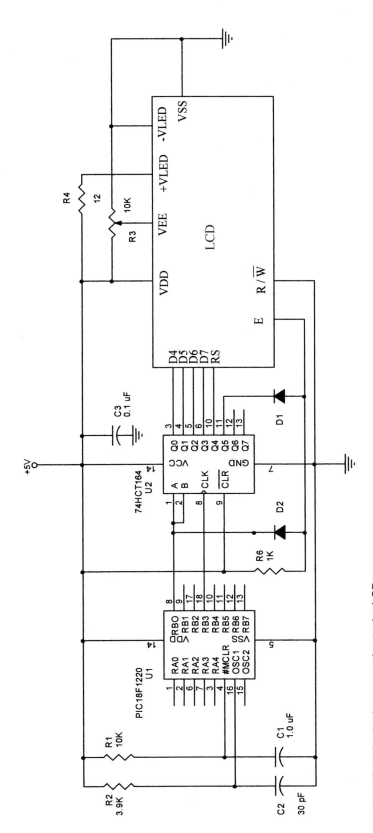

FIGURE 7-13 2-wire connection to the LCD.

```
void clockShiftReg (void)
{
      PORTBbits.RB3 = 1;                // pulse clock on RB3
      PORTBbits.RB3 = 0;
}

void clearShiftReg (void)
{
      char a;
      PORTBbits.RB0 = 0;                // data line low
      for (a = 0; a < 6; a++)           // send 6 clocks
           clockShiftReg ();
}

void sendNibble (char nib, char RS)
{
      char a;
      clearShiftReg ();                 // clear shift register
      PORTBbits.RB0 = 1;                // send E
      clockShiftReg ();
      PORTBbits.RB0 = RS;               // send RS
      clockShiftReg ();
      for (a = 0; a < 4; a++)
      {
           PORTBbits.RB0 = nib & 1;
           clockShiftReg ();
           nib >>= 1;
      }
}

void sendLCDdata (char data, char RS)
{
      sendNibble (data >> 4, RS);       // send left nibble
      sendNibble (data, RS);            // send right nibble
}

void initLCD (void)
{
      Delay1KTCYx (10);                 // wait 20 ms (see text)

      sendLCDdata (0x20, 0);            // send 0x20
      Delay1KTCYx (3);                  // wait 6 ms
      sendLCDdata (0x20, 0);            // send 0x20
      Delay10TCYx (1);                  // wait 100 us
      sendLCDdata (0x20, 0);            // send 0x20
      sendLCDdata (0x28, 0);            // send 0x28
      sendLCDdata (0x01, 0);            // send 0x01
      Delay1KTCYx (1);                  // wait 2 ms
      sendLCDdata (0x0C, 0);            // send 0x0C
      sendLCDdata (0x06, 0);            // send 0x06
}

void main (void)
{
      ADCON1 = 0x0F;                    // select all digital signals
      TRISB = 0;                        // Port B is output
      PORTB = 0;                        // all Port B pins = 0
      initLCD ();                       // initialize LCD

      // display data here
}
```

C18 Support for the LCD Display

The C language compiler has support for the LCD display in the xlcd.h header file with the functions listed in Table 7-2. This table describes a complete set of functions for the LCD, including changing the data in the character generator so custom characters can be displayed.

In order to use the xlcd.h and its functions, three time delays must be provided: DelayFor18TCY (for a delay of 18 instruction cycles), DelayPorXLCD (for a delay of 15 ms),

TABLE 7-2 Functions for the LCD display based on the HD44780 controller.

Function	Example	Note
BusyXLCD (void)	while (BusyXLCD());	Tests to see if the LCD is busy
OpenXLCD (unsigned char init)	OpenXLCD (FOUR_BIT & LINES_5X7);	Initializes the LCD for 4 data bits and multiple lines of 5 × 7 characters FOUR_BIT for 4 data lines EIGHT_BIT for 8 data lines LINE_5X7 for a single 5 × 7 line LINE _5X10 for a single 5 × 10 line LINES_5X7 for multiple 5 × 7 lines
putcXLCD (char data)	See WriteDataXLCD	Same as WriteDataXLCD
putsXLCD (char *buffer)	putsXLCD (buffer);	Displays the contents of a data memory buffer that contains a null-terminated string
putrsXLCD (const rom char *buffer)	putrsXLCD (buffer);	Displays the contents of a program memory buffer that contains a null-terminated string
unsigned char ReadAddrXLCD (void)	addr = ReadAddrXLCD();	Reads the display RAM address from the LCD
char ReadDataXLCD (void)	data = ReadDataXLCD();	Reads the data byte from the LCD
SetCGRamAddr (unsigned char addr)	SetCGRamAddr (0x40);	Addresses the @ symbol in the character generator
SetDDRamAddr (unsigned char addr)	SetDDRamAddr (0x80);	Addresses a data character position
WriteCmdXLCD (unsigned char cmd)	WriteCmdXLCD (BLINK_ON);	Writes a command to the LCD DON display on DOFF display off BLINK_ON cursor blinks BLINK_OFF cursor does not blink CURSOR_OFF cursor off SHIFT_RIGHT_CUR SHIFT_LEFT_CUR SHIFT_DISP_RIGHT SHIFT_DISP_LEFT
WriteDataXLCD (char data)	WriteDataXLCD ('A');	Write a data byte to the LCD

and DelayXLCD (for a delay of 5 ms). If the header is left unchanged, the RB4 pin is used for the E signal, the RB5 pin is used for the RS signal, the RB6 pin is used for the RW signal, and the RB0 through RB3 pins are used as the data connections for 4-bit operation to the display. A simple example illustrating the use of the xlcd header appears in Example 7-13. Because the R/W pin must be connected for these functions to operate correctly, which requires an additional I/O pin, these functions are rarely used.

EXAMPLE 7-13

```
/*
 * Serial LCD example written for a PIC18F1220
 */

#include <p18cxxx.h>
#include <delays.h>
#include <xlcd.h>
```

```
/* Set configuration bits
 * - set HS oscillator (4 MHz)
 * - disable watchdog timer
 * - disable low-voltage programming
 * - disable brownout reset
 * - enable master clear
 */

#pragma config OSC = HS
#pragma config WDT = OFF
#pragma config LVP = OFF
#pragma config BOR = OFF
#pragma config MCLRE = ON

// ROM DATA

near rom char str1[] = "I am PIC18F";
near rom char str2[] = "How are you?";

#pragma code

// main program

void DelayFor18TCY (void)              // for a 4-MHz clock
{
     Delay10TCYx (2);                  // 20 us
}
void DelayPORXLCD (void)
{
     Delay1KTCYx (15);                 // delay of 15 ms
}
void DelayXLCD (void)
{
     Delay1KTCYx (5);                  // delay of 5 ms
}

void main (void)
{
     ADCON1 = 0x0F;                    // select all digital signals

// initialize LCD

     OpenXLCD (FOUR_BIT &              // open LCD
               LINES_5X7);
     while (1)
     {
          SetDDRamAddr (0x80);   // line 1
          putrsXLCD (str1);
          SetDDRamAddr (0xC0);   // line 2
          putrsXLCD (str2);
     }

}
```

Vacuum Fluorescent Displays

Vacuum fluorescent displays (VFDs) are most often found in alarm clocks and automobile clock/radio displays. These displays are typically characterized by their blue-green color that is visible in many different lighting situations, although other color VFDs are available. The VFD was the first electronic display device available and predates the LED. These units are (extraneous) common in many consumer electronic devices, and operate for up to 80,000 hours.

Figure 7-14 illustrates the internal construction of the VFD. The VFD functions similar to a triode vacuum tube, except that the anode voltage of the VFD is 10 to 15 volts instead of hundreds of volts used in vacuum tube triodes. The filament (cathode) is connected to 1.5 V and the current through it causes it to heat and generate a cloud of electrons. If a positive voltage is applied to an anode, the electrons are attracted to the anode and when they strike it they produce a

FIGURE 7-14 Internal construction of the vacuum fluorescent display (VFD).

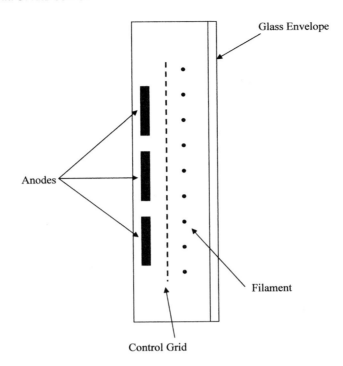

Glass Envelope

Anodes

Filament

Control Grid

photon of light that is visible. The anode is coated with a phosphor that determines the color of the light emitted. The control grid, between the anode and the filament, can be connected to a small negative voltage to repel some of the electrons to reduce the brightness of the display. Usually a switch is connected to the control grid to connect it to either ground or a negative 1 V. If grounded, the display shines bright and if connected to −1 V, the display is dimmed.

Because this device works well with 12 V and is fairly bright, most automobiles use it for the odometer and the clock and radio display. As far as programming, the device is available in a serial version that is programmed similar to the LCD display, so an example is not shown here.

7-3 CONTROLLING MOTORS

This section discusses controlling motors. A few basic motor types are controlled by embedded systems, including stepper motors and DC motors. Interfacing and software examples of each motor type is presented in this section.

Stepper Motors

A device often interfaced to an embedded system is the **stepper motor**. A stepper motor is considered a digital motor because it is moved in discrete steps as it traverses through 360°. A common stepper motor is geared to move approximately 15° per step to 1° per step in high-precision stepper motor. In all cases, these steps are achieved through many magnetic poles and/or gearing. Notice that two opposing coils are energized simultaneously in Figure 7-15. This is called **full-stepping** and causes the armature to move to 0°, 90°, 180°, and 270°. In order to save power, a single coil could be energized at a time, but reduces the amount of force.

Figure 7-15 shows a four-coil stepper motor that uses an armature with a single pole. The stepper motor is shown four times with the armature (permanent magnetic) rotated to four discrete

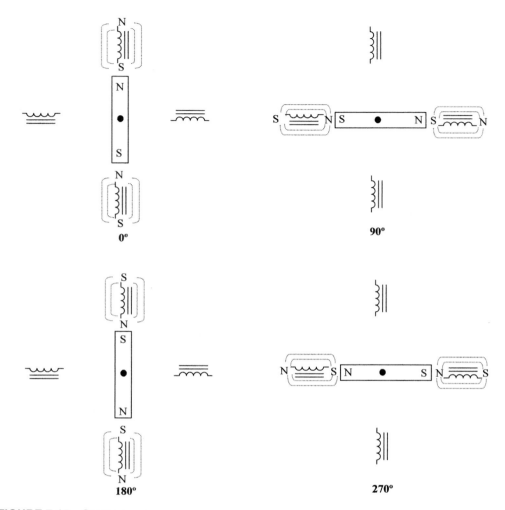

FIGURE 7-15 Stepper motor.

places. This is accomplished by energizing the coils, as shown. This is an illustration of full stepping. The stepper motor is driven by using N-channel MOSFETs to provide a large current to each coil. The MOSFET amplifiers could be replaced with Darlington amplifiers. These amplifiers contain internal damper diodes to shunt any inductive kickback voltage generated by the collapse of the magnetic field around the MOSFET.

A circuit that drives this stepper motor is illustrated in Figure 7-16, with the four coils shown in place. This circuit uses the microcontroller to provide it with the drive signals that are used to rotate the armature of the motor in either the right-hand or left-hand direction.

A simple procedure that drives the motor (assuming that Port B is programmed as an output port) is listed in Example 7-14 in C language. This function is called with parameter steps holding the number of steps and direction of the rotation. If the value of steps is greater than 0x8000, the motor spins in the right-hand direction; if the value of steps is less than 0x8000, it spins in the left-hand direction. For example, if the number of steps is 0x0003, the motor moves in the left-hand direction three steps and if the number of steps is 0x8003, it moves three steps in the right-hand direction. The leftmost bit of steps is removed and the remaining 15 bits contain the number of steps. Notice that the procedure uses a time delay of 1.0 ms. This time delay is required to allow the stepper-motor armature time to move to its next position.

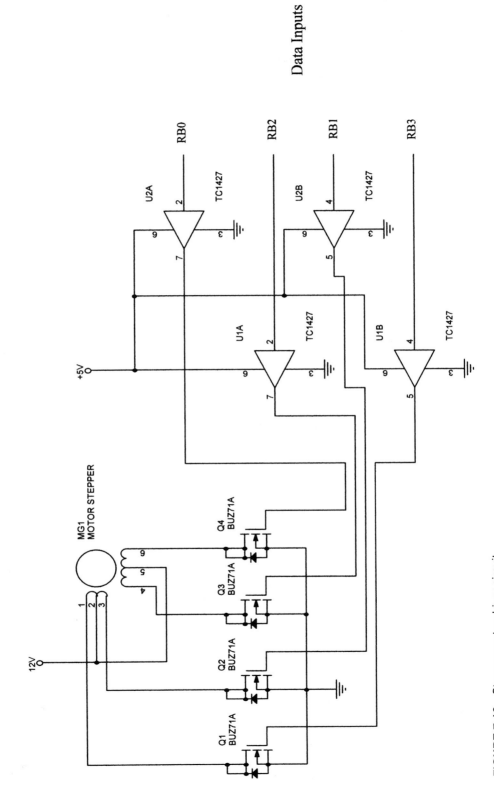

FIGURE 7-16 Stepper motor driver circuit.

EXAMPLE 7-14

```
char currentPosition = 0x33;

void moveStepper (int steps)
{
        unsigned int a;
        if ((steps & 0x8000) == 0x8000)
        {
                for (a = 0; a < steps & 0x7FFF; a++)
                {
                        _asm RLNCF currentPosition, 1, BANKED _endasm
                        PORTB = currentPosition;
                }
        }
        else
        {
                for (a = 0; a < steps; a++)
                {
                        _asm RRNCF currentPosition, 1, BANKED _endasm
                        PORTB = currentPosition;
                }
        }
}
```

The current position is stored in memory location currentPosition, which must be initialized with 0x33, 0x66, 0xEE, or 0x99. This allows a simple RRNCF (step right) or RLNCF (step left) instruction to rotate the binary bit pattern for the next step. When using assembly language in a C language program, all fields of the instruction must be included as illustrated in Example 7-14. Why was assembly language used to rotate the number in memory location currentPosition? These are only shift instructions in C language, so accomplishing the rotate in C is much more complex, uses much more memory, and is time consuming.

Stepper motors also operate in the half-step mode, which allows eight steps per sequence. This is accomplished by using the full-step sequence described, with a half step obtained by energizing one coil interspersed between the full steps. Half-stepping allows the armature to be positioned at 45°, 135°, 225°, and 315° in addition to the full-step positions. The half-step position codes are 0x11, 0x22, 0x44, and 0x88. A complete sequence of eight steps interspersing half and full-steps is: 0x11, 0x33, 0x22, 0x66, 0x44, 0xCC, 0x88, and 0x99. This sequence is either output from a lookup table or generated with software.

The type of stepper motor illustrated in Figure 7-15 is a 6-wire stepper motor. Also available is a 4-wire stepper motor that uses series-connected coils with four wires. Its main disadvantage is that a dual polarity supply or an H-bridge circuit must be used for control.

DC Motors

A DC motor is either an on–off device or its speed and direction are controlled. Control is accomplished by changing the voltage across the motor or by changing the duration of pulse applied to the motor. The later technique is explained here because it is the most common method of controlling motor speed today. This section explores controlling DC motors using the pulse-width modulator (PWM) located within the microcontroller.

Figure 7-17 shows the drivers required to drive a DC motor both as a unidirectional device and a bidirectional device. This illustration uses MOSFET drivers because the cost is similar to bipolar junction transistors and fewer parts are required with a MOSFET interface. The bidirectional driver is optically isolated from the H-bridge motor driver. This is a common practice to prevent problems and isolate the microcontroller from the motor driver. The control input to the H-bridge turns on a pair of opposing MOSFETs to drive the motor. The threshold voltage of most power MOSFETs is approximately 3.0 V. To turn a MOSFET on, the input is driven more positive than the threshold voltage and to turn it off, the voltage is driven below the threshold voltage. Because MOSFETs are

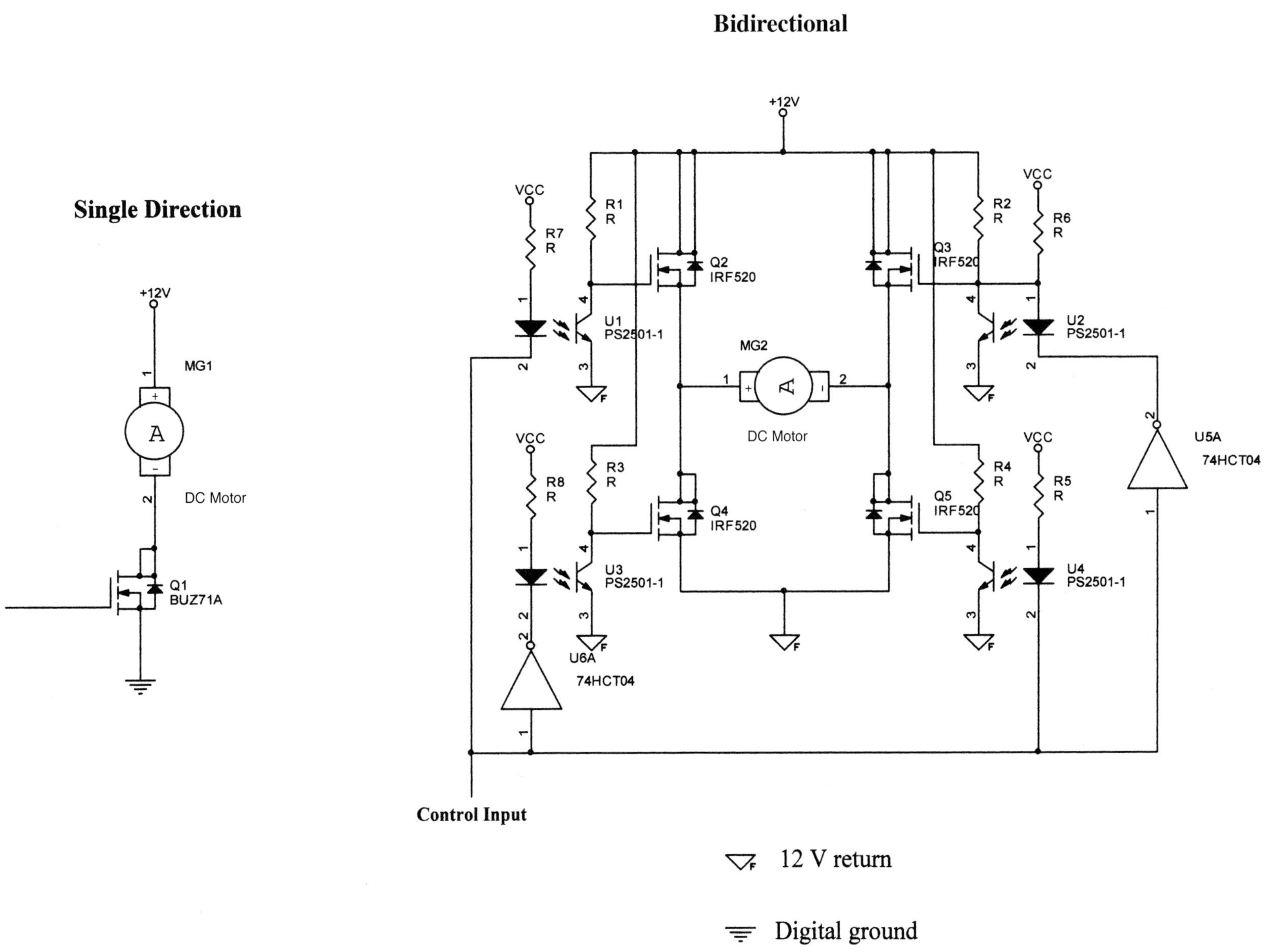

Bidirectional

Single Direction

FIGURE 7-17 Single-phase DC motor drivers.

231

voltage-controlled devices, one would assume that very little gate current flows. This is true in a power MOSFET for a single device, but what Figure 7-17 does not show is that thousands of MOS-FETs are placed in parallel inside the package to drive large current. Because the gate of a MOSFET is a capacitor, and when capacitors are placed parallel the capacitance adds, the input have a fairly large capacitance. To charge or discharge the gate capacitance can require a considerable amount of current. For example, it is fairly typical to require 1.5 mA of gate current to switch 1A with a power MOSFET, but to switch 12A requires 150 mA of gate current. Drivers for large current applications are therefore needed to interface the microcontroller with a power MOSFET.

To vary the speed of a motor, the pulse width modulator in the microcontroller is used to deliver a varying duty cycle waveform to the driver. The PWM uses Timer 2 and operates as a single output, a pair of outputs as a half H-bridge controller, or four outputs for controlling a full H-bridge. Various family members have one or more PWMs to drive more than a single motor. The signal to the motor should be above the hearing range or a motor can sing at the modulating frequency. The basic frequency of the PWM should therefore be at least 5 KHz. (Most motors will not mechanically vibrate at 10 KHz, which is what produces the sound). It should also be less than about 30 KHz or problems can occur if FCC acceptance it needed for a product. In most cases an 8-bit PWM is sufficient to drive most single-direction DC motors. This allows for 256 different speeds. If a bidirectional motor is controlled, this allows 128 speeds in each direction, which may or may not be enough speeds for complete control. To generate more speed steps, a wider PWM is required. Luckily, the PWM in the microcontroller is 10 bits wide, which allows for up to 1,024 speed steps. This is probably more than is required for most applications.

Chapter 6 introduced timers, but did not explain the operation of Timer 2. The block diagram of Timer 2 is illustrated in Figure 7-18. Timer 2 is an 8-bit counter labeled TMR2 in the

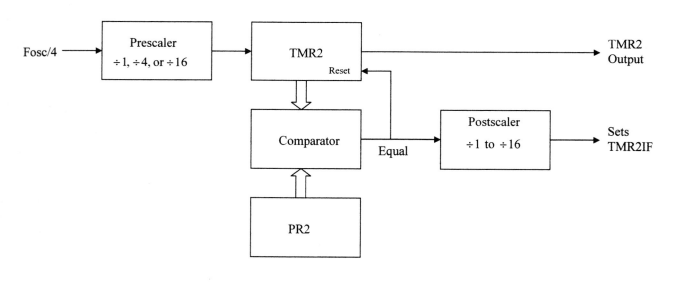

T2CON

	TOUTPS3	TOUTPS2	TOUTPS1	TOUTPS0	TMR2ON	T2CKPS1	T2CKPS0

TOUTPS3–TOUTPS0 = Postscaler $(0000 = \div1, 0001 = \div2$ to $1111 = \div16)$
TMR2ON = Timer 2 on/off $(0 = \text{off}, 1 = \text{on})$
T2CKPS1–T2CKPS0 = Prescale $(00 = \div1, 01 = \div4, 1X = \div16)$

FIGURE 7-18   Block diagram of Timer 2 and Timer 2 control register.

illustration. The input to the timer unit is the system clock divided by 4 and this is scalable by a factor of 1, 4, or 16 by the prescaler as programmed by Timer 2 control register (T2CON) bits T2CKPS1 and T2CKPS0. For example, if the system uses a 4-MHz oscillator and a prescaler of 4 is selected, the timer will increment once every 4 μs $\left(\dfrac{4\,\text{MHz}}{16} = 250\,\text{KHz}\right)$. The period register (PR2) determines the count that will reset TMR2. Timer 2 increments on each of its input clocks, then if PR2 is equal to TMR2 the timer resets on the next clock. The input to TMR2 is divided by PR2 + 1. For example, if the prescaler is set to divide by 4, the input clock is 4 MHz, and PR2 is programmed with a 9, the timer 2 output frequency is 25 KHz.

Timer 2 has two outputs: one is used for the PWM circuit inside of the microcontroller, and the other sets the interrupt flag bit (TMR2IF) to cause a Timer 2 interrupt. The signal to the interrupt flag is divided by the level set in the post-scaler, which is programmed to divide the Timer 2 signal by any whole number from 1 to 16.

Before looking at the software to control the PWM and generate a pulse-width modulation signal to drive the motor, the operation of the PWM must be understood. The CCP unit (capture/compare/PWM) provides a pulse-width modulator with up to four outputs. In the capture mode, the CCP captures the contents of Timer 1 or Timer 3 when an event occurs on the RB3 pin. The event is a rising edge, falling edge, every fourth rising edge, or every 16th rising edge. This is used to measure the time between edges as defined by Timer 1 or Timer 3. The compare mode of Timer 1 or Timer 3 is compared against the CCPR1 (16-bit) register. If a match occurs, the RB3 pin is driven high, low, toggles, or an interrupt is requested. This mode is useful for generating a signal after a certain number of clocks have elapsed. The PWM mode, which is the mode of the CCP unit of interest here, uses Timer 2 and the compare feature of the CCP to generate a pulse-width modulated output signal on the P1A pin, which is RB3 on most versions of the microcontroller. This is an output to drive a unidirectional motor. If a half- or full-H bridge is used to drive a bidirectional motor, then the P1A, P1B or P1A, P1B, P1C, and P1D pins are used. Some versions of the microcontroller have more than one PWM.

Figure 7-19 illustrates the block diagram of the PWM for 8-bit operation and an output showing a 50% duty cycle. (The **duty cycle** is the high time divided by the low time.) Timer 2 determines the period of the waveform and CCPR1L selects the high time of the output, which determines the duty cycle of the PWM output. When TMR2 (Timer 2) counts up to the count preset by PR2, the output is set and the contents of CCPR1L are loaded into CCPR1H, and Timer 2 is cleared to zero. When Timer 2 counts up to the predefined count as programmed into CCPR1L, the output clears. For example, to generate a 50% duty cycle if PR2 is programmed with 100, load CCPR1L with 50. This will set the output each time TMR2 clears, and then when TMR2 hits 50, the output clears and remains cleared until the count hits 100, where it sets again. A 25% duty cycle is obtained, for example, when PR2 is 100 and CCPR1L is 25.

Suppose that a DC motor is connected to the P1A pin (RB3) as illustrated in Figure 7-20. The motor is driven in one direction only so the motor driver circuit is not complex. This interface uses a 4-MHz crystal for timing and two pushbuttons connected to PA0 and PA1 for speed control. One is labeled UP and the other is labeled DOWN. The UP pushbutton increases the speed of the motor and the down pushbutton decreases its speed. If the PWN generates hundreds of output duty cycles, a lot of work is needed to change the speed from stopped to full, so Example 7-15 uses only 10 different speeds. The motor in this example could be the fan motor in a ceiling fan.

EXAMPLE 7-15

```
PWM period = (PR2 + 1) x TMR2PS x 4 x Tosc

PWM period = (10 + 1) x 4 x 4 x 1/4 μs
           = 11 x 4 x 1 μs
           = 44 μs
```

FIGURE 7-19   PWM in the PIC18 and the output waveform.

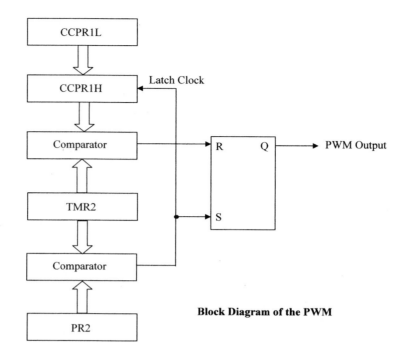

Block Diagram of the PWM

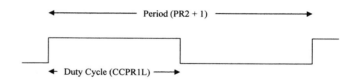

PWM Output Waveform

PWM frequency $= \dfrac{1}{44\mu s} = 22.7$ KHz

(Note: TMR2PS is the Timer 2 prescaler value)

The system software, listed in Example 7-16, programs the PWM for a period as determined by the equation in Example 7-15. In this example, the PWM frequency is 22.7 KHz. The Timer 2 clock is always the oscillator frequency divided by 4. In this case the Timer 2 clock is 1 MHz, because a 4-MHz oscillator is used in the system. Because the Timer 2 prescaler is adjusted to divide by 4, Timer 2 operates at 250 KHz. Timer 2 is programmed to reset after it hits a count of 10 (PR2). Timer 2 counts from zero to 10 and then resets by dividing the 250-KHz clock input by 11, $\left(\dfrac{250 \text{ KHz}}{11} = 22.7 \text{ KHz} \right)$. This frequency is not audible.

The initial count for CCPR1L is 11. This causes the output pin of the PWM (RB3) to be placed at a logic one because the output of the CCP comparator never clears the internal RS flip-flop. Likewise, if CCPR1L is programmed with a zero, the output becomes a constant logic zero because the output latch is constantly reset. The logic 1 input to the motor driver connected to

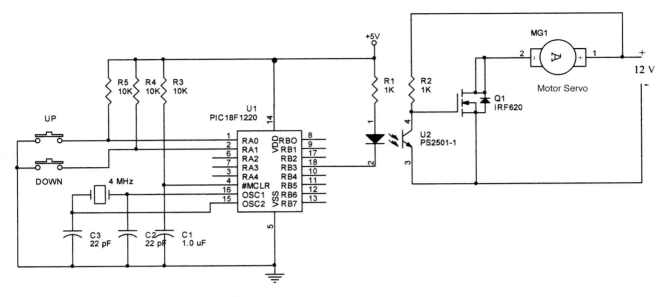

FIGURE 7-20 Driving a unidirectional DC motor.

RB3 causes the MOSFET driver to turn off because there is no input voltage. This stops the motor. If the UP pushbutton is pressed, the content of CCPR1L is decremented to 10. This causes the output of the PWM to generate a waveform that is high for 10 counts and low for 1 count. The MOSFET driver is on for 1 count and off for 10. Figure 7-21 shows a few waveforms for a few different counts.

FIGURE 7-21 A few waveforms of Figure 7-19 and the software in Example 7-14.

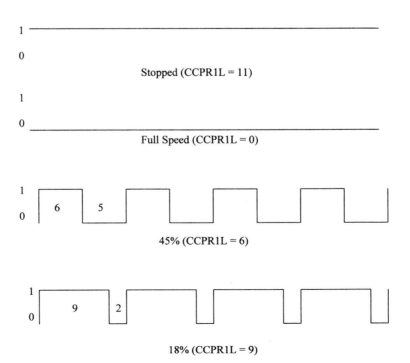

EXAMPLE 7-16

```
/*
 * Motor speed control for a PIC18F1220
 */

#include <p18cxxx.h>
#include <delays.h>

/* Set configuration bits
 * - set HS oscillator
 * - disable watchdog timer
 * - disable low-voltage programming
 * - disable brownout reset
 * - enable master clear
 */

#pragma config OSC = HS
#pragma config WDT = OFF
#pragma config LVP = OFF
#pragma config BOR = OFF
#pragma config MCLRE = ON

#pragma code

// main program

void Switch (char bitP)            // debounce switches
{
     do                            // wait for release (bits == 3)
     {
          while ((PORTA & bitP) != bitP);
          Delay1KTCYx (15);
     }while ((PORTA & bitP) != bitP);

     do                            // wait for press (bits != 3)
     {
          while ((PORTA & bitP) == bitP);
          Delay1KTCYx (15);
     }while ((PORTA & bitP) == bitP);
}

void main (void)
{
     ADCON1 = 0x0F;               // select all digital signals
     TRISB = 0;                   // Port B is output
     TRISA = 3;                   // Port A bits 0 and or 1 are input

     T2CON = 1;                   // select a prescaler of divide by 4
     CCP1CON = 0x0C;              // CCP module off, active high output
     TMR2 = 0;                    // clear Timer 2
     PR2 = 10;                    // Timer 2 clears on a 10
     CCPR1L = 11;                 // stop fan (11)
     T2CONbits.TMR2ON = 1;        // start Timer 2 and PWM

     while (1)
     {
          Switch (3);   // check and debounce switches
          if (PORTAbits.RA0 == 0)      // UP pressed
               if (CCPR1L != 0)
                    CCPR1L--;
          else
               if (CCPR1L != 11)       // DOWN pressed
                    CCPR1L++;
     }
}
```

In this software, the CCP1CON (CCP control register) is programmed to select the operation of the CCP module. Figure 7-22 illustrates the contents of the control register for the CCP module. In the example, CCP1CON is programmed with a 0x0C, which selects a single output

FIGURE 7-22 CCP control
register.

CCP1CON

P1M1	P1M0	DC1B1	DC1B0	CCP1M3	CCP1M2	CCP1M1	CCP1M0

P1M1—P1M0 PWM Output Configuration Bits
00 = single output P1A
01 = full-bridge output forward
10 = half-bridge output
11 = full-bridge output reverse

DC1B1—DC1B0 PWM Extension Bits
The least significant bits of a 10 bit PWM

CCP1M3—CCP1M0 Mode Select bits
0000 = reset CCP
0001 = not used
0010 = compare mode, toggle on match
0011 = not used
0100 = capture mode, every falling edge
0101 = capture mode, every rising edge
0110 = capture mode, every 4th rising edge
0111 = capture mode, every 16th rising edge
1000 = compare mode, set on match
1001 = compare mode, clear on match
1010 = compare mode, generate interrupt on match
1011 = compare mode, trigger special event
1100 = PWM mode, P1A, P1C active high, P1B, P1D active-high
1101 = PWM mode, P1A, P1C active high, P1B, P1D active-low
1110 = PWM mode, P1A, P1C active-low, P1B, P1D active-high
1111 = PWM mode, P1A, P1C active-low, P1B, P1D active-low

(P1A) that is active high. The various modes of operation and the signals available at the P1A through P1D pins are illustrated in Figure 7-23. The P1A and P1C outputs are meant to activate one of the pull-up MOSFETS in an H bridge, whereas the signal to move the motor is available on the P1D or P1B output pins for a full bridge output.

Suppose a motor must be controlled with a joystick-type of device that contains a potentiometer. The potentiometer selects a voltage as the joystick handle is moved from one extreme to another. Because the microcontroller contains an ADC, the output of the joystick is used as the input to the ADC, generating a digital number that corresponds to the setting of the joystick. The ADC generates an 8-bit value that is used as an input to the PWM to control the speed and direction of a motor connected to the microcontroller as a MOSFET-controlled H-bridge. Such a system is depicted in Figure 7-24. In this circuit, four IRF540N handle up to 33 amps of current, so a fairly large motor can be driven from this circuit. Optical isolators are again used to isolate the motor driver from the microcontroller. Drivers for the MOSFETs are required for large current motors.

The full-bridge signals are used to control the motor. In the forward direction, P1A is a logic one (RB3), which turns on MOSFET Q3 and P1C (RB6) is a logic zero, which turns off MOSFET Q1. The forward signal from the PWM is applied to P1D (RB7) and to Q2. In the forward mode the pair of Q2 and SQ3 are active, which allows current to flow from the + terminal on the motor to the − terminal spinning it in the forward direction. In the reverse direction Q3 is turned off and Q1 is turned on and the signal from the PWM appears on P1B (RB2), which controls Q4. In the reverse direction Q1 allows current to flow through the motor from the − terminal to the + terminal through Q4 spinning the motor in the reverse direction.

FIGURE 7-23 Half- and full-bridge outputs of the PWM unit.

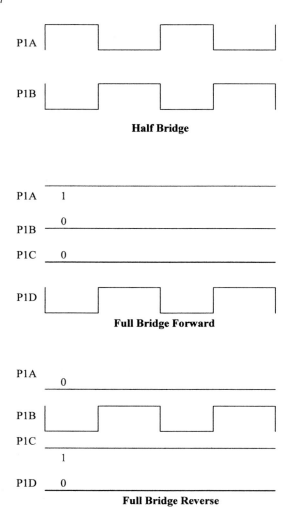

The joystick is connected to the RA0 pin or analog input zero of the ADC. The software in Example 7-17 reads the voltage off of the joystick and generates a digital input to the PWM to control the speed and direction of the motor.

EXAMPLE 7-17

```c
/*
 * Joystick-controlled motor written for a PIC18F1220
 */

#include <p18cxxx.h>

/* Set configuration bits
 * - set HS oscillator
 * - disable watchdog timer
 * - disable low-voltage programming
 * - disable brownout reset
 * - enable master clear
 */

#pragma config OSC = HS
#pragma config WDT = OFF
#pragma config LVP = OFF
#pragma config BOR = OFF
#pragma config MCLRE = ON
```

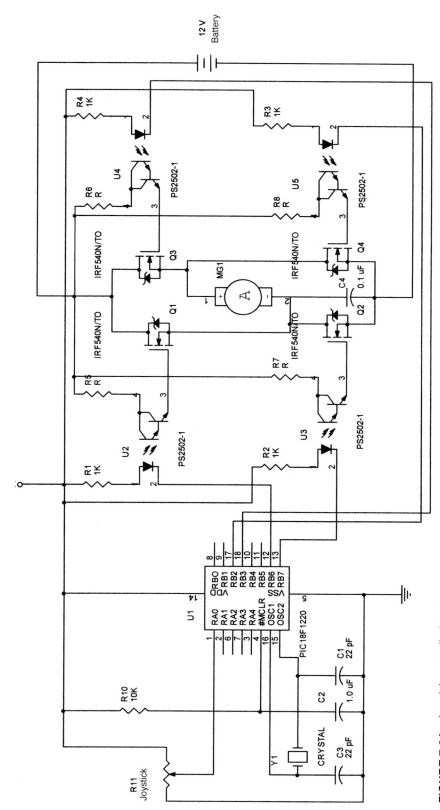

FIGURE 7-24 Joystick controlled motor.

239

```
#pragma code

// main program

char getJoy (void)
{
      ADCON0bits.GO = 1;    //start a conversion
      while (ADCON0bits.GO == 1);   // wait for completion
      return (ADRESL >> 2 | (ADRESH << 6);
}

void main (void)
{
      ADCON0 = 1;                   // select input AN0, enable ADC
      ADCON1 = 0x0e;                // AN0 is analog, VDD and VSS are references
      ADCON2 = 0x8C;                // convert using 1 MHz
      TRISA = 1;                    // Port A bit 1 = output
      TRISB = 0;                    // Port B is output

      T2CON = 0;                    // select a prescaler or divide by 1
      CCP1CON = 0x4C;               // CCP module off, full, active high output
      TMR2 = 0;                     // clear Timer 2

      PR2 = 0xff;                   // Timer 2 clears on 255;

      while (1)
      {
            CCPR1L = getJoy()    // read joystick; change speed
      }
}
```

7-4   RELAYS, SOLENOIDS, AND SENSORS

This section explains how to interface and control a variety of devices, including relays, solenoids, and various sensors.

Relays

Relays are used to switch several signals or voltages at a time, and have many applications. Relays are available as electromechanical devices or solid-state devices. Electromechanical relays have a life of 10,000 to 100,000 cycles. The solid state relay has a much longer lifespan.

To interface an electromechanical relay to a microcontroller usually requires a driver because the relay coil current is normally greater that the 3 mA (logic one) available at an output pin. Figure 7-25 illustrates a reed relay connected to RB0 of the microcontroller. In this example

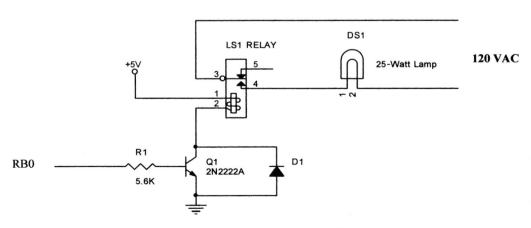

FIGURE 7-25   Interfacing a small reed relay to a microcontroller.

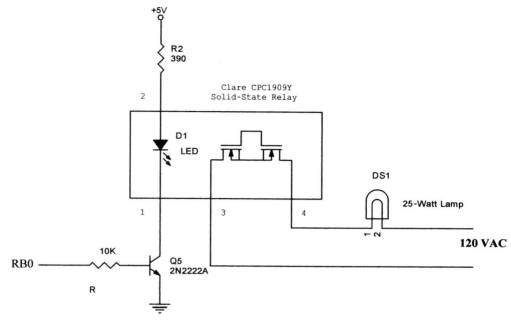

FIGURE 7-26 Driving a solid state relay.

the reed relay is used to control a 25-watt light bulb connected to 120 VAC. It also illustrates a simple driver transistor used to enable the relay. This relay requires a coil current of 60 mA. The gain of the 2N2222A is 100 minimum, so the base current is $\dfrac{60\text{ mA}}{100}$ or 0.6 mA. Because the minimum logic one voltage from the microcontroller port pin is 4.2 V, the voltage across the base resister is 3.5 V. The base resister value is $\dfrac{3.5\text{ V}}{0.6\text{ mA}}$, or 5833 Ω. The nearest standard resister value is 5.6 KΩ. The diode protects the transistor from the inductive kickback voltage produced when the magnetic field collapses around the relay coil. Without it, the transistor is destroyed the first time the relay is turned off because the voltage drop across the transistor is approximately 12,000 V, (60 mA $\times$ 20 KΩ, which is the off resistance of a transistor).

Figure 7-26 illustrates the same application as Figure 7-25, except that a solid state relay is used to switch the 25-watt lamp on and off. Here a Clare 10A solid-state relay is used to illuminate the lamp. The input current to the LED inside of this solid-state relay must be at least 10 mA, so a transistor switch is needed to amplify the current from the microcontroller. The voltage across the LED is 1.2 V. The base resister in this example produces much more current than required to enable the solid-state relay.

Solenoids

Solenoids are controlled similarly to relays, except a solenoid typically has no contacts. Solenoids are used to move objects a short distance. For example, a solenoid is used in a cash register to open the drawer by pulling a catch about 5 mm (1/4″). Once the catch is pulled out of the way, a spring propels the drawer open. The solenoid in a cash register requires a current of about 0.5 A to operate at 12 V. To ensure that the drawer moves past the catch, the solenoid is energized for approximately 500 ms. The driver circuit for the solenoid is illustrated in Figure 7-27 and the software, which uses a Timer 3 from the microcontroller and an interrupt, is illustrated in Example 7-18. The timer is chosen so that

FIGURE 7-27 Solenoid
driver.

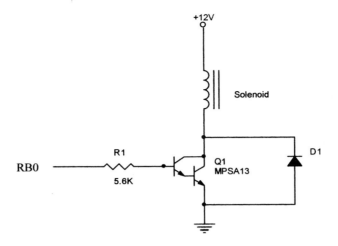

the microcontroller can do other things during the 500 ms required to energize the solenoid. The MPSA13 Darlington amplifier has a gain of 10,000 and a maximum continuous current of 500 mA.

EXAMPLE 7-18

```
#include <timers.h>

void OpenDrawer (void)              // open the drawer
{
      WriteTimer0 (64560);          // program Timer 0 to divide by 976

      OpenTimer0 (TIMER_INT_ON &    // Timer 0 interrupt on
                 T0_16BIT &         // Timer 0 is 16 bits
                 T0_SOURCE_INT &    // Timer 0 clock is internal
                 T0_PS_1_256);      // Timer 0 prescaler is 256

      PORTBbits.RB0 = 1;            // set RB0
}

void Timer0Int (void)               // interrupt service procedure
{
      INTCON.TMR0IF = 0;            // clear the interrupt
      PORTBbits.RB0 = 0;            // clear RB0
      CloseTimer0();                // kill Timer 0
}
```

Example 7-18 uses two functions. The software required to enable the interrupt structure of the microcontroller is not shown. One function opens the drawer by placing a logic one on the RB0 pin and starting Timer 0 so it causes an interrupt in 500 ms. The other is the interrupt service procedure for Timer 0, which places a logic zero on the RB0 pin and stops the timer. In this example it is assumed that an RC clock is used at a frequency of 2 MHz. Because the clock is divided by 4 before being applied to the Timer 0 prescaler, the input to the timer after the prescaler is 500 KHz (2 μs) divided by 256, or 1953 Hz (512 μs). If the timer is set to go off in 976 clocks, it will fire the Timer 0 interrupt in 499.7 ms after it is started or opened. The WriteTimer0 (64560) programs Timer 0 so it causes an interrupt on overflow in 976 clocks. It can also be written as WriteTimer0 (−976), which may be a little clearer. The interrupt enable it not shown here and neither is the interrupt service procedure; only the function called when the Timer 0 interrupt occurs.

In this example, the timers.h include file is used to provide the functions that control the timer. When the timer interrupts the microcontroller, the RB0 pin is placed back at a logic zero level and the Timer 0 interrupt is closed until the drawer is once again opened by calling Open-Drawer. This process generates a pulse at RB0 of 500 ms in duration. More information on the timers.h is provided in Chapter 8 on interrupt-processed I/O.

Why wasn't a time delay used to time the firing of the solenoid? If a time delay is used, nothing else in the system functions for the duration of the time delay. In other words, the system is not available for 500 ms while the time delay operates. For time delays of a few milliseconds this presents no problem, but for long delays, the system may need to do other things. A timer and an interrupt is often the best approach for delays of more than a few milliseconds.

Sensors

There are many types of sensors. Some are presented here and others are presented in later chapters. Not all sensors are explained because there are so many different types. This textbook shows just an illustrative sample of them.

One common type of sensor is a temperature sensor. Figure 7-28 illustrates the LM70 from National Semiconductor interfaced to the microcontroller. The device itself is the temperature sensor and is small enough to mount just about anywhere. The LM70 has a chip select input to enable it for a read or a write, a SI/O pin for serial data, and a SC input that functions as a serial clock signal. To read the temperature, place a logic zero on the chip select input and then pulse the SC input. After pulsing the SC input, read a bit of the temperature from the SI/O pin. The only reason to be able to transmit to the LM70 is to read the manufacturer product code, which is

FIGURE 7-28 Temperature sensor.

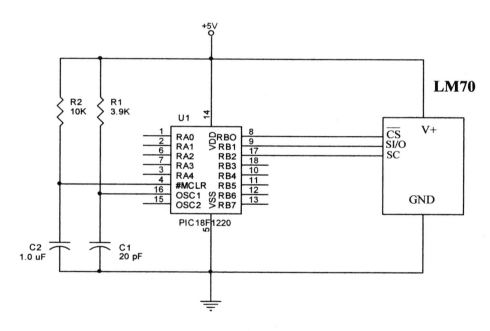

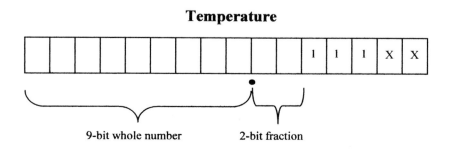

not normally required in most systems. The first data bit is available when the chip select input is grounded; the second and subsequent data bits appear 70 ns after the falling edge of the SC input.

The data appear on the SI/O pin a bit at a time beginning with the most significant bit of the 16-bit temperature. The temperature itself is read, as illustrated in Figure 7-27, to produce an 11-bit temperature that is in two's complement form if negative. Even though the rightmost five bits contain no data, they must be read each time that a temperature is received from the LM70. The software to retrieve a temperature is listed in Example 7-19. The LM70 reads only Celsius temperature, but the example returns the Fahrenheit temperature as a rounded signed integer from the function getTemp.

EXAMPLE 7-19

```
/*
 * Get temperature example written for a PIC18F1220
 */

#include <p18cxxx.h>

/* Set configuration bits
 * - set RC oscillator
 * - disable watchdog timer
 * - disable low-voltage programming
 * - disable brownout reset
 * - enable master clear
 */

#pragma config OSC = RC
#pragma config WDT = OFF
#pragma config LVP = OFF
#pragma config BOR = OFF
#pragma config MCLRE = ON

#pragma code

// main program

#define SC PORTBbits.RB2
#define CS PORTBbits.RB0
#define SIO PORTBbits.RB1

void sendClock (void)
{
      SC = 0;         // SC = 0
      SC = 1;         // SC = 1
}

int getTemp (void)
{
      int temp;
      int b;
      char a
      CS = 0;                               // #CS = 0;

      for (a = 0; a < 16; a++)              // get 16 bits
      {
            temp <<= 1;
            temp != SIO >> 1;
            sendClock();
      }

      a = 0;
      CS = 1;                               // #CS = 1

      if ((temp & 0x8000) == 0x8000)        // check sign
      {
            a = 1;
            temp = -temp;                   // make positive
      }
```

```
        b = temp;
        temp >>= 7;
        if ((b & 1) == 1)                // round result
        temp++;

        temp = (9 * temp) / 5 + 32;     // make Fahrenheit

        if (a)
                temp = -temp;
        return temp;                     // return with temperature
}
void main (void)
{
        ADCON1 = 0x0f;                   // all port pins digital
        TRISB = 0x02;                    // Port B programmed
        PORTB = 0x05;                    // #CS = 1 and SC = 1

        // do other stuff here
}
```

Rotary Shaft Encoders

Rotary encoders are devices that usually contain **optical sensors** that report the position of a shaft. These sensors come in a variety of sizes from 2 bits, which indicate positions at 90° increments of the shaft, to encoders that produce thousands of pulses per revolution. Some produce a binary output and some produce only pulses on a signal line. Shaft encoders produce gray code or natural binary code at the outputs if encoded. Gray code follows the sequence 000, 001, 011, 111, 110, 100 for a 3-bit encoder. Notice that only one bit changes at one time from one code to the next. The 3-bit gray code has six states, the 4-bit gray code has eight states, and so forth.

Figure 7-29 illustrates a shaft encoder connected to a microcontroller. Here a US Digital S4 shaft encode is interfaced. This encoder uses a 4-wire interface cable that contains a +5 V (pin 1), ground (pin 3), signal A (pin 2), and signal B (pin 4). The signal A (least significant bit) and B lines produce the codes 00, 01, 11, 10 (gray code) as the shaft is rotated in the forward direction and 10, 11, 01, 00 as the shaft is rotated in the reverse direction.

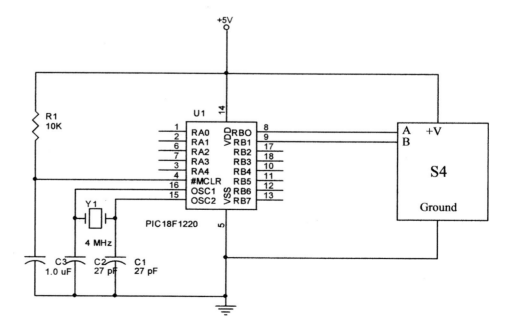

FIGURE 7-29   Rotary shaft encoder connected to a PIC18F1220.

Suppose that the shaft encoder is used to measure the speed of a machine. The machine rotates at up to 4000 RPM. It rotates up to 67.7 times per second and the shaft encoder outputs change up to 267 times a second. In order to measure this speed, the inputs must be sampled at least every 3.75 ms. A sample program (see Example 7-20) tests the RB0 and RB1 inputs every 1.024 ms to detect these changes. An interrupt is used to perform the checks and determine whether the motor is rotating clockwise or counter-clockwise and also to post the speed in RPMs in an integer memory location called speedRPM. The speed is accurate from 15 RPM to 4000 RPM. If a slower speed must be resolved, then the interrupt frequency rate must be reduced from once every 1.024 ms. For example, if it is reduced to 10.24 ms, the resolution is 1.5 RPM to 4000 RPM.

EXAMPLE 7-20

```c
/*
 * Speed and direction example written for a PIC18F1220
 */

#include <p18cxxx.h>
#include <timers.h>

/* Set configuration bits
 * - set HS oscillator
 * - disable watchdog timer
 * - disable low-voltage programming
 * - disable brownout reset
 * - enable master clear
 */

#pragma config OSC = HS                 // assumes a 4-MHz clock
#pragma config WDT = OFF
#pragma config LVP = OFF
#pragma config BOR = OFF
#pragma config MCLRE = ON

void MyHighInt (void);                  // prototypes for interrupts

#pragma interrupt MyHighInt             // MyHighInt is an interrupt
#pragma code high_vector=0x08           // high_vector is the vector at 0x08

void high_vector (void)
{
      _asm GOTO MyHighInt _endasm
}

// data memory data

int speedRPM;                   // speed in RPM
char direction;                 // direction 1 = forward
                                //           0 = reverse

char lastCount;                 // most recent position must be here
int timer;                      // elapsed count

// high-prioity interrupt
#pragma code

void Timer0 (void)
{
      char temp = PORTB & 3;

      WriteTimer0 (0xFC);          // Timer 0 to 252

      OpenTimer0 (TIMER_INT_ON &   // Timer 0 interrupt on
                  T0_8BIT &                 // Timer 0 is 16 bits
                  T0_SOURCE_INT &           // Timer 0 clock is internal
                  T0_PS_1_256);             // Timer 0 prescaler is 256
      if (lastCount != temp)
      {
            switch (lastCount)
```

```
                        {
                                case 0:
                                {
                                        if (temp == 1)
                                                direction = 1;
                                        else
                                                direction = 0;
                                        break;
                                }
                                case 1:
                                {
                                        if (temp == 2)
                                                direction = 1;
                                        else
                                                direction = 0;
                                        break;
                                }
                                case 2:
                                {
                                        if (temp == 0)
                                                direction = 1;
                                        else
                                                direction = 0;
                                        break;
                                }
                                case 3:
                                {
                                        if (temp == 2)
                                                direction = 1;
                                        else
                                                direction = 0;
                                        break;
                                }
                        }
                        speedRPM = timer / 1.024 * 15; // number of 90-degree interrupts
                        timer = 0;
                }
                else
                        timer++;
}

void MyHighInt (void)
{
        if (INTCONbits.TMR0IF == 1)
                Timer0();
}

// main program

void main (void)
{
        ADCON1 = 0x0f;                  // all port pins digital
        TRISB = 0x00;                   // Port B programmed

        lastCount = timer = 0;

        WriteTimer0 (0xFC);             // Timer 0 to 252

        OpenTimer0 (TIMER_INT_ON &      // Timer 0 interrupt on
                    T0_8BIT &                   // Timer 0 is 16 bits
                    T0_SOURCE_INT &             // Timer 0 clock is internal
                    T0_PS_1_256);               // Timer 0 prescaler is 256

        RCONbits.IPEN = 1;                      // IPEN = 1
        INTCONbits.GIEH = 1;                    // enable high-priority interrupts

        // do other stuff here
}
```

FIGURE 7-30 Flow sensor interfaced to the microcontroller.

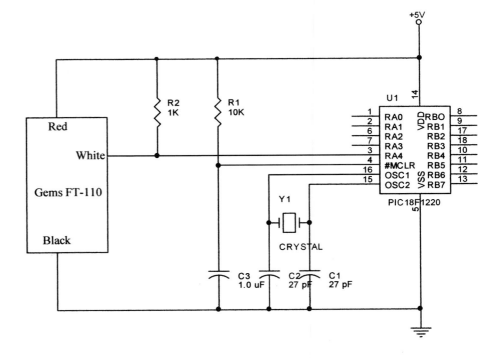

Flow Sensor

Flow sensors produce pulses as liquids flow through them. Suppose that fuel flow and consumption need to be measured in a system. This is accomplished by using a flow sensor such as the Gems FT-110 flow sensor. This device (model 173934) produces 8300 pulses per gallon of flow. The measurable flow rate is from .07 to 5.3 gallon per minute. Figure 7-30 illustrates the connection of this flow sensor to the microcontroller. This is a simple device to interface and requires only a single pull-up resistor and three wires.

The software is listed in Example 7-21. The software maintains the instantaneous flow rate through the sensor in floating-point format at location flowRate. The number in this location is the gallons per minute of flow rate through the sensor updated 10 times per second.

EXAMPLE 7-21

```
/*
 * Flow example written for a PIC18F1220
 */

#include <p18cxxx.h>
#include <timers.h>

/* Set configuration bits
 * - set HS oscillator
 * - disable watchdog timer
 * - disable low-voltage programming
 * - disable brownout reset
 * - enable master clear
 */

#pragma config OSC = HS
#pragma config WDT = OFF
#pragma config LVP = OFF
#pragma config BOR = OFF
#pragma config MCLRE = ON
```

```
void MyHighInt (void);                  // prototypes for interrupts

#pragma interrupt MyHighInt        // MyHighInt is an interrupt
#pragma code high_vector=0x08      // high_vector is the vector at 0x08

void high_vector (void)
{
      _asm GOTO MyHighInt _endasm
}

// data memory data

float flowRate;

// high-prioity interrupt

#pragma code

void Timer1 (void)                      // every 1/10 of a second
{
      PIR1bits.TMR1IF = 0;              // reenable Timer 1
      flowRate = ReadTimer0() * 10.0 * 60 / 8300;
      WriteTimer0 (0);                  // reset count
}

void MyHighInt (void)
{
      if (PIR1bits.TMR1IF == 1)
            Timer1();
}

// main program

void main (void)
{
      ADCON1 = 0x0f;                     // all port pins digital
      TRISA = 0x10;                      // Port A programmed

      WriteTimer0 (0);                   // Timer 0 to 0

      OpenTimer0 (TIMER_INT_OFF &             // Timer 0 interrupt on
                  T0_16BIT &                  // Timer 0 is 16 bits
                  T0_SOURCE_EXT &             // Timer 0 clock is RA4
                  T0_EDGE_FALL &              // pin RA4 negative edge
                  T0_PS_1_1);                 // Timer 0 prescaler is 1

      WriteTimer1 (5303);                     // Timer 1 to 5303

      OpenTimer1 (TIMER_INT_ON &              // Timer 1 interrupt on
                  T1_16BIT_RW &               // Timer 1 is 16 bits
                  T1_SOURCE_INT &             // Timer 1 clock is internal
                  T1_PS_1_8 &                 // Timer 1 prescaler is 8
                  T1_OSC1EN_OFF);

      RCONbits.IPEN = 1;                      // IPEN = 1
      INTCONbits.GIEH = 1;                    // enable high-priority interrupts

      // do other stuff here

}
```

Infrared Remote Control Devices

Infrared remote controls are extremely common and have many uses with embedded applications. The sensor and emitter are usually a simple infrared LED, or multiple LEDs for distances greater than a few meters. The detector is usually an infrared photodiode, phototransistor, or photo-Darlington. The signal is often transmitted at the relatively low rate of 36 KHz for many remote control devices, which makes the interface especially suited to a microcontroller. The standard frequencies in popular use are 36 KHz, 38 KHz, and 40 KHz.

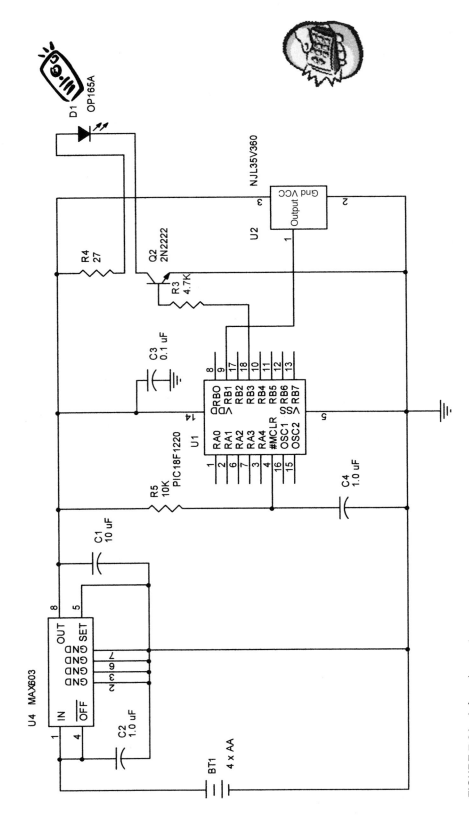

FIGURE 7-31 Infrared repeater.

A useful device that illustrates the principle of infrared sensors is a remote control range-extender circuit constructed with the microcontroller. This device requires both the infrared emitter and sensor, and illustrates some basic serial programming. The circuit uses the inexpensive PIC18F1220 and a few components. The power supply here is battery operated using four AA batteries and a MAX603 regulator to provide a steady 5 V to the circuit. This MAX603 integrated circuit uses an internal change pump and MOSFET to generate the 5 V output voltage from any input voltage between 2.7 V and 11.5 V. This is an excellent component to use for battery-powered equipment. Figure 7-31 illustrates the infrared range extender.

Data are sent in bursts of 36 KHz pulses using bi-phase code sometimes called Manchester code. Data consists of 14 bits that include two logic 1 start bits, followed by a control bit that toggles each time that a button is pressed on the transmitter. After the control bit is a 5 bit address. Finally, the last six bits are the command sent to a device. A bit is 1.728 ms wide and every 30 ms a code is retransmitted if a button on the remote control is held down. The Manchester code is sent so that a logic one changes the output to a zero from a one ↑ at mid-bit, and a logic zero changes the output from a logic one to a logic zero ↓ at mid-bit. The signal sent to the infrared diode is 36 KHz if the Manchester code is a logic one. Our system does not have to generate these codes, just detect them, receive them, and resend them. Figure 7-32 illustrates the format for the RC5 remote control code used by virtually all electronic components.

The received signal is applied to RB1, which is also an interrupt input, because when a signal is not received for an extended period, the sleep mode is entered to conserve power. Recall from Chapter 6 that the sleep mode is exited by any interrupt. Once a signal is received, software is used to measure the input and relay the command through to the transmitter. The PWM unit, as discussed with motor speed control, is used to generate the 36-KHz signal for the transmitting LED. Example 7-22 shows the calculation of the output frequency of the PWM. As seen form the calculation, the nearest frequency is 37 KHz using a 4-MHz internal clock frequency, but this is close enough that the system will function correctly. Here the value loaded into the PM2 register is 26.

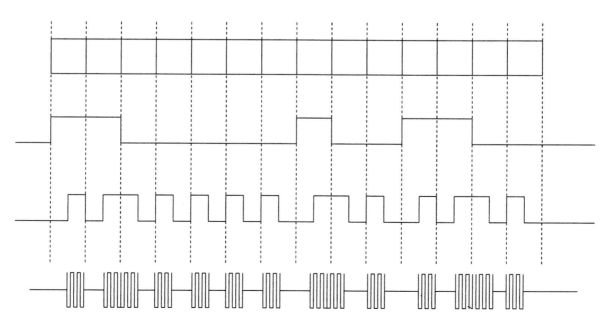

FIGURE 7-32 RC5 remote-control code.

EXAMPLE 7-22

```
PWM period = (PR2 + 1) x TMR2PS x 4 x Tosc

PWM period = (26 + 1) x 1 x 4 x 1/4 µs
           = 27 x 1 x 1 µs
           = 27 µs
```

$$\text{PWM frequency} = \frac{1}{27\ \mu s} = 37\ KHz$$

```
(Note: TMR2PS is the Timer 2 prescaler value)
```

A complete software listing appears in Example 7-23 for this application. Notice that when the input to RB1 changes, it causes an interrupt that wakes up the microcontroller. In the interrupt service procedure, the 36-KHz output to the LED is turned on, and as long as the input from the detector occurs once every 35 µs (as counted out by Timer 0), the output remains at 36 KHz. Timer 0 is cleared every half period of the input signal from the detector. If the Manchester code and the output is examined in Figure 7-31, the longest period of time that the signal drops to zero is one clocking period or 27 µs. The software detects if the signal stops for 35 µs, which should be long enough to detect when the signal stops changing. When the input stops changing, the 36-KHz output is turned off, the interrupt is re-enabled, and the return from the interrupt takes the program back to the main infinite while loop where the system goes back to sleep to conserve power.

EXAMPLE 7-23

```
/*
 * Remote control range extender
 */

#include <p18cxxx.h>

/* Set configuration bits
 * - set internal oscillator
 * - disable watchdog timer
 * - disable low-voltage programming
 * - disable brownout reset
 * - enable master clear
 */

#pragma config OSC = INTIO2
#pragma config WDT = OFF
#pragma config LVP = OFF
#pragma config BOR = OFF
#pragma config MCLRE = ON

// ********* DATA MEMORY VARIABLES ************

void MyHighInt (void);                    // prototypes for interrupts

#pragma interrupt MyHighInt              // MyHighInt is an interrupt
#pragma code high_vector=0x08            // high_vector is the vector at 0x08

void high_vector (void)
{
    _asm GOTO MyHighInt _endasm
}

#pragma code

// Interrupt service procedure
//    occurs for any positive edge on INT1 (RB1)
//

void MyHighInt (void)
{
    INTCON3bits.INT1IF = 0;              // clear interrupt request
    char state = PORTBbits.RB1;
    T2CONbits.TMR2ON = 1;                // start 36-KHz output
```

```
        do
        {
            TMR0L = 0;                          // clear Timer 0
              while (state == PORTBbits.RB1 &&
                    TMR0L < 20);                // wait for change or time-out
              state = PORTBbits.RB1;
        }
        while (TMR0L < 20);

        T2CONbits.TMR2ON = 0;                   // 36 KHz turned off
}

void main (void)
{
        OSCCON = 0x63                           // internal oscillator is 4 MHz
        ADCON1 = 0x0F;                          // ports are digital

        TRISB = 2;                              // program ports
        PORTB = 0;

        T2CON = 0;                              // select a prescaler of 1
        CCP1CON = 0x0C;                         // CCP module off, active high output
        TMR2 = 0;                               // clear Timer 2
        PR2 = 26;                               // Timer 2 clears on 10;
        CCPR1L = 14;                            // set for square wave (50% duty cycle)

        T0CON = 0xC8;                           // Timer 0 counts in µs

        RCONbits.IPEN = 0;                      // only high-priority interrupt on

        INTCON2bits.INTEDG1 = 1;                // make INT1 positive edge-triggered
        INTCON3bits.INT1IE = 1;                 // enable INT1
        INTCONbits.GIE = 1;                     // enable interrupts

        while (1)
        {
            Sleep();                            // just go to sleep
        }
}
```

Suppose that this system also needed to function and decode signals from an external remote control? The input signal is detected in the interrupt service procedure, but neither the high or low times are measured. In the software in Example 7-23, Timer 0 can be used to time the input signal, and also locate when it started and stopped to decode the Manchester code sent from the remote control. This portion of the software is not listed here but will be developed for the other applications, which use Manchester code, in later chapters.

Sensing Gas

Many sensors are available for measuring the amount of certain gasses in the atmosphere. This section presents one of them to measure the carbon dioxide content. Many other sensors are available that measure oxygen, methane, alcohol, hydrogen, hydrogen sulfide, and so forth.

Suppose that an instrument is needed to measure the concentration of carbon dioxide in the air. Many carbon dioxide sensors are available, for example, the TGS-2442 from Figaro Corporation. The device contains a heater and a sensor. The heater activates the sensor whose resistance changes with the concentration of carbon dioxide. To monitor the amount of carbon dioxide, the heater is activated for 14 ms and the resistance is measured 5 ms into the heating cycle. This heating monitoring cycle is repeated every second while a measurement is taken. This sensor can measure the concentration of CO_2 from 30 to 1000 parts per million (ppm).

The heater requires 5 V at a current of 200 mA for 14 ms out of every 1000 ms, so the average is only 2.8 mA of current. The temperature sensor produces a voltage that can be monitored by the ADC in the microcontroller, and displayed as the count of the concentration of CO_2 or used in any other manner. Figure 7-33 illustrates the sensor and circuit used to monitor the concentration of CO_2.

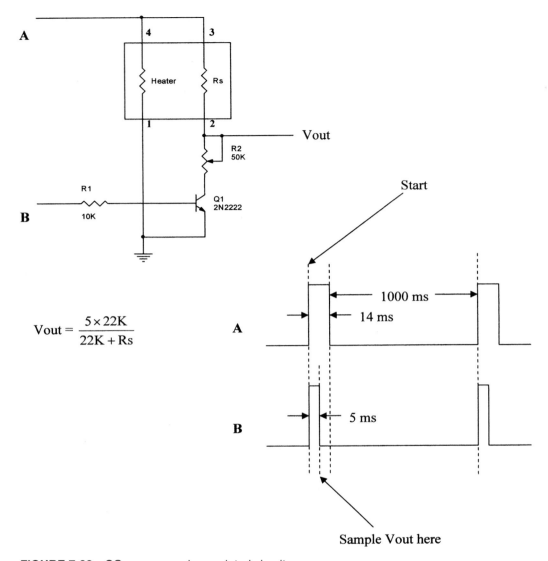

$$Vout = \frac{5 \times 22K}{22K + Rs}$$

FIGURE 7-33 CO_2 sensor and associated circuitry.

The output voltage obtained at Vout is determined by the variable resistance of the sensor RS. The TG-2442A1 version has an RS value of 6.81 KΩ–21.5 KΩ. These devices must be calibrated for the reading to be accurate, which is why the 50 KΩ potentiometer is used in the collectors of the 2N2222.

Figure 7-34 shows the sensor interfaced to a PIC18F1220 microcontroller. Also included in the circuit is a 1×16 LCD display to indicate the CO_2 content in ppm. For this to be accurate, the sensor is placed in a known concentration of CO_2 and the potentiometer is adjusted for the correct reading on the LCD. The sensor is linear so once adjusted, all the readings will be accurate. The sensor also requires a 2-day burn-in period before it is adjusted.

The software for the system is illustrated in Example 7-24. The multiplication factor used to display the ppm generates a full-scale reading of 500, but needs to be adjusted to match the actual sensor. These adjustments are accomplished with the calibration potentiometer.

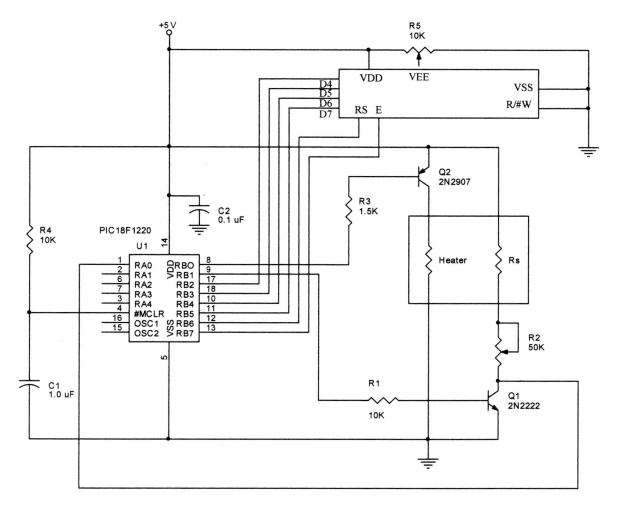

FIGURE 7-34 CO_2 sensor interface.

EXAMPLE 7-24

```
/*
 * Priority of multiple interrupts
 */

#include <p18cxxx.h>
#include <delays.h>

/* Set configuration bits
 * - set RC oscillator
 * - disable watchdog timer
 * - disable low-voltage programming
 * - disable brownout reset
 * - enable master clear
 */

#pragma config OSC = INTIO2
#pragma config WDT = OFF
#pragma config LVP = OFF
#pragma config BOR = OFF
#pragma config MCLRE = ON
```

```c
// ********* DATA MEMORY VARIABLES ************

void MyHighInt (void);              // prototypes for interrupts

#pragma interrupt MyHighInt         // MyHighInt is an interrupt
#pragma code high_vector=0x08       // high_vector is the vector at 0x08

void high_vector (void)
{
      _asm GOTO MyHighInt _endasm
}

#pragma code

char str1[] = "CO2 = ";
char str2[] = " ppm";

// Interrupt service procedure
//     occurs each second
//
void sendLCDdata (char data, char rs)
{
PORTB = (PORTB & 0xC3)       | ((data >> 2) & 0x3C); // send left
      PORTBbits.RB6 = rs;         // set RS
      PORTBbits.RB7 = 1;         // pulse E
      PORTBbits.RB7 = 9;
      Delay10TCYx(1);            // delay 40 us
      PORTB = (PORTB & 0xC3) | ((data << 2) & 0x3C); // send right
      PORTBbits.RB6 = rs;         // set RS
      PORTBbits.RB7 = 1;         // pulse E
      PORTBbits.RB7 = 0;
      Delay10TCYx(1);            // delay 40 us
}

void initLCD (void)
{
      Delay1KTCYx (5);                   // wait 20 ms (see text)

      sendLCDdata (0x20, 0);             // send 0x30
      Delay1KTCYx (2);                   // wait 8 ms

      sendLCDdata (0x20, 0);             // send 0x30
      Delay10TCYx (5);                   // wait 100 us

      sendLCDdata (0x20, 0);             // send 0x30
      sendLCDdata (0x08, 0);             // send 0x28
      sendLCDdata (0x01, 0);             // send 0x01
      Delay100TCYx (5);                  // wait 2 ms
      sendLCDdata (0x0C, 0);             // send 0x0C
      sendLCDdata (0x06, 0);             // send 0x06
}

void string (char *str, char position)
{
      int ptr = 0;
      sendLCDdata (position, 0);         // send position
      while (str[ptr] != 0)
            sendLCDdata(str[ptr++], 1); // send character
}

void MyHighInt (void)
{
      float analog;
      PIR1bits.TMR1IF = 0;                 // re-enable interrupt
      TMR1L = 0xEE;                        // preload Timer 1 with -31, 250
      TMR1H = 0x85;                        // for an interrupt per second
      PORTBbits.RB0 = 0;                   // power sensor and heater
      PORTBbits.RB1 = 1;
      Delay1KTCYx(1);                      // wait 4 ms
      Delay100TCYx(2);                     // wait .8 ms
      Delay10TCYx(5);                      // wait .2 ms
                                           // take sample
```

```
        ADCON0bits.GO = 1;              // start a conversion
        while (ADCON0bits.GO == 1);                      // wait for completion
        analog = (ADRESL + (ADRESH << 8)) * 0.489;
                                                          // scaled to 500 max ppm
        PORTBbits.RB1 = 0;
        Delay1KTCYx (2);               // wait 8 ms
        PORTBbits.RB0 = 1;
        string (str1, 0x80);           // display "CO2 = "
        str2[0] = analog / 100 + 0x30;
        analog = (int) analog % 100;
        str2[1] = (int) analog / 10 + 0x30;
        str2[2] = (int) analog % 10 + 0x30;
        string (str2, 0x86);
}

void main (void)
{
        OSCCON = 0x43;                 // internal oscilator is set to 1 MHz
        ADCON1 = 0x7E;                 // select input AN0, enable ADC
        ADCON1 = 0x0e;                 // AN0 is analog, VDD and VSS are references
        ADCON2 = 0x84;
        TRISA = 1;                     // Port A bit 1 = input
        TRISB = 0;                     // Port B = output
        PORTB = 1;                     // Turn off sensor

        initLCD();

        T1CON = 0xF9;                  // Timer 1 runs at 32 us per count
        TMR1L = 0xEE;                  // preload Timer 1 with -31, 250
        TMR1H = 0x85;                  //    for an interrupt per second

        RCONbits.IPEN = 0;                  // only high-priority interrupt on

        PIE1bits.TMR1IE = 1;                // enable Timer 1 interrupt
        INTCONbits.GIE = 1;                 // enable interrupts

        while (1)
        {
                // nothing
        }
}
```

7-5   SUMMARY

1. Switches bounce for approximately 10 ms and must be debounced for proper operation in many applications. Debouncing is best handled with software.
2. Keypads are often connected to a microcontroller for inputting various types of information. Keypads are most often constructed in the form of a matrix where pressing a key connects a row to a column.
3. Light-emitting diodes that are interfaced to a microcontroller often require a driver circuit to produce enough current to light the LED. Numeric displays use LEDs in a 7-segment configuration that uses 7-segment code.
4. Liquid crystal displays (LCD) are often connected to a microcontroller to display both numeric and alphabetic data. Various sizes of LCD display panels are available from 1-line to 4-line displays that have from 16 to 40 characters on a display line.
5. Vacuum fluorescent displays (VFDs) are bright blue-green displays found in many consumer products to display both alphabetic and numeric data.
6. Motors in the form of DC or stepper motors are often used to position or move devices, and are commonly interfaced to the microcontroller.
7. Motors are often driven using MOSFET power transistors. A common configuration of a MOSFET driver is the H-bridge.

8. The PWM unit in the microcontroller is often used to control the speed and direction of a DC motor. By varying the duty cycle of a signal at the PWM output, the speed of a motor is changed because the average current through the motor changes.

9. Relays, both electromechanical and solid state, are used to switch high voltages or many contacts simultaneously. Most relay circuits require a driver circuit to energize them because of the amount of current required.

10. Solenoids are electromechanical devices that are used to move a device a short distance. Solenoids must be energized for a specific length of time, which require a software time delay. Solenoids also require large currents and a driver circuit.

11. Sensors are devices that convert motion or some other physical event into an electric signal that is processed by the microcontroller. These include rotational motion, pressure, temperature, liquid flow, and so forth.

12. The RC5 code is most often used with infrared remote controls to send signals at 36 KHz.

7-6   QUESTIONS AND PROBLEMS

1. What problems can occur with mechanical switches?
2. How long will a mechanical switch bounce?
3. What type of software is used to eliminate bouncing from the contacts of a switch?
4. Formulate a function call using the Switch function of Example 7-1 to debounce two switches connected to Port A bits 0 and 7.
5. What is the difference between a SPST and a SPDT switch?
6. Why is a pull-up resistor needed when connecting a switch to a microcontroller?
7. Describe the construction of a keypad matrix.
8. Modify the software in Example 7-2 so that a 4×6 keypad can be used with the software.
9. Because the software in Example 7-2 uses Port A as an input, how many rows maximum could be connected to Port A on a PIC18F1220?
10. Suppose that a white LED is interfaced to Port B bit RB0 on a PIC microcontroller. The LED requires a voltage of 2.5 V and a current of 30 mA to fully illuminate. Using a 2N2222 transistor, design this interface.
11. Repeat Question number 10 for a bright red LED that requires 2.0 V and 20 mA of current.
12. Connect a pair of red LEDs to Port B bits RB0 and RB1. The LEDs require 10 mA of current and 1.65 V. Develop a function for the PIC that alternately flashes the two LEDs at the rate of 2 Hz for 10 seconds when the function flashLeds is invoked. Assume that the clock is 4 MHz.
13. Tricolor LEDs are available. These devices have four connections: (1) a common anode connection, (2) a red LED cathode, (3) a green LED cathode, and (4) a blue LED cathode. Interface this device to the PIC pins RB0, RB1, and RB2. The red and green sections require 1.65 V at 10 mA and the blue section requires 2.5 V at 20 mA. Develop a function that flashes random colors at the rate of 3 Hz for 20 seconds whenever the input to RA0 is a logic one. The random colors are the numbers 000 through 111 sent to the three Port B bits. Assume that the clock is 4 MHz.
14. When LEDs are multiplexed they are individually on for only part of the time, but they appear to be lit all of the time. Why?
15. What are the minimum and maximum flash rates for multiplexed displays?
16. When six LED displays are multiplexed, the peak current for each display must be boosted by what factor?
17. What sets the flash rate in Example 7-6?

18. The LCD display panel has a VEE connection used to set the contrast. What is normally connected to VEE?
19. What is the purpose of the RS input to an LCD display panel?
20. What is the purpose of the E input to an LCD display?
21. Describe the process required to initialize the LCD display.
22. How is the cursor moved on the LCD display panel?
23. What are the display addresses of line 1 for a 24×1 LCD display?
24. Develop a function called cursor that moves the cursor to any LCD display position available. The prototype for the function is void cursor(char position). Assume that E is connected to RB4, RS is connected to RB5, and 4-bit data is connected to RB0–RB3.
25. Given a 2×16 LCD display, develop a function that erases only line number 1 of the display. (Outputting a space will erase a character.)
26. What command is sent to the LCD display to clear the entire screen? Write a function to clear the screen called Clear.
27. A vacuum fluorescent display functions as a _____ vacuum tube.
28. What is the main advantage of the VFD?
29. What color is normally displayed by a VFD?
30. Why is a stepper motor called a stepper motor?
31. What is the main advantage of a stepper motor?
32. Explain the difference between a full step and a half step in a stepper motor.
33. What is an H-bridge?
34. Explain how varying the pulse width to a DC motor can change the speed of the motor.
35. What does the term "singing motor" indicate and how is it prevented?
36. What is a kickback or damper diode and what does it prevent?
37. How fast will a DC motor connected to an H-bridge rotate if a pulse width modulated signal has a 50% duty cycle?
38. Approximately how fast will a DC motor spin if it is driven with a single MOSFET and the PWM input to it is 50%?
39. Suppose a relay is interfaced to the microcontroller and it requires 100 mA of current. Design a driver circuit for the relay using a 2N2222 transistor driver.
40. Interface a solenoid to the microcontroller assuming that the solenoid requires 600 mA of current at 12 V. Develop software to energize the solenoid for 700 ms each time the function Fire is called. Assume the clock frequency is 4 MHz and the solenoid is interfaced to RB3.
41. What is the gray code as used with a rotary shaft encoder?
42. Write the gray code for a 4-bit shaft encoder.
43. What is the purpose of a rotary shaft encoder?
44. The LM70 temperature senor resolves tempers to _____ degrees Celsius.
45. The LM70 temperature sensor presents a _____ bit binary number for the temperature.
46. Explain how the software in Example 7-19 rounds the temperature.
47. A flow sensor is used to indicate what information?
48. Go to the Internet and list three sensors that are not explained in the text. Also describe the output signal generated by the sensor.
49. Is a sensor available to detect the amount of oxygen in the environment? (Again, scour the Internet).
50. Because many gas sensors are available, would it be possible to design a device that would detect just about anything that has a characteristic odor? If so, list at least three examples of what might be detected.
51. How many bits comprise one code in the RC5 code?
52. What frequency is used for transmitting an infrared signal using the RC5 code?

CHAPTER 8

Interrupts

Interrupts were introduced earlier in this book, but because of their importance, an entire chapter is devoted to using interrupts in a variety of applications. Almost all systems use interrupts. Interrupts enable an application to efficiently share the use of the microcontroller, and without them many applications would be extremely difficult to implement. However, if an error occurs, interrupts are difficult to troubleshoot or debug.

Upon completion of this chapter you will be able to:

1. Efficiently use interrupts to program applications for the microcontroller.
2. Use both the low- and high-priority interrupts in a program.
3. Program the USART as an interrupting device.
4. Program using the bit change interrupts.
5. Understand how a small network functions using interrupts.
6. Program and use barcode devices.

8-1   INTERRUPTS REVISITED

The microcontroller has two interrupt levels: the high-priority interrupt and the low-priority interrupt. The priority levels take precedence only if both interrupts occur simultaneously. When both are active, the high-priority interrupt is acted upon first by the microcontroller. Priority also takes effect if a low-priority interrupt is executing wherein the high-priority interrupt will interrupt the low-priority interrupt. Some applications use only the high-priority interrupt, whereas other applications use both interrupt levels. The choice of which level or levels to use is up to the software designer, but in general high-priority interrupts are used for events that must be serviced quickly, and low-priority interrupts are used for less urgent events. The IPEN bit in the reset control register (RCON) selects either the high-priority interrupt (IPEN = 0) or the high- and low-priority interrupts (IPEN = 1).

For example, suppose that a system is using a timer, which interrupts the microcontroller once every 100 ms to update a real-time clock. A second interrupt is used to monitor an external event from a flow sensor. The signal from the flow sensor occurs at an interval of between 1 ms to 40 ms. Because the signal from the flow sensor occurs more often than the real-time clock interrupt, the signal from the flow sensor is given a high priority and the real-time clock is given a low priority. Generally, events that occur more frequently are assigned a high priority, but not always. The high-priority interrupt must usually be serviced first because data is lost if it is not serviced before other interrupts.

As mentioned in Chapter 6, the high-priority interrupt vectors to program memory address location 0x0008, and the low-priority interrupt vectors to program memory address location 0x0018 for their respective interrupt service procedures. Because the amount of memory at these vector locations is limited, the vector normally contains a GOTO instruction to transfer control to the interrupt service procedure. Today most applications are written in C language because the amount of time required for coding an application is significantly reduced. The shell for implementing the interrupt structure in C appears in Example 8-1. The placement of the interrupt vectors is critical and much of the software must appear in the order listed. The code that follows the two interrupt vectors must be delineated with the #pragma code statement. The main function in this example enables both interrupts after selecting both priority interrupts with RCON bit IPEN. If IPEN is cleared, only the high-priority interrupt is enabled. In other words, either both interrupt levels or only the high-priority interrupt level can be enabled.

EXAMPLE 8-1

```
/*
 * Interrupt example
 */

#include <p18fxxx.h>

/* Set configuration bits
 *   - set RC oscillator
 *   - disable watchdog timer
 *   - disable low-voltage programming
 *   - disable brownout reset
 *   - enable master clear
 */

#pragma config OSC = RC
#pragma config WDT = OFF
#pragma config LVP = OFF
#pragma config BOR = OFF
#pragma config MCLRE = ON

// ********** INTERRUPT PROTOYPES AND VECTORS ********************

void MyHighInt (void);            // prototypes for the interrupt
void MyLowInt (void);             // service procedures

#pragma interrupt MyHighInt       // MyHighInt is an interrupt
#pragma code high_vector=0x08     // high_vector is at 0x0008

void high_vector (void)           // the high-prioity vector
{
     _asm GOTO MyHighInt _endasm      // goto high software
}

#pragma interruptlow MyLowInt     // MyLowInt is an interrupt
#pragma code low_vector=0x18      // low vector is at 0x0018

void low_vector (void)            // the low-prioity vector
{
     _asm GOTO MyLowInt _endasm       // goto low software
}

// ***************** CODE ***********************

//    the program code must be placed here after the
//    interrupt vectors are defined
//

#pragma code             // start code here
```

```
void MyHighInt (void)
{
      // high-priority interrupt goes here
}

void MyLowInt (void)
{
      // low-priority interrupt goes here
}

// main program

void main  (void)
{
      RCONbits.IPEN = 1;         // IPEN = 1 to enable priority interrupts
      INTCONbits.GIEH = 1;       // enable high-priority interrupt
      INTCONbits.GIEL = 1;       // enable low-priority interrupt

      // interrupts are now on so do other software

}
```

Interrupt Service Procedure

The software in Example 8-1 is all that is needed to install and enable both the low- and high-priority interrupts. The interrupt service procedures that handle the low- and high-priority interrupts are not included in this example; only the shells of the interrupt service procedures are included.

The interrupt service procedure, sometimes called the **interrupt handler**, must correctly service an interrupt. Otherwise, the system will hang or lock up, making troubleshooting very difficult or impossible. Each interrupting device in the microcontroller has three control bits associated with it:

- Interrupt priority bit (IP)
- Interrupt enable bit (IE)
- Interrupt flag bit (IF)

The interrupt priority bit (IP) selects either the high- or low-priority interrupt for the device. A logic one in the IP bit for a given interrupt input selects the high-priority interrupt, and a logic zero selects the low-priority interrupt. The interrupt enable bit (IE) enables the interrupting device when placed at a logic one. The interrupt flag bit (IF) is set to indicate that the device has caused an interrupt. The IF bit is polled by the software in the interrupt service procedure to determine which device caused the interrupt. The interrupt flag must be cleared by the interrupt handler when the interrupt is serviced. If not cleared inside of the interrupt handler, another interrupt will never occur.

Suppose that Timer 0, Timer 1, and the INT1 interrupt input pin are used in an application. The Timer 0 interrupt is the high-priority interrupt and the Timer 1 and INT1 input are the low-priority interrupts. Example 8-2 illustrates the software needed to implement this structure. This example does not show that the interrupt service procedures are terminated with a return from interrupt instruction (RETFIE). Proper termination with the RETFIE instruction is assured by the #pragma interrupt and #pragma interruptlow statements in the program. These #pragma statements identify the functions used to process both the high- and low-priority interrupts. The RETFIE instruction (return from function with interrupts enabled) automatically reenables future interrupts at the end of an interrupt function. The contents of the interrupt control registers are described in Chapter 6.

EXAMPLE 8-2

```
/*
 * A high- and low-priority interrupt example
 */

#include <p18cxxx.h>
```

```
/* Set configuration bits
 *   - select high-speed crystal oscillator
 *   - disable watchdog timer
 *   - disable low-voltage programming
 *   - disable brownout reset
 *   - enable master clear
 */

#pragma config OSC = HS
#pragma config WDT = OFF
#pragma config LVP = OFF
#pragma config BOR = OFF
#pragma config MCLRE = ON

// ********** INTERRUPT PROTOYPES AND VECTORS *******************

void MyHighInt (void);                 // prototypes for the interrupt
void MyLowInt (void);                  // service procedures

#pragma interrupt MyHighInt           // MyHighInt is an interrupt
#pragma code high_vector=8            // high_vector is at 0x0008

void high_vector (void)                // the high-prioity vector
{
      _asm GOTO MyHighInt _endasm         // goto high software
}

#pragma interruptlow MyLowInt         // MyLowInt is an interrupt
#pragma code low_vector=0x18          // low vector is at 0x0018

void low_vector (void)                 // the low-prioity vector
{
      _asm GOTO MyLowInt _endasm          // goto low software
}

// ******************* CODE ***********************************

#pragma code                          // start code here

// high-priority interrupt

void MyHighInt (void)
{
      INTCONbits.TMR0IF = 0;          // clear Timer 0 request

      // process Timer 0 interrupt
}

// low-priority interrupt

void MyLowInt (void)
{
      if (INTCON3bits.INT1IF == 1)    // is it INT1?
      {
            INTCON3bits.INT1IF = 0;   // clear request

            // process INT1 interrupt
      }
      else if (PIR1bits.TMR1IF == 1)  // is it Timer 1?
      {
            PIR1bits.TMR1IF = 0;      // clear request

            // process Timer 1 interrupt
      }
}

// main program

void main (void)
{
      // program Timer 0, Timer 1, and INT1 here

      INTCON2bits.TMR0IP = 1;         // Timer 0 is high priority
      INTCONbits.TMR0IE = 1;          // enable Timer 0 interrupt
```

```
IPR1bits.TMR1IP = 0;              // Timer 1 is low priority
PIE1bits.TMR1IE = 1;              // enable Timer 1 interrupt

INTCON3bits.INT1IP = 0;           // INT1 is low priority
INTCON3bits.INT1IE = 1;           // enable INT1

RCONbits.IPEN = 1;                // IPEN = 1 to enable priority interrupts
INTCONbits.GIEH = 1;              // enable high-priority interrupt
INTCONbits.GIEL = 1;              // enable low-priority interrupt

// interrupts are now on so do other software

}
```

To illustrate a sample system, suppose that a bank requires a time and temperature display and has commissioned us to develop such a display. To develop this system, a large display is needed so 5-inch high LED displays (Liteon 50801HRB) are obtained for the display portion of the system. Four of these LED displays are used for the time and four are used for the temperature. Figure 8-1 shows the configuration of this display, illustrating a time of 11:03 and a temperature of 104°F. (It must be a hot day.) Single 8-mm LEDs are used for displaying the degree symbol and the colons between the hours and minutes. The eight displays are numbered D0–D7 so that a system schematic can be constructed.

Because the displays are multiplexed and require 15 I/O pins on the microcontroller, the PIC18F1220 is not suitable for this application. Thus, a larger microcontroller is chosen to implement the system. Here the PIC18F2220 is chosen using the SPDIP 28-pin package. The Liteon five-inch 7-segment LED displays require 4 V per segment at 30 mA of current. The 8 mm LEDs require 2 V at 20 mA of current.

The inputs to the microcontroller (see Figure 8-2) are a pushbutton to set the time and the LM70 temperature sensor. These devices use the RA0–RA2 pins on the microcontroller. The segment driver is the ULN2003 (Darlington amplifiers) connected to Port C bits RC0–RC6. The anode switches are a pair of high-side drivers (754410) from Texas Instruments. The ULN2003 is an

FIGURE 8-1 Display for time and temperature.

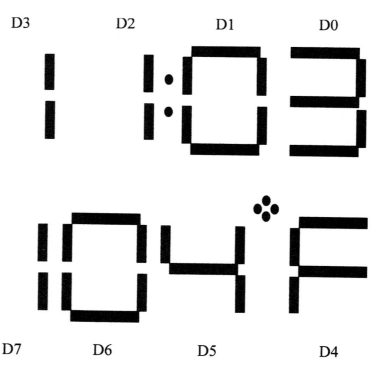

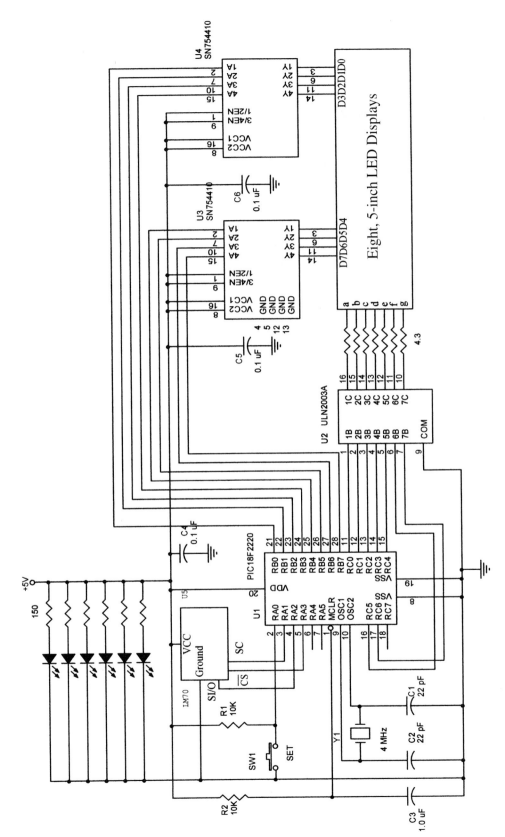

FIGURE 8-2 Circuit for time and temperature.

inverting amplifier, so the 7-segment code is an active high code and the anode switches are noninverting drivers, so a logic one selects a display. The anode switches use all eight bits of Port B.

The software for the application is written to use a single interrupt from Timer 1 to control the entire system. The Timer 1 interrupt increments the clock. Once per minute, the Timer 1 interrupt service procedure retrieves the temperature from the LM70 and also displays the time of day. The main program in this system does nothing but read the pushbutton switch in case the time must be changed. The way the software for setting the time is written, one button is used to set both the hours and minutes. (Maybe the digital watch manufacturers will adopt this technique for setting the time).

The sequence of operation for the set button is:

1. If the pushbutton is pressed, the set time mode is entered and the rightmost display of the temperature shows the letter H for hours.
2. If the pushbutton is pressed within 3 seconds, the hours increment for each press of the pushbutton.
3. If the pushbutton is not pressed for more than 3 seconds, the set minutes mode is entered and the rightmost display of the temperature shows the letter M for minutes.
4. If the button is pressed within 3 seconds, the minutes increment and if not, the set seconds mode is entered and the rightmost display of the temperature shows the letter S for seconds.
5. If the button is pressed within 3 seconds, the seconds increment and if not, the clock resumes normal operation and displays the new the time and temperature.

Example 8-3 illustrates the complete software for this system. The system is written in C language, because assembly language would create a very large listing file and be very difficult to read and debug. The total memory used by this system is 1044 words (only about half the amount available) for the program memory and 432 bytes of data memory. This software is mostly pieced together from other examples in earlier chapters. The biggest change in the software is the Switch function. Because the bit position is known, it is not transferred to Switch in the call. Another change to Switch is the way it times out after 3 seconds. If a return 0 occurs, the pushbutton was pressed and if a return 1 occurs, the button was not touched for at least 3 seconds. These changes are essential to using the pushbutton to set the time of day.

EXAMPLE 8-3

```
/*
 * Time and temperature example
 */

#include <p18cxxx.h>
#include <timers.h>
#include <delays.h>

/* Set configuration bits
 *  - set RC oscillator
 *  - disable watchdog timer
 *  - disable low-voltage programming
 *  - disable brownout reset
 *  - enable master clear
 */

#pragma config OSC = HS
#pragma config WDT = OFF
#pragma config LVP = OFF
#pragma config BOR = OFF
#pragma config MCLRE = ON

void MyHighInt (void);          // prototypes for the interrupt
void MyLowInt (void);           // service procedures
void timeTemp (void);
int getTemp (void);
```

```
#pragma interrupt MyHighInt          // MyHighInt is an interrupt
#pragma code high_vector=0x08        // high_vector is at 0x0008

void high_vector (void)              // high-priority vector
{
      _asm GOTO MyHighInt _endasm          // goto high software
}

#pragma interruptlow MyLowInt        // MyLowInt is an interrupt
#pragma code low_vector=0x18         // low vector is at 0x0018

void low_vector (void)               // low-prioity vector
{
      _asm GOTO MyLowInt _endasm           // goto low software
}
// program memory data

rom near char look7[] = // 7-segment lookup table
{
      0x3F,        // 0    active high signals
      0x06,        // 1
      0x5B,        // 2
      0x4F,        // 3
      0x66,        // 4
      0x6D,        // 5
      0x7D,        // 6
      0x07,        // 7
      0x7F,        // 8
      0x6F         // 9
};

// data memory variables

char tenths;                         // time storage
char seconds;
char minutes;
char hours;
char displayRAM[8];                  // display information
char setTimeFlag;                    // set time flag
char select;                         // display selection code
char pointer;                        // display pointer

#pragma code                         // start code here

void MyHighInt (void)
{
      INTCONbits.TMR0IF = 0;         // clear Timer 0 request
      PORTB = select;
      PORTC = displayRAM[pointer++];
      select <<= 1;
      if (pointer == 8)
      {
            pointer = 0;
            select = 1;
      }
}

void MyLowInt (void)
{
      PIR1bits.TMR1IF = 0;           // clear Timer 1 request
      WriteTimer1(53000);            // reload count minus bias (36)
      tenths++;                      // increment clock
      if (tenths == 10)
      {
            tenths = 0;
            seconds++;
            if (seconds == 60)
            {
                  seconds = 0;
                  minutes++;
                  if (minutes == 60)
```

```
                              {
                                      minutes = 0;
                                      hours++;
                                      if (hours == 13)
                                              hours = 1;
                              }
                      }
              }
              if ((tenths|seconds) == 0)              // once per minute
                      timeTemp();
      }


      void timeTemp (void)
      {
              int temp;
              if (setTimeFlag == 0)
                      displayRAM[4] = 0x71;           // F for temperature
              displayRAM[0] = look7[minutes % 10];
              displayRAM[1] = look7[minutes / 10];
              displayRAM[2] = look7[hours % 10];
              displayRAM[3] = look7[seconds / 10];
              temp = getTemp();
              displayRAM[7] = 0;
              if (temp < 0)
              {
                      temp = -temp;
                      displayRAM[7] = 0x40;
              }
              else
              if (temp >= 100)
              {
                      temp -= 100;
                      displayRAM[7] = 6;
              }
              displayRAM[4] = look7[temp % 100];
              displayRAM[5] = look7[temp / 10];
      }


      void sendClock (void)
      {
              PORTAbits.RA1 = 0;                      // SC = 0
              PORTAbits.RA1 = 1;                      // SC = 1
      }


      int getTemp (void)
      {
              int temp = 0;
              char a;
              PORTAbits.RA3 = 0;                      // #CS = 0;
              for (a = 0; a < 16; a++)                // get 16 bits
              {
                      temp <<= 1;
                      temp |= PORTAbits.RA2 >> 1;
                      sendClock();
              }
              a = 0;
              if ((temp & 0x8000) == 0x8000)          // check sign
              {
                      a = 1;
                      temp = -temp;                   // make positive
              }
              temp >>= 7;
              if (STATUSbits.C = 1)                   // round result
                      temp++;
              temp = (9 * temp) / 5 + 32;             // make Fahrenheit
              if (a)
                      temp = -temp;
              PORTAbits.RA3 = 1;                      // #CS = 1
```

```
        return temp;                        // return with temperature
}
int Switch (void)
{
        int delay = 1500;           // for a 3-second delay
        do                          // wait for release
        {
                while ((PORTA & 1) != 1);
                Delay1KTCYx(15);

        }while ((PORTA & 1) != 1);

        do                          // wait for press
        {
                =while ((PORTA & 1) == 1)
                {
                        Delay1KTCYx(2);
                        delay--;
                        if (delay == 0)
                                return 1;   // if timed out
                }
                Delay1KTCYx (15);

        }while ((PORTA & 1) == 1);
        return 0;                   // if pushbutton pressed
}

// main program

void main  (void)
{
        ADCON1 = 0x0F;                  // all digital
        TRISA = 0x05;                   // program Port A
        TRISB = 0;                      // program Port B
        TRISC = 0;                      // program Port C
        PORTB = 0;                      // all displays off
        PORTA = 0x0A;                   // #CS and SC = 1

        tenths = seconds = minutes = setTimeFlag = pointer = 0;
        select = 1;
        hours = 12;

        timeTemp();                     // initialize display

        INTCON2bits.TMR0IP = 1;         // Timer 0 is high priority
        IPR1bits.TMR1IP = 0;            // Timer 1 is low priority

        WriteTimer0 (0);
        OpenTimer0 (TIMER_INT_ON &      // every 1024 us
                    T0_8BIT &
                    T0_SOURCE_INT &
                    T0_PS_1_4);

        WriteTimer1 (53036);            // every 100 ms
        OpenTimer1 (TIMER_INT_ON &
                    T1_8BIT_RW &
                    T1_SOURCE_INT &
                    T1_PS_1_8);

        RCONbits.IPEN = 1;              // IPEN = 1 to enable priority interrupts
        INTCONbits.GIEH = 1;            // enable high-priority interrupt
        INTCONbits.GIEL = 1;            // enable low-priority interrupt

        while (1)                       // main loop
        {
                while (Switch() == 1);  // wait for switch press
                setTimeFlag = 1;        // do not display F
                displayRAM[4] = 0x76;   // display H
                timeTemp();
                while (Switch () == 0)
                {
                        hours++;
```

```
                          if (hours == 13)
                                hours = 1;
                                timeTemp();
                 }
                 displayRAM[4] = 0x6D;    // display S
                 timeTemp();
                 while (Switch () == 0)
                 {
                         minutes++;
                         if (minutes == 60)
                                minutes = 0;
                         timeTemp();
                 }
                 setTimeFlag = 0;
                 timeTemp();
          }
    }
```

This software uses two interrupts: a high-priority interrupt that multiplexes the displays and a low-priority interrupt that occurs every 100 ms to advance the clock. The low-priority interrupt also changes the contents of the display RAM once per minute when the time changes by calling the timeTemp function.

This sample system uses two timers that are important to the function of the clock and the display. Timer 0 is used to multiplex the displays and every time it overflows it causes a high-priority interrupt. Timer 0 is set up so it uses a prescaler of 4, which divides the timer input of 1 MHz (the system clock is 4 MHz) down to 250 KHz (4 µs). The timer register is written with a 0 so that it causes an interrupt for every 256 of the 4 µs input clocks, or once every 1024 µs. This causes the flash rate of each display to be approximately 125 Hz.

Timer 1 is used to cause a low-priority interrupt once every 100 ms by using a prescaler of 8 to produce a clock rate of 8 µs as the input of Timer 1. Timer 1 is programmed with a count of 53,036, which is 12,500 clock pulses before it overflows. The 12,500 times 8 µs is 100 ms, so the low-priority interrupt occurs every 100 ms. The clock divides the 100 ms by 10 to produce a second's count and the remainder of the time. The clock is a 12-hour clock with AM or PM indication, which is common for this type of clock. Note that as an alternate to using a count of 53,036, a -12,500 can be used and probably should be used because it is easier to understand. A 53,036 is a 16-bit 0xCF2C and a -12,500 is also a 0xCF2C so either can be used in a program.

Example 8-3 uses the timers.h header file to simplify programming the timer. The timers.h header file contains four functions for each timer: OpenTimer, CloseTimer, ReadTimer, and WriteTimer. This example uses only the OpenTimer and WriteTimer functions. For a complete list of these functions and all the details, refer to the Microchip Website and the MPLAB–C18 library's PDF file and to Appendix B, which is a condensed listing of the library functions.

8-2   USART AND INTERRUPTS

The USART (universal synchronous/asynchronous receiver/transmitter) generates and receives either synchronous or asynchronous serial data. As with most devices in the microcontroller, this device also uses interrupts during normal operation to improve the efficiency of a system. The USART can be operated without interrupts if needed, but in most cases the interrupt operation is more efficient. Also available in the C18 library is a software UART (universal asynchronous receiver/transmitter) that functions with asynchronous serial data. The UART is explained in later sections of this book.

Serial Data

Serial data is either synchronous or asynchronous. Synchronous data is sent with a clock pulse for synchronization, and asynchronous data is sent without a clock pulse. Most data today is asynchronous data. In addition to these two basic serial data types, the data can be sent in simplex, half-duplex, or full-duplex (usually called just duplex) modes. **Simplex** is information that is always sent in one direction such as the radio signal from a satellite transmitter. No information is ever transmitted to the satellite, it's only received. **Half-duplex** is information that can be sent or received, but in only one direction at a time. An example of a half-duplex system is CB (citizen's band) radio communications or the USB (universal serial bus) interface in a computer. In CB radio communications, a transmitted message is ended with the word "over" to inform the listener it's their turn to transmit. In the USB system, which is a half-duplex system, the protocol requires that a listener device waits for the serial line to become available before transmitting information. In a **full-duplex** system, information is transmitted and received simultaneously. An example of a full-duplex system is the telephone system.

Figure 8-3 shows how asynchronous data appears when transmitted by the USART or UART. A frame of serial asynchronous data is 10 bits in most cases and starts with a start bit and ends with at least one stop bit, although there may be as many stop bits as needed to separate information. A logic zero is often called a **space** and a logic one is often called a **mark**. The data, which is between the start and stop bits, is usually eight bits with the least significant bit sent first. Data can be successfully received because the transmission rate or bit time is known. Most systems are tolerant to a $\pm5\%$ variation in the bit time. For example, if data are transmitted at 9600 bps (bits per second), 960 characters or bytes are transmitted per second. Each character requires 10 bit times. The data are transmitted at 960 Bps (bytes per second). The bit rate (often called the baud rate) is 1/9600 or 104.2 μs per bit. Note that the bit rate is the Baud rate and not the data rate. The data rate is 960 Bps and the bit rate or Baud rate is 9600 bps. The bit rate is named in honor of **Émile Baudot** who developed a code called Baudot code used with telegraphy in the 1800s. The Baudot code uses a start bit, 5 data bits, and one and one-half stop bits, and was popular until the late 1960s for use with national news providers. Today, standard asynchronous serial data almost always uses a start bit, 8 data bits, and 1 stop bit.

Software could easily be used to generate this type of data, except it would consume most of the microcontroller's execution time. Therefore, a circuit called a USART is included in the microcontroller to transmit and receive both synchronous and asynchronous serial data. The UART was the first commercially available LSI (large scale integration) circuit produced in 1968 by Texas Instruments.

Synchronous data, which is not very common today, contains a signaling line and a clock line. Figure 8-4 illustrates the form of synchronous serial data. It does not contain any start or stop pulse, which increases the data rate because instead it is using 10 bit times for a byte of data and only 8 bit times are used per byte. This type of data was dropped from common usage because of the extra wire for the clock. Comparing Figure 8-3 with Figure 8-4, the same data are sent (A followed by B), but synchronous data takes less time to transmit the same information. What Figure 8-4 does not show is that synchronous data is normally transmitted in packets. A packet contains a fixed number of bytes, preceded by a byte that indicates the start of the packet followed by a byte that indicates the end of the packet. The time required by these two bytes reduces the data rate of the information. This text concentrates mainly on asynchronous serial data.

Controlling the USART

The USART within the PIC18 contains a programmable baud rate generator to set the bit rate of the data. The baud rate generator is programmed by placing a value in a register called SPBRG, which programs an internal timer to divide the system clock by the number loaded into SPBRG. Example 8-4 shows the equations used to calculate the value of the number loaded into SPBRG to generate

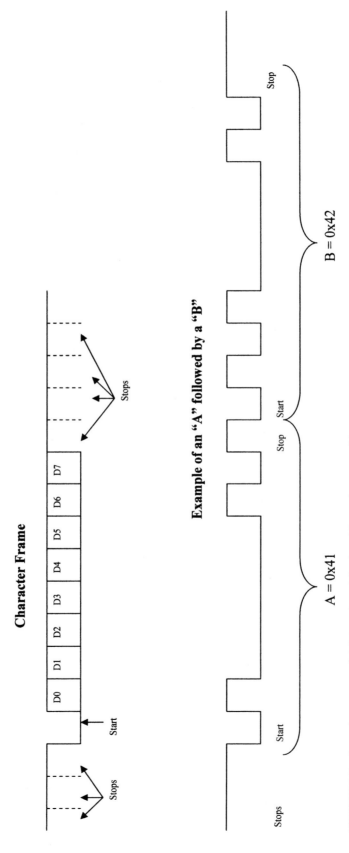

FIGURE 8-3 Asynchronous serial data using eight data bits and one stop bit.

FIGURE 8-4 Synchronous data transmission.

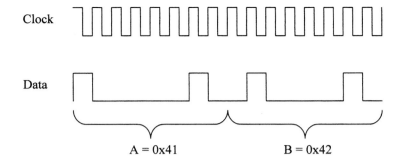

A = 0x41   B = 0x42

a specific baud rate. Because the baud rate timer is an 8-bit timer, the largest number that can be programmed into it is 0xFF (255) or, if zero is used, a 256. These three equations are used to calculate the value of SPBRG for varying conditions such as the clock rate and desired baud rate. The only difference between the equations are the constants 64 (called low-speed), 16 (called high-speed), or 4 (for synchronous operation), which sets an internal prescaler on the baud rate timer.

EXAMPLE 8-4

```
Asynchronous baud rate low speed (X is programmed into SPBRG)
```

$$x = \left(\frac{Fosc^*}{Baud\ Rate} \right) \div 64 - 1$$

```
Asynchronous baud rate high speed (X is programmed into SPBRG)
```

$$x = \left(\frac{Fosc}{Baud\ Rate} \right) \div 16 - 1$$

```
Synchronous baud rate (X is programmed into SPBRG)
```

$$x = \left(\frac{Fosc}{Baud\ Rate} \right) \div 4 - 1$$

```
*Note: Fosc = clock frequency
```

For example, to operate the USART at an asynchronous baud rate of 9600, either the low-speed or high-speed equation is used to determine the value loaded into SPBRG. Using a 16-MHz clock, the low-speed value is 25 and the high-speed value is 103. To select the high speed, the BRGH bit in the TXSTA register is set to a logic one. The USART uses pin RC6 for transmit data (TX) and pin RC7 for received data (RX) on the PIC18F2220 microcontroller. Other family members can use different pins, so consult the data sheet for the selected device. Figure 8-5 illustrates the PIC18F2220 connected to a serial interface with the DS275 line driver circuit used to convert from TTL levels to standard RS-232 levels. Standard RS-232C logic levels are defined as 3 V to 25 V for a logic zero and −3 V to −25 V for a logic one. In practice and using the DS275, the voltages are ±12 V. The DB9 connector (9-pin D-type connector) is used with the PC serial COM port. The connection shown is for a standard serial cable (not a null modem cable). Both the connector on this circuit and the connector on the back of a PC are female connectors. The cable to connect the two must be a male-to-male cable.

Software for this interface is listed in Example 8-5. The serial interface is programmed to operate asynchronously with 8 data bits at a baud rate of 9600. This is usually the default baud rate on a PC. The OpenUSART function programs the USART to operate, in this case, with both receiver and transmitter interrupts enabled with the BRGH bit set to high. The calculated value for SPBRG is 25 for a baud rate of 9600. This sample program responds to two commands sent

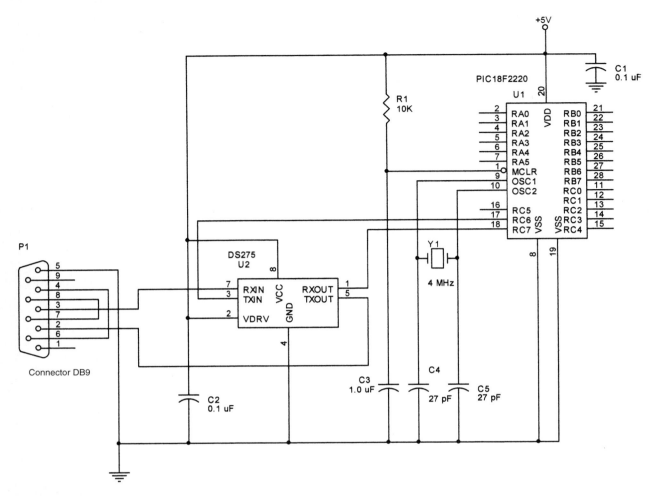

FIGURE 8-5 USART interfaced to a stand DB9 connector for RS-232C data.

from the PC: the letter C for connect and the letter G for goodbye. If the microcontroller receives the letter C, it returns a character string back to the PC with the word "Hello" as a standard ASCII character string. Standard ASCII character strings terminate with a carriage return (0x0D) followed by a line feed (0x0A). If the microcontroller receives the letter G from the PC, it sends the ASCII string "Goodbye" back to the PC. This software can be the basis for communications between the PC and the microcontroller. The microcontroller uses two queues to buffer the communications between the PC and itself. It also uses interrupts to handle both received and transmitted data.

EXAMPLE 8-5

```
/*
 * USART interrupt example
 */

#include <p18cxxx.h>
#include <usart.h>

/* Set configuration bits
 *  - set high-speed oscillator
```

```
 *   - disable watchdog timer
 *   - disable low-voltage programming
 *   - disable brownout reset
 *   - enable master clear
 */

#pragma config OSC = HS
#pragma config WDT = OFF
#pragma config LVP = OFF
#pragma config BOR = OFF
#pragma config MCLRE = ON

void MyHighInt (void);                                  // prototypes for the interrupt
void MyLowInt (void);                                   // service procedures

#pragma interrupt MyHighInt                             // MyHighInt is an interrupt
#pragma code high_vector=0x08                           // high_vector is at 0x0008

void high_vector (void)                                 // high-priority vector
{
      _asm GOTO MyHighInt _endasm                       // goto high software
}

#pragma interruptlow MyLowInt                           // MyLowInt is an interrupt
#pragma code low_vector=0x18                            // low vector is at 0x0018

void low_vector (void)                                  // low-priioty vector
{
      _asm GOTO MyLowInt _endasm                        // goto low software
}

// data memory data

char inQueue[16];
char outQueue[16];
char inPi;
char inPo;
char outPi;
char outPo;

#pragma code                                            // start code here

int outInQueue (void)
{
      int temp;
      if (inPi == inPo)
            return 0x100;                               // if empty
      temp = inQueue[inPo];                             // get data
      inPo = (inPo + 1) & 0x0F;
      return temp;
}

int inInQueue (char data)
{
      if (inPi == ((inPo + 1) & 0x0F))
            return 0x100;                               // if full
      inQueue[inPi] = data;
      inPi = (inPi + 1) & 0x0F;
      return 0;
}

int outOutQueue (void)
{
      int temp;
      if (outPi == outPo)
            return 0x100;                               // if empty
      temp = outQueue[outPo];                           // get data
      outPo = (outPo + 1) & 0x0F;
      return temp;
}
```

```
int inOutQueue (char data)
{
      if (outPi == ((outPo + 1) & 0x0F))
            return 0x100;                         // if full
      outQueue[outPi] = data;
      outPi = (outPi + 1) & 0x0F;
      PIE1bits.TXIE = 1;                          // transmitter on
      return 0;
}

void MyHighInt (void)
{
      int temp;
      if (PIR1bits.RCIF == 1)
      {
            PIR1bits.RCIF = 0;                    // clear interrupt
            inInQueue (RCREG);                    // store received data in queue
      }
      else if (PIR1bits.TXIF == 1 && PIE1bits.TXIE == 1)
      {
            PIR1bits.TXIF = 0;                    // clear interrupt
            temp = outOutQueue();
            if (temp == 0x100)
            {
                  while (TXSTAbits.TRMT == 0);
                  PIE1bits.TXIE = 0;              // transmitter off
            }
            else
                  TXREG = temp;                   // send data
      }
}

void MyLowInt (void)
{
      // nothing in this example
}

// main program

void main (void)
{
      int temp;
      ADCON1 = 0x0F;                             // all digital
      TRISC = 0x80;                              // program Port C

      IPR1bits.TXIP = 1;                         // select high priority
      IPR1bits.RCIP = 1;

      inPo = inPi = outPo = outPi = 0;           // setup queues

      OpenUSART (USART_TX_INT_ON &               // USART operates at 9600 baud
            USART_RX_INT_ON &
            USART_ASYNCH_MODE &
            USART_EIGHT_BIT &
            USART_CONT_RX &
            USART_BRGH_HIGH,
            25);

      RCONbits.IPEN = 1;                         // IPEN = 1 to enable
                                                 // priority interrupts
      INTCONbits.GIEH = 1;                       // enable high-priority interrupt
      INTCONbits.GIEL = 0;                       // enable low-priority interrupt

      while (1)
      {
            temp = outInQueue();                 // get PC data from InQueue
            if (temp != 0x100)                   // if data found
            {
                  if (temp == 'C')               // if code C (connect)
                  {
                        inOutQueue('H');         // send H
                        inOutQueue('e');         // send e
```

```
                        inOutQueue('l');              // send l
                        inOutQueue('l');              // send l
                        inOutQueue('o');              // send o
                        inOutQueue(13);               // send carraige return
                        inOutQueue(10);               // send line feed
                }
                else if (temp == 'G') // if code is G (goodbye)
                {
                        inOutQueue('G');              // send G
                        inOutQueue('o');              // send o
                        inOutQueue('o');              // send o
                        inOutQueue('d');              // send d
                        inOutQueue('b');              // send b
                        inOutQueue('y');              // send y
                        inOutQueue('e');              // send e
                        inOutQueue(13);               // send carraige return
                        inOutQueue(10);               // send line feed
                }
        }
    }
}
```

For this software to function with the PC, an application for the PC is needed. The hyperterminal application can be used, or a short Visual C++ application can be written. Figure 8-6 shows a screen shot of a Visual C++ application written to test the microcontroller and the program in Example 8-5.

The Visual C++ software, with or without an Active-X control, required to accomplish this task appears in Example 8-6. This software works in any version of Windows, from Windows 98 to Windows XP. All that is needed is Visual Studio of any 32-bit version (Visual C version 5.0 or newer) and the knowledge to develop a Visual C++ application. This application [Example 8-6(a)] uses the mscomm active-x object to control the serial interface and is written in Visual Studio.net 2003. Some modification may be needed for the software to function in other versions of Visual

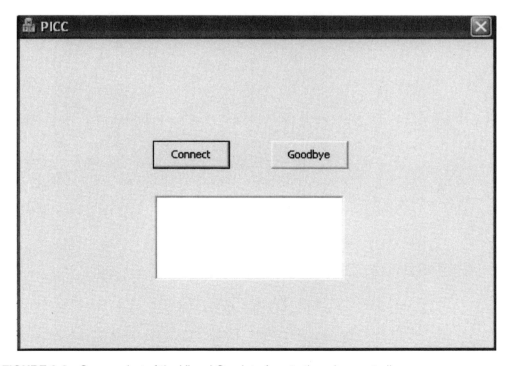

FIGURE 8-6   Screen shot of the Visual C++ interface to the microcontroller.

Studio because Microsoft keeps changing certain aspects of the package. This mscomm active-x object is either in your Visual Studio package or available for download from the Internet. The speed and so forth are set up in the property window for control in Visual C. The software listed is only what is added to a simple dialog application that uses Button1 for Connect, Button2 for Goodbye, and List1 for the list box where the replies from the microcontroller are displayed. To add the button handlers in Visual Studio.net 2003, click on the yellow lightning bolts. The two communications port functions are added to the dialog class by using the add function wizard. For this program to function correctly, the microcontroller must be powered before this Visual C program is executed. Also make sure the COM1 port is connected to the microcontroller.

Example 8-6(b) shows how to access the Windows kernel directly to communicate with the COM port. If the active-x control is available, use it because the software is a little shorter, and the kernel software is probably better because it will likely work in all future versions of Windows.

EXAMPLE 8-6(A)

```
void OpenComPort()
{
        mscomm1.put_CommPort(1);            // select COM1
        mscomm1.put_InputLen(1);
        mscomm1.put_RThreshold(1);          // do OnComm for each received char
        mscomm1.put_PortOpen(true);         // open COM1
}

void CPICCDlg::OnBnClickedButton1()    // connect button
{
        int temp;
        CString message;
        WriteComPort ("C");                 // send a "C"
}

void CPICCDlg::OnBnClickedButton2()    // goodbye button
{
        int temp;
        CString message;
        WriteComPort ("G");                 // send a "G"
}

void CPICCDlg::WriteComPort (CString data)
{
        mscomm1.put_Output(COleVariant(data));   // send string
}

void CPICCDlg::OnCommMscomm1()
{
        CString temp;
        if (comm.get_CommEvent() == 2)    // 2 = received char event
        {
                do
                {
                        temp += comm.get_Input();
                }while (comm.get_InBufferCount() > 0);
        }
        Lis1.AddString (temp);
}
```

EXAMPLE 8-6(B)

```
HANDLE hPort;      // define the handle
DCB dcb;           // define the control block

void CPICCDlg::OpenCom (CString port, DWORD baudrate)
{
        COMMTIMEOUTS Timeout;       // timeout structure
        hPort = CreateFile (        // create the handle to COM port
                port,
                GENERIC_WRITE | GENERIC_READ,
```

```
                    0,
                    NULL,
                    OPEN_EXISTING,
                    0,
                    NULL);
         GetCommState (hPort,&dcb);        // Set COM port state
         dcb.BaudRate = baudrate;          // baud rate
         dcb.ByteSize = 8;                 // 8 data bits
         dcb.Parity = NOPARITY;            // no parity
         dcb.StopBits = ONESTOPBIT;        // 1 stop
         SetCommState (hPort,&dcb);
         GetCommTimeouts (hPort, &Timeout);    // SET COM port timeouts
         Timeout.ReadIntervalTimeout = 500;
         Timeout.ReadTotalTimeoutConstant = 500;
         Timeout.ReadTotalTimeoutMultiplier = 100;
         Timeout.WriteTotalTimeoutConstant = 500;
         Timeout.WriteTotalTimeoutMultiplier = 100;
         SetCommTimeouts (hPort, &Timeout);
}

void CPICCDlg::SendPort (CString str)
{
         DWORD byteswritten;
         WriteFile (hPort, str, 1, &byteswritten, 0);
         return;
}

CString CPICCDlg::ReadPort (void)
{
         char mes[100];
         DWORD transferred;
         DWORD error;
         COMSTAT status;
         do
         {
                 ClearCommError (
                 hPort,                          // handle to Com port
                 &error,                         // pointer to error codes
                 &status);                       // pointer to status
                 mes[status.cbInQue] = 0;
                 ReadFile (hPort, &mes, status.cbInQue, &transferred, 0);
         } while (status.cbInQue == 0);   // until data is received
         return mes;
}
```

8-3   INTERRUPT ON CHANGE

The interrupt on change inputs are Ports B bits RB4 through RB7. If the interrupt on change is enabled and Port B is programmed as an input port, an interrupt occurs when any of these four pins change. This feature can be useful in many applications. If any of these pins change, an interrupt on change interrupt occurs, so do not forget to test these four pins in the interrupt service procedure to determine exactly which pin changed if more than one pin is used in an application.

Chapter 5 discussed the DCC using software and polling of an input pin to measure the width of pulses. A better way to measure a pulse width is by using the interrupt on change interrupt and a timer. By doing so, the microcontroller is free to do other things while measuring the pulse width of signals. In Chapter 5 we learned that the DCC signal uses a logic one signal that has a high and low pulse width of between 52 μs and 64 μs each. A logic zero data bit is much wider and has a high and low pulse width of between 90 μs and 10,000 μs. Suppose that the circuit in Figure 8-7 is connected to the track in a DCC system. The 754410 drivers supply up to 1 A of current at 12 V to each of the output pins when activated by a control signal that is sent

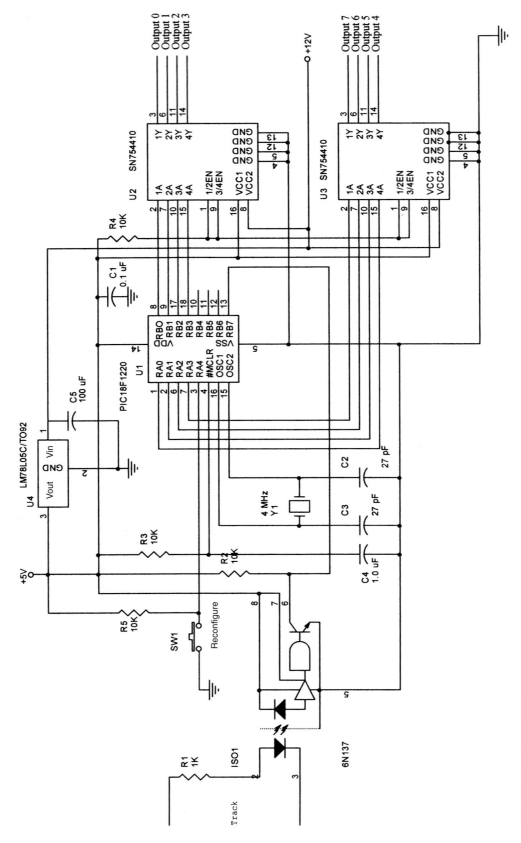

FIGURE 8-7 Octal DCC accessory controller.

down the track and detected by the microcontroller. Eight output pins control any combination of eight separate devices or four bidirectional devices such as switch machines.

As in Chapter 5, the track signal is electrically isolated from the circuit with an optical isolator. The signal from the track appears in Figure 5-4, and after conditioning, it is a TTL signal applied to pin RB7 of the microcontroller. The software written for this circuit appears in Example 8-7. Each time RB7 changes, an interrupt occurs. It is very important that PORTB is read before clearing the RBIF flag bit in the interrupt service procedure. If PORTB is not read before the RBIF bit is cleared, the change condition remains true and the RBIF bit will immediately set after clearing the RBIF. In the interrupt service procedure, a zero-to-one transition clears Timer 0 and so does a one-to-zero transition. The difference is that when the input changes from one to zero the contents of Timer 0 are examined. Timer 0 counts in one-fourth μs intervals so, if the count is between 52 * 4 or 208 (52 μs) and 64 * 4 or 256 (64 μs), the variable location data is changed to 0x01 and the location flag is set indicating the valid data. Likewise, if Timer 0 contains a number between 90 * 4 or 360 (90 μs) and 10,000 * 4 or 40,000 (10 ms), the variable location data is cleared to zero and the flag is set indicating a valid logic zero. This system measures only the high time of the input, so it produces a true image of the data received from the track. Chapter 5 measures both the high and low timer.

Figure 5-6 illustrates the valid data sent to control a device. The software in Example 8-7 detects the valid data in the main programs infinite while loop. This program must be able to detect the two types of packets: one controls the eight outputs and the other programs the device. Programming is limited to setting the address of the accessory controller. The accessory address is stored at data EEPROM locations CV513 (control variable 513) and CV521, where CV513 holds the low-order portion of the device address (A5–A0) and CV521 holds the most significant three bits of the device address (A8–A6). The three most significant address bits are inverted. That is, for accessory address 0x074, the value in CV513 is 0x34 and the value in CV521 is 0x06. The initial values in CV513 and CV521 are 0x00. This means that the accessory device address is initially 0x1C0 or 448 decimal. This address is changed by sending CV commands to the system for the command control unit. The accessory device can also be reset to the factory setting of CV513 and CV521 = 0x00. Resetting is accomplished by holding the pushbutton down while powering the system. The aaa bits in the command depicted in Figure 5-6 are used to select one of the eight outputs of the system. Here 000 is Output 0, 001 is Output 1, and so forth. The C bit is the data sent to the output. In this system if C = 1, the output becomes a source of 12 V with a current of up to 1 A and if C = 0, the output supplies no current. These commands are generated by the command control unit, which is purchased from a company such as Digitrax or Lenz. It can also be designed using another microcontroller-based system.

EXAMPLE 8-7

```
/*
 * DCC accessory decoder example
 */

#include <p18cxxx.h>
#include <timers.h>

/* Set configuration bits
 *  - set HS oscillator
 *  - disable watchdog timer
 *  - disable low-voltage programming
 *  - disable brownout reset
 *  - enable master clear
 */

#pragma config OSC = HS
#pragma config WDT = OFF
#pragma config LVP = OFF
```

```
#pragma config BOR = OFF
#pragma config MCLRE = ON

void MyHighInt (void);          // prototypes for the interrupt
void MyLowInt (void);           // service procedures

#pragma interrupt MyHighInt     // MyHighInt is an interrupt
#pragma code high_vector=0x08   // high_vector is at 0x0008

void high_vector (void)         // high-priority vector
{
      _asm GOTO MyHighInt _endasm     // goto high software
}

#pragma interruptlow MyLowInt   // MyLowInt is an interrupt
#pragma code low_vector=0x18    // low vector is at 0x0018

void low_vector (void)          // low-prioity vector
{
      _asm GOTO MyLowInt _endasm       // goto low software
}

// data memory

char flag;                      // data available flag
char data;

// ******** DATA EEPROM DATA ADDRESS ASSIGNMENTS ********

#define CV513   0               // low address
#define CV521   1               // high address

#pragma code                    // start code here

// Reads a data EEPROM location from address

char eeRead (char address)
{
      EECON1bits.EEPGD = 0;
      EEADR = address;
      EECON1bits.RD = 1;
      return EEDATA;
}

// Write a data EEPROM location at address with data

void eeWrite (char address, char data)
{
      INTCONbits.GIEH = 0;
      INTCONbits.GIEL = 0;
      EECON1bits.EEPGD = 0;
      EECON1bits.WREN = 1;
      EEADR = address;
      EEDATA = data;
      EECON2 = 0x55;
      EECON2 = 0xAA;
      EECON1bits.WR = 1;
      while (PIR2bits.EEIF == 0);
      PIR2bits.EEIF = 0;
      EECON1bits.WREN = 0;
      INTCONbits.GIEH = 1;
      INTCONbits.GIEL = 1;
}

void MyHighInt (void)
{
      if (INTCONbits.RBIF == 1)
      {
            int temp = PORTB;       // must read PORTB to clear change
            INTCONbits.RBIF = 0;    // clear interrupt
            if (PORTBbits.RB7 == 0)
            {
```

```
                                    if (ReadTimer0() == 0) // initial
                                    {
                                            OpenTimer0 (TIMER_INT_OFF &
                                                                TO_16BIT &
                                                                TO_SOURCE_INT &
                                                                TO_PS_1_256);
                                    }
                                    else if (ReadTimer0() < 4 * 52)
                                            WriteTimer0(0);      // invalid, reset Timer 0
                                    else if (ReadTimer0() >= 4 * 52 &&
                                                    ReadTimer0() <= 4 * 64)
                                    {
                                            WriteTimer0(0);      // good 1, reset Timer 0
                                            data = 1;
                                            flag = 1;
                                    }
                                    else if (ReadTimer0() >= 4 * 90 &&
                                                    ReadTimer0() <= 4 * 10000)
                                    {
                                            WriteTimer0 (0);     // good 0, reset Timer 0
                                            data = 0;
                                            flag = 1;
                                    }
                                    else            // bad data
                                            WriteTimer0 (0);    // reset Timer 0
                            }
                    else
                            WriteTimer0 (0);                // reset on a one transition
            }
}

void MyLowInt (void)
{
        // nothing in this example
}

char wait4Bit (void)
{
        while (flag == 0);       // wait for a bit
        flag = 0;
        return data;
}

char getByte (void)             // receive a byte
{
        char temp = 0;
        int a;
        for (a = 0; a < 8; a++)
        {
                temp <<= 1;
                temp |= wait4Bit();
        }
        return temp;
}

// main program

void main (void)
{
        char check;
        char count;
        char packetBytes[6];
        int a;
        char mask;

        ADCON1 = 0x0F;          // all digital
        TRISA = 0;              // program direction
        TRISB = 0x80;
        PORTA = PORTB = flag = 0;
```

```c
        if (PORTAbits.RA4 == 0)   // if reconfigure
        {
                eeWrite (CV513, 0);
                eeWrite (CV521, 0);
        }

        INTCON2bits.RBIP = 1;     // high priority
        INTCONbits.RBIE = 1;      // enable bit change interrupt

        WriteTimer0 (0);          // intitilaize Timer 0

        RCONbits.IPEN = 1;        // IPEN = 1 to enable priority interrupts
        INTCONbits.GIEH = 1;      // enable high-priority interrupt
        INTCONbits.GIEL = 0;      // enable low-priority interrupt

        while (1)
        {
                count = 0;
                while (count != 10)       // find preamble of at least 10 ones
                {
                        if (wait4Bit() == 1)
                                count++;
                        else
                                count = 0;
                }

                while (wait4Bit() == 1);   // wait for additional ones

                // past the preamble at this point

                count = 0;
                do                                    // get all packet bytes
                {
                        packetBytes[count++] = getByte();
                }

                while (wait4Bit == 0);

                // got all the packet bytes count = number of them

                for (a = 0; a < count; a++)
                        check ^= packetBytes[a];

                if (check == 0)              // good checksum
                {
                        if ((packetBytes[0] & 0x3F) == eeRead(CV513) &&
                        (((packetBytes[1] >> 4) & 0x07) ^ 0x07)
                                == eeRead (CV521))   // address match
                        {
                                if (count == 3) // if command
                                {
                                        mask = 1;
                                        mask <<= packetBytes[1] & 3;
                                        if ((packetBytes[1] & 4) == 0 &&
                                                (packetBytes[1] & 8) == 0)

                                                PORTB &= mask ^ 0xFF;
                                        else if ((packetBytes[1] & 4) == 0 &&
                                                (packetBytes[1] & 8) == 8)
                                                PORTB |= mask;
                                        else if ((packetBytes[1] & 4) == 4 &&
                                                (packetBytes[1] & 8) == 0)
                                                PORTA &= mask ^ 0xFF;
                                        else
                                                PORTA |= mask;
                                }
                                else if (count == 5)  // if program CV
                                {
                                        if ((packetBytes[1] & 3) == 2  &&
                                                packetBytes[2] == 0) // if CV513
                                                eeWrite(CV513, packetBytes[3]);
```

```
                                                  else if ((packetBytes[1] & 3) == 2 &&
                                                          packetBytes[2] == 8) // if CV521
                                                          eeWrite(CV521, packetBytes[3]);
                                          }
                                  }
                          }
                  }
          }
```

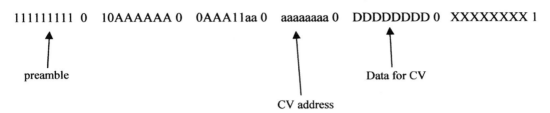

FIGURE 8-8   Accessory command for programming a CV register.

The formats for programming the accessory address in CV513 (address 0x200) and CV521 (address 0x208) are shown in Figure 8-8. These packets are five bytes in length instead of three bytes for the accessory decoder command. The CV addresses start at CV1 (0x000), so CV513 is programmed at address 512 (0x200).

8-4   SAMPLE INTERRUPT SYSTEMS

This section presents a few interrupt-processed systems to use as case studies or just to emulate. The more systems that are examined, the more proficient you become at programming and interfacing.

Example System 1

The first system is the one first presented in Chapter 2 that logged information from ID cards and transmitted the information on a serial link to a central data logging computer system. The device contains a serial interface, an LCD display, and a keypad, and could function either as a time clock to record the hours worked by an employee or as an attendance meter in a school to track student attendance. Whatever its use, the system needs interrupts for: its real-time clock, processing the numbers read from an ID card, and processing the data on the serial link. This system has many interrupts and is a worthwhile system to explore.

Now that you have a conceptual view of the system, the components must be selected before the system is designed. The following components are needed:

1. A microcontroller (type to be decided later).
2. An LCD display with 2 lines of display and 40 characters per line. Two lines are needed to display employee numbers and names and also messages to an employee.
3. An optical card reader. A magnetic card reader could be chosen, but then special equipment needs to be purchased to create magnetic ID cards. Optical ID cards can be created on a modern printer and the only special equipment needed is a laminating machine.
4. A simple serial interface that uses a twisted pair of wires that is compatible with a PC. The lines on this interface may need to be quite long so the USB is unacceptable because it allows for wire lengths of only a few meters. The choice of the interface is discussed later in this section.

5. A keypad. Because this system must be easy to use without any special code, a telephone-style keypad is not recommended because it does not have enough keys. The number of keys and their functions will be discussed later in this section.
6. A power source. If a power source is not available, use two twisted pairs of wire, one for the power supply to the unit and one for the serial data. Two pairs of telephone wire or CAT5 (category 5) wire might be a good choice for the cable. CAT5 cable is commonly available and contains four pairs of wire, which is more than needed. The connectors are RJ-45 connectors.

An 18-pin PIC probably does not have enough pins to support all these connections. A 28-pin microcontroller is therefore needed. The LCD requires six connections, and the keyboard probably requires nine connections, assuming that a 5×4 keypad is sufficient. A few pins for the serial connection, and a few pins for the optical card reader and the 28-pin microcontroller is enough to handle this system. The Microchip website recommends the PIC18F2580, which has 32K of program memory, 1536 bytes of SRAM, and 256 bytes of EEPROM, which should be enough memory for this system. Suppose that ID numbers are limited to 10 digits/alphabetic characters in length. If 412 bytes of SRAM are used as a backup storage for storing ID numbers, space for 41 numbers are available. On the rare occasion that the link to the PC is down, numbers might need to be stored in the system temporarily. The remaining bytes of SRAM should be more than enough to handle any task that needs to be performed.

The optical card reader selected for the system is the Opticon LCO SR scanner module, which has a TTL level output signal. The output is on a 5-pin 240° DIN connector with pin 1 connected to 5 V, pin 2 is the TLL output, and pin 3 is ground. Figure 8-9 illustrates the entire system schematic including the card reader.

The CAT5 cable uses 7 of the 8 wires unless there are repeaters in a system, then one additional line is needed. The outermost wires (pins 1, 2 and 7, 8) carry power from the PC head end power station to each of up to 16 ID card readers connected in the system. More that 16 drops can be used, if a repeater is placed in the line. The system uses the RS-422 interface standard, which has a maximum cable length of 4000 feet, almost a mile. The differential data signal is carried on the center wires (pins 4 and 5). The remaining connections, except for pin 3, which is sent from the power unit to turn the line around in the repeaters, are not used. This system is wired as a half-duplex system, which means that only one device can transmit at a time, while all the other devices on the system receive. The MAX 1483 is the line driver/receiver chosen for this system. Whether it sends or receives data is determined by the DR/RE connection. If this connection is placed at a logic one, the line driver becomes active and the unit sends data back toward the PC or head end. If the DR/$\overline{\text{RE}}$ connection is placed at a logic zero, the line receiver becomes active and the system receives data from the PC. The RA6 pin controls the direction of the MAX 1483.

Figure 8-10 depicts the function of the PC and each drop or ID card system attached to this single cable. Actually, there is one additional component in the system that is explained in Example System 2, located later in this section. The head end also requires another system to communicate between the PC and this RS-422 network. This other system, called a power station, provides power to the network and also interfaces the PC to the network. A minimal system requires two parts: the power station that connects to the PC, and the ID card reader system. This system is scalable to almost any size, but the software design limits it to 1,000 ID card readers. That should be enough units for just about any application.

Each section of the system can have up to 15 ID card readers plus a repeater. What does not appear here is that if repeaters are needed, another signal line is needed to control the direction of the repeater. This will be explained with the power station later in this section. The PC connected to the power station is normally in the transmit mode, and all of the repeaters and ID card readers are normally in the receiver mode. When the PC polls an ID card reader, the line is turned around

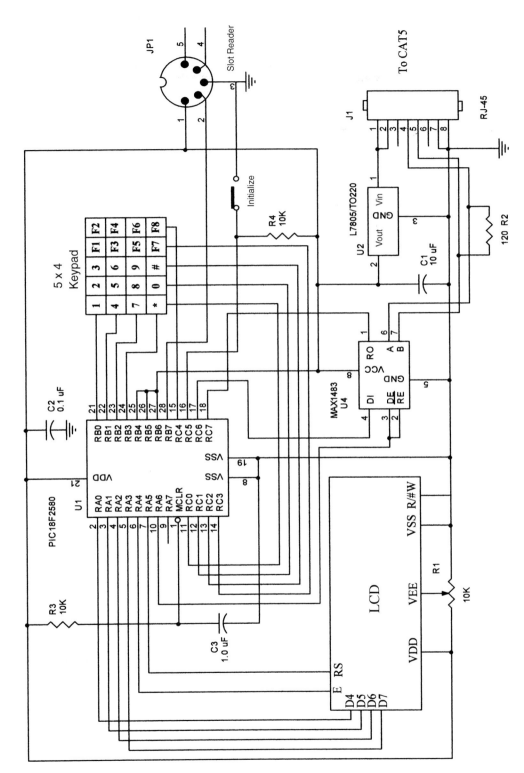

FIGURE 8-9 ID card system.

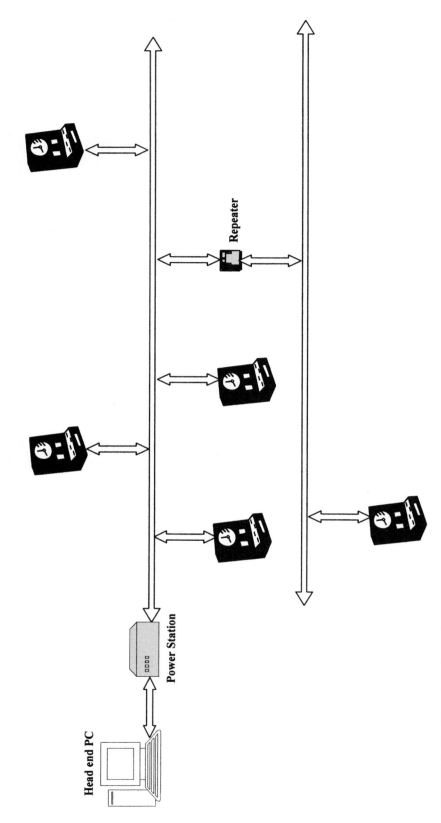

FIGURE 8-10 ID card reader system.

TABLE 8-1 Commands for the ID card reader system.

Byte	Command	Comment
0x06	**ACK**	Acknowledge receipt
0x21	**NAK**	Negative acknowledge receipt
0x17	**TIME HH MM SS MM DD YY YY crc**	Broadcast time and date from PC to ID card readers (YY YY) sent as 20 06 (No ACK/NAK sent)
0x18	**READ address crc**	Ask an ID card reader for its data Address is (LL HH) (ACK/NAK sent)
0x19	**SEND packet crc**	Send an ID card number to PC packet
0x1a	**SETID address newaddress crc**	Program ID number Addresses are (LL HH) (No ACK.NAK sent)
0x1b	**SETIDL length crc**	Set ID number length (No ACK/NAK)

and the PC becomes receptive to data and one of the ID card readers drives the bus. This is a half-duplex operation.

In order to handle data transfers, a set of commands is needed for this network and also something to control the flow of data if an error occurs. This system uses the ACK/NAK flow control technique to validate data and control data flow. Table 8-1 lists the commands available in this system.

Is this type of network reliable? It is if some safeguards are inserted in the system. A hardware failure will crash the system, but that is practically impossible to prevent unless some redundancy is incorporated into the system. When any ID card reader is powered, its bus transceiver (MAX 1483) is set so it receives data. If an ID card reader hangs, the watchdog timer (WDT) reboots it and places the ID card reader into receive mode. That prevents any problems with the ID card readers. The PC also needs to detect if an ID card reader crashed in the middle of a transmission. That is possible by using a WDT timer in the PC software as well. The WDTs prevent system hang-ups.

Each unit in the system has an address. The PC has address 0x0000, and each ID card reader has an address of 0x0001 when it is first installed into the system. The software in the PC periodically polls that system to see if any new ID card readers are attached and assigns them a unique ID number. The PC operator can enter the location of the unit into its databank at a later time. The software in the PC should poll the network for a new ID card reader about once every minute. That way, there is little chance that two readers will be installed at the same time and both receive the same address. It could happen, so the software in the ID card reader allows the address to be changed at the ID card reader. Note that the ID number is stored in the EEPROM so that if the power unit fails, the ID card reader still contains the ID number. It would probably be worthwhile to have a spare power unit for the system.

The software for the ID card reader appears in Example 8-9. Because this is the software for an entire system, it is longer than other applications thus far presented. One other change must be made to the linker file to allow this program to execute: If an array is larger than 256 bytes, the linker file must be modified to be able to address the array. In this example, 412 bytes of memory are used to store ID numbers locally. In order to accomplish this, the linker file is modified as shown in Example 8-8. Two statements are added and a few are removed. The underlined statements were added. One statement defines the queue as the name "big" and the other statement defines the SECTION as RAM. What does not appear here is that banks grp4 and grp5 have been removed. If the linker file in Example 8-9 is viewed, #pragma udata is where the array dataQueue is stored.

EXAMPLE 8-8

```
// $Id: 18f2580.lkr,v 1.2 2004/08/22 23:40:54 curtiss Exp $
// File: 18f2580.lkr
// Sample linker script for the PIC18F2580 processor

LIBPATH .

FILES c018i.o
FILES clib.lib
FILES p18f2580.lib

CODEPAGE       NAME=vectors     START=0x0          END=0x29          PROTECTED
CODEPAGE       NAME=page        START=0x2A         END=0x7FFF
CODEPAGE       NAME=idlocs      START=0x200000     END=0x200007      PROTECTED
CODEPAGE       NAME=config      START=0x300000     END=0x30000D      PROTECTED
CODEPAGE       NAME=devid       START=0x3FFFFE     END=0x3FFFFF      PROTECTED
CODEPAGE       NAME=eedata      START=0xF00000     END=0xF000FF      PROTECTED

ACCESSBANK     NAME=accessram   START=0x0          END=0x5F
DATABANK       NAME=gpr0        START=0x60         END=0xFF
DATABANK       NAME=gpr1        START=0x100        END=0x1FF
DATABANK       NAME=grp2        START=0x200        END=0x2FF
DATABANK       NAME=grp3        START=0x300        END=0x3FF

DATABANK       NAME=big         START=0x400        END=0x5FF         PROTECTED

DATABANK       NAME=sfr13       START=0xD00        END=0xDFF         PROTECTED
DATABANK       NAME=sfr14       START=0xE00        END=0xEFF         PROTECTED
DATABANK       NAME=sfr15       START=0xF00        END=0xF5F         PROTECTED
ACCESSBANK     NAME=accesssfr   START=0xF60        END=0xFFF         PROTECTED

SECTION        NAME=CONFIG      ROM=config

SECTION        NAME=queue       RAM=big

STACK SIZE=0x100 RAM=grp3
```

This system uses bar code **code 128**, a very common bar code introduced in 1981 for encoding alphabetic and numeric data. The advantage of this bar code is that Windows and a trutype font set can print ID cards without any special software package except Microsoft Word. There are actually three bar code 128 codes: A, B, and C. The code 128 A version has a partial ASCII code set, version code 128 B has a full ASCII code set, and version code 128 C has only the numbers 0 through 9. Figure 8-11 illustrates an example of this bar code. Refer to Table 8-2 for a complete list of the code 128 bar codes. The pattern for each code is listed as b and s, where b is a bar and s is a blank space. A bb is a double wide bar and an ss is a double wide space. Each character in the code is comprised of three bars and three spaces of varying widths.

A bar code is always sent with a start (start A in our example) followed by the characters in the body of the code. After the body of the code is a modulo103 check byte and then the stop code. Figure 8-12 show the message WOW3. The check code is generated as shown. The sensor in the card reader produces a logic zero for a dark bar and a logic one for a light bar. The program in Example 8-9 uses the interrupt on change feature to detect the bar code and convert it to a number that is stored in the queue for transmission to the network when requested.

FIGURE 8-11 Example of a code 128 bar code.

PIC18F2580

TABLE 8-2 Code 128 bar codes.

Code A	Code B	Code C	Value	Pattern
Space	Space	00	0	bbsbbssbbss
!	!	01	1	bbssbbsbbss
"	"	02	2	bbssbbssbbs
#	#	03	3	bssbssbbsss
$	$	04	4	bssbsssbbss
%	%	05	5	bsssbssbbss
&	&	06	6	bssbbssbsss
'	(	07	7	bssbbsssbss
(	)	08	8	bsssbbssbss
)	*	09	9	bbssbssbsss
*	*	10	10	bbssbsssbss
+	+	11	11	bbsssbssbss
,	,	12	12	bsbbssbbbss
-	-	13	13	bssbbsbbbss
.	.	14	14	bssbbssbbbs
/	/	15	15	bsbbbssbbss
0	0	16	16	bssbbbsbbss
1	1	17	17	bssbbbssbbs
2	2	18	18	bbssbbbssbs
3	3	19	19	bbssbsbbbss
4	4	20	20	bbssbssbbbs
5	5	21	21	bbsbbbssbss
6	6	22	22	bbssbbbsbss
7	7	23	23	bbbsbbsbbbs
8	8	24	24	bbbsbssbbss
9	9	25	25	bbbssbsbbss
:	:	26	26	bbbssbssbbs
;	;	27	27	bbbsbbssbss
<	<	28	28	bbbssbsbss
Equal	Equal	29	29	bbbssbbssbs
>	>	30	30	bbsbbsbbsss
?	?	31	31	bbsbbsssbbs
@	@	32	32	bbsssbbsbbs
A	A	33	33	bsbsssbbsss
B	B	34	34	bsssbsbbsss
C	C	35	35	bsssbsssbbs
D	D	36	36	bsbbsssbsss
E	E	37	37	bsssbbsbsss
F	F	38	38	bsssbbsssbs
G	G	39	39	bbsbsssbsss
H	H	40	40	bbsssbsbsss
I	I	41	41	bbsssbsssbs
J	J	42	42	bsbbsbbbsss
K	K	43	43	bsbbsssbbbs
L	L	44	44	bsssbbsbbbs
M	M	45	45	bsbbbsbbsss
N	N	46	46	bsbbbsssbbs
O	O	47	47	bsssbbbsbbs
P	P	48	48	bbbsbbbsbbs
Q	Q	49	49	bbsbsssbbbs
R	R	50	50	bbsssbsbbbs
S	S	51	51	bbsbbbsbsss
T	T	52	52	bbsbbbsssbs

TABLE 8-2 (*continued*)

Code A	Code B	Code C	Value	Pattern
U	U	53	53	bbsbbbsbbbs
V	V	54	54	bbbsbsbbsss
W	W	55	55	bbbsbsssbbs
X	X	56	56	bbbsssbsbbs
Y	Y	57	57	bbbsbbsbsss
Z	Z	58	58	bbbsbbsssbs
[	[	59	59	bbbsssbbsbs
\	\	60	60	bbbsbbbbsbs
]	]	61	61	bbssbsssssbs
^	^	62	62	bbbbsssbsbs
_	_	63	63	bsbssbbssss
NUL	'	64	64	bsbssssbbss
SOH	a	65	65	bssbsbbssss
STX	b	66	66	bssbssssbbs
ETX	c	67	67	bsssbsbbss
EOT	d	68	68	bssssbssbbs
ENQ	e	69	69	bsbbssbssss
ACK	f	70	70	bsbbssssbss
BEL	g	71	71	bssbbsbssss
BS	h	72	72	bssbbsssssbs
HT	i	73	73	bsssbbsbss
LF	j	74	74	bsssbbssbss
VT	k	75	75	bbssssbssbs
FF	l	76	76	bbssbsbssss
CR	m	77	77	bbbbsbbbsbs
SO	n	78	78	bbsssssbsbss
SI	o	79	79	bsssbbbbsbs
DLE	p	80	80	bsbssbbbbss
DC1	q	81	81	bssbsbbbbss
DC2	r	82	82	bssbsbbbbs
DC3	s	83	83	bsbbbbssbss
DC4	t	84	84	bssbbbbsbss
NAK	u	85	85	bssbbbbssbs
SYN	v	86	86	bbbbsbssbss
ETB	w	87	87	bbbbssbsbss
CAN	x	88	88	bbbbssbssbs
EM	y	89	89	bbsbbsbbbbs
SUB	z	90	90	bbsbbbbsbbs
ESC	{	91	91	bbbbsbbbsbs
FS			92	92	bsbsbbbbsss
GS	}	93	93	bsbsssbbbbs
RS	~	94	94	bsssbsbbbbs
US	DEL	95	95	bsbbbbsbsss
FNC3	FNC3	96	96	bsbbbbsssbs
FNC2	FNC2	97	97	bbbbsbsbsss
Shift	Shift	98	98	bbbbsbsssbs
Switch Code C	Switch Code C	99	99	bsbbbsbbbbs
Switch Code B	FNC4	Switch Code B	100	bsbbbbsbbbs
FNC4	Switch Code A	Switch Code A	101	bbbsbsbbbbs
FNC1	FNC1	FNC1	102	bbbbsbsbbbs
START Code A	START Code A	START Code A	103	bbsbsbbbbs
START Code B	START Code B	START Code B	104	bbsbssbssss
START Code C	START Code C	START Code C	105	bbsbssbbbss
STOP	STOP	STOP	106	bbsssbbbbsbsbb

FIGURE 8-12 Numbers and bar code generated for a WOW3.

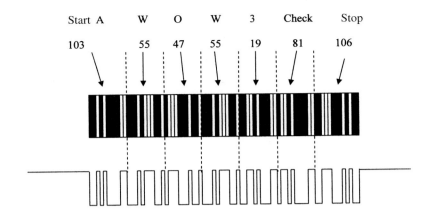

Check = 1(55) + 2 (47) + 3(55) + 4(19) = 390

TTL signal

390 divided by 103 = 3 remainder of 81 where the 81 is the check

EXAMPLE 8-9

```
// Sample system program
//

#include <p18cxxx.h>
#include <timers.h>
#include <delays.h>
#include <usart.h>
#include <string.h>

/* Set configuration bits
 *   - set RC oscillator
 *   - disable watchdog timer
 *   - disable low voltage programming
 *   - disable brownout reset
 *   - enable master clear
 */

#pragma config OSC = IRCIO67
#pragma config WDT = ON
#pragma config WDTPS = 256        // once per second
#pragma config LVP = OFF
#pragma config BOR = OFF
#pragma config MCLRE = ON

// ********* DATA MEMORY VARIABLES ************

#pragma udata queue               // big buffers (see text)
char dataQueue[312];              // ID queue
char cardBuffer[200];             // buffer for card number

#pragma udata

char inQueue[16];                 //USART queues
char outQueue[16];
char inPi;
char inPo;
char outPi;
char outPo;

int dataQueuePtr;
char dataQueueBusy;

char cardBufferPtr;
char cardBufferTimeOut;
int oneBits;
```

```
short long time;          // time
char day = 1;             // date
char month = 0;
int year = 2225;

char flag;                // data available flag
char dataL;
char dataH;

char alpha;               // alpha numbers mode
char dirty;               // dirty flag
char idNumberCount;
char idBuffer[10];

char state;               // packet buffer
char packetBuffer[10];
char packetPtr;

// ********** PROGRAM MEMORY STATIC DATA ***************
rom near char *mes1 = "Welcome--just swipe your card.";
rom near char *mes2 = "Alpha A-I";
rom near char *mes3 = "Numeric";
rom near char *mes4 = "ID Number = ";
rom near char *mes5 = "                              ";
rom near char *mes6 = "Alpha J-R";
rom near char *mes7 = "Alpha S-Z";
rom near char *mes8 = "ID Number is too short! Reenter it.";

rom near char *monthName[] =
{
      "January ",
      "February ",
      "March ",
      "April ",
      "May ",
      "June ",
      "July ",
      "August ",
      "September ",
      "October ",
      "November ",
      "December "
};

rom near char lookupKey[] =
{
      1,  4,  7, 10,      // left column
      2,  5,  8,  0,
      3,  6,  9, 11,
      12, 13, 14, 15,
      16, 17, 18, 19      // right column
};

rom near int code128[] =   // code 128 A lookup table
{
   0b11011001100,         // space  0
   0b11001101100,         // !      1
   0b11001100110,         // "      2
   0b10010011000,         // #      3
   0b10010001100,         // $      4
   0b10001001100,         // %      5
   0b10011001000,         // &      6
   0b10011000100,         // `      7
   0b10001100100,         // (      8
   0b11001001000,         // )      9
   0b11001000100,         // *      10
   0b11000100100,         // +      11
   0b10110011100,         // ,      12
   0b10011011100,         // -      13
```

```
0b10011001110,        // .          14
0b10111001100,        // /          15
0b10011101100,        // 0          16
0b10011100110,        // 1          17
0b11001110010,        // 2          18
0b11001011100,        // 3          19
0b11001001110,        // 4          20
0b11011100100,        // 5          21
0b11001110100,        // 6          22
0b11101101110,        // 7          23
0b11101001100,        // 8          24
0b11100101100,        // 9          25
0b11100100110,        // :          26
0b11101100100,        // ;          27
0b11100110100,        // <          28
0b11100110010,        // =          29
0b11011011000,        // >          30
0b11011000110,        // ?          31
0b11000110110,        // @          32
0b10100011000,        // A          33
0b10001011000,        // B          34
0b10001000110,        // C          35
0b10110001000,        // D          36
0b10001101000,        // E          37
0b10001100010,        // F          38
0b11010001000,        // G          39
0b11000101000,        // H          40
0b11000100010,        // I          41
0b10110111000,        // J          42
0b10110001110,        // K          43
0b10001101110,        // L          44
0b10111011000,        // M          45
0b10111000110,        // N          46
0b10001110110,        // O          47
0b11101110110,        // P          48
0b11010001110,        // Q          49
0b11000101110,        // R          50
0b11011101000,        // S          51
0b11011100010,        // T          52
0b11011101110,        // U          53
0b11101011000,        // V          54
0b11101000110,        // W          55
0b11100010110,        // X          56
0b11101101000,        // Y          57
0b11101100010,        // Z          58
0b11100011010,        // [          59
0b11101111010,        // \          60
0b11001000010,        // ]          61
0b11110001010,        // ^          62
0b10100110000,        // _          63
0b10100001100,        // nul        64
0b10010110000,        // soh        65
0b10010000110,        // stx        66
0b10000101100,        // etx        67
0b10000100110,        // eot        68
0b10110010000,        // enq        69
0b10110000100,        // ack        70
0b10011010000,        // bel        71
0b10011000010,        // bs         72
0b10000110110,        // ht         73
0b10000110010,        // lf         74
0b11000010010,        // vt         75
0b11001010000,        // ff         76
0b11110111010,        // cr         77
0b11000010100,        // so         78
0b10001111010,        // si         79
0b10100111100,        // dle        80
0b10010111100,        // dc1        81
```

```
        0b10010011110,          // dc2         82
        0b10111100100,          // dc3         83
        0b10011110100,          // dc4         84
        0b10011110010,          // nak         85
        0b11110100100,          // syn         86
        0b11110010100,          // etb         87
        0b11110010010,          // can         88
        0b11011011110,          // em          89
        0b11011110110,          // sub         90
        0b11110110110,          // esc         91
        0b10101111000,          // fs          92
        0b10100011110,          // gs          93
        0b10001011110,          // rs          94
        0b10111101000,          // us          95
        0b10111100010,          // fnc3              96
        0b11110101000,          // fnc2              97
        0b11110100010,          // shift       98
        0b10111011110,          // switch C    99
        0b10111101110,          // switch B    100
        0b11101011110,          // fnc4              101
        0b11110101110,          // fnc1              102
        0b11010111100,          // start A     103
        0b11010010000,          // start B     104
        0b11010011100,          // start C     105
        0b1100011101011         // stop              106
};

// ******** DATA EEPROM DATA ADDRESS ASSIGNMENTS ********

#define timeMode 0      // 0 = 12 hr; 1 = 24 hr
#define unitAddressL 1
#define unitAddressH 2
#define passWord0 3
#define passWord1 4
#define passWord2 5
#define passWord3 6
#define passWord4 7
#define idLength 8

// ********* INTERRUPT DEFINITIONS ************

// ********* PROTOTYPES ***********

void MyHighInt (void);          // prototypes for the interrupt
void MyLowInt (void);           // service procedures
void timeTemp (void);
int getTemp (void);
void DoClock (void);
int GetInQueue (void);
int SaveInQueue (char data);
int GetOutQueue (void);
int SaveOutQueue (char data);
void abortCard (void);

// ********** SETUP INTERRUPT VECTORS ***************
// *** Special note for PIC18F2580
// ***          requires the high-priority interrupt to be
// *** defined as interrupt low (see errata for this
// *** microcontroller).

#pragma interruptlow MyHighInt
#pragma code high_vector=0x08   // high_vector is at 0x0008

void high_vector (void)         // high-prioity vector
{
        _asm GOTO MyHighInt _endasm     // goto high software
}

#pragma interruptlow MyLowInt   // MyLowInt is an interrupt
#pragma code low_vector=0x18    // low vector is at 0x0018
```

```
void low_vector (void)              // low-prioity vector
{
      _asm GOTO MyLowInt _endasm          // goto low software
}

#pragma code                        // start code here

// High-priority interrupt service procedure
//    -- Interrupt on Change for ID card slot reader

void MyHighInt (void)
{
      if (INTCONbits.RBIF == 1)
      {
            int temp = PORTB;          // must read PORTB to clear change
            INTCONbits.RBIF = 0;       // clear interrupt
            if (ReadTimer0() == 0)     // initial
            {
                  OpenTimer0 (TIMER_INT_OFF &
                              T0_16BIT &
                              T0_SOURCE_INT &
                              T0_PS_1_256);            // 128 us period
                  cardBufferPtr = 0;
                  cardBufferTimeOut = 10;
            }
            else
            {
                  cardBuffer[cardBufferPtr++] = ReadTimer0();
                  cardBuffer[cardBufferPtr++] = ReadTimer0() >> 8;
                  cardBuffer[cardBufferPtr] = PORTBbits.RB7;
                  cardBuffer[cardBufferPtr++ ^= 0x01;      // invert bit
                  if (cardBufferPtr >= (idLength + 3) * 6)
                        abortCard();
            }
      }
}

// Low-priority interrupt service procedure
//     -- USART receiver (highest)
// -- USART transmitter
// -- Timer 1 (RTC) (lowest)

void MyLowInt (void)
{
      int temp;
      if (PIR1bits.RCIF == 1)     // is it USART receiver?
      {
            PIR1bits.RCIF = 0;         // clear interrupt
            SaveInQueue (RCREG);       // store received data in queue
      }
      else if (PIR1bits.TXIF == 1 && PIE1bits.TXIE == 1)// is it transmitter?
      {
            PIR1bits.TXIF = 0;         // clear interrupt
            temp = GetOutQueue();      // get queue data
            if (temp == 0x100)
            {
                  while (TXSTAbits.TRMT == 0);
                  PIE1bits.TXIE = 0;   // transmitter off
            }
            else
                  TXREG = temp;        // send data
      }
      else if (PIR1bits.TMR1IF == 1)    // is it 100 ms RTC?
      {
            if (cardBufferTimeOut == 0 && ReadTimer0() != 0)
                  abortCard();    // if timed out
            else if (cardBufferTimeOut != 0)
                  cardBufferTimeOut--;
            DoClock();                 // crank the clock
      }
}
```

```
// ************** FUNCTIONS **************

// Card slot reader time out

void abortCard (void)
{
      CloseTimer0();
      cardBufferTimeOut = 0;
      WriteTimer0(0);
      if (cardBufferPtr >= (idLength + 3) + 6)
            flag = 1;
}

// Read a data EEPROM location from address

char eeRead (char address)
{
      EECON1bits.EEPGD = 0;
      EEADR = address;
      EECON1bits.RD = 1;
      return EEDATA;
}

// Write a data EEPROM location at address with data

void eeWrite (char address, char data)
{
      INTCONbits.GIEH = 0;
      INTCONbits.GIEL = 0;
      EECON1bits.EEPGD = 0;
      EECON1bits.WREN = 1;
      EEADR = address;
      EEDATA = data;
      EECON2 = 0x55;
      EECON2 = 0xAA;
      EECON1bits.WR = 1;
      while (PIR2bits.EEIF == 0);
      PIR2bits.EEIF = 0;
      EECON1bits.WREN = 0;
      INTCONbits.GIEH = 1;
      INTCONbits.GIEL = 1;
}

// Read from InQueue, 0x100 is empty

int GetInQueue (void)
{
      int temp;
      if (inPi == inPo)
            return 0x100;       // if empty
      temp = inQueue[inPo];     // get data
      inPo = (inPo + 1) & 0x0F;
      return temp;
}

// Save data in InQueue, 0x100 is full

int SaveInQueue (char data)
{
      if (inPi == ((inPo + 1) & 0x0F))
            return 0x100;       // if full
      inQueue[inPi] = data;
      inPi = (inPi + 1) & 0x0F;
      return 0;
}

// Read from OutQueue, 0x100 is empty

int GetOutQueue (void)
{
      int temp;
```

```
                if (outPi == outPo)
                        return 0x100;                   // if empty
                temp = outQueue[outPo];                 // get data
                outPo = (outPo + 1) & 0x0F;
                return temp;
        }

        // Save data in OutQueue, 0x100 is full

        int SaveOutQueue (char data)
        {
                if (outPi == ((outPo + 1) & 0x0F))
                        return 0x100;                   // if full
                inQueue[outPi] = data;
                        outPi = (outPi + 1) & 0x0F;
                PIE1bits.TXIE = 1;                      // transmitter on
                return 0;
        }

        // Send LCD byte data, with RS = rs

        void SendLCDdata (char data, char rs)
        {
                PORTA = data >> 4;                      // send left nibble
                PORTAbits.RA5 = rs;                     // set RS
                PORTAbits.RA4 = 1;                      // pulse E
                PORTAbits.RA4 = 0;
                Delay10TCYx(8);                         // wait 40 us
                PORTA = data & 0x0F;                    // send right nibble
                PORTAbits.RA5 = rs;                     // set RS
                PORTAbits.RA4 = 1;                      // pulse E
                PORTAbits.RA4 = 0;
                Delay10TCYx(8);                         // wait 40 us
        }

        // Initialize LCD

        void InitLCD (void)                             // intialize LCD
        {
                Delay1KTCYx(40);                        // wait 20 ms
                SendLCDdata(0x20, 0);                   // send 0x20
                Delay1KTCYx(12);                        // wait 6 ms
                SendLCDdata(0x20, 0);                   // send 0x20
                Delay10TCYx(20);                        // wait 100 us
                SendLCDdata(0x20, 0);                   // send 0x20
                SendLCDdata(0x08, 0);                   // send 0x28
                SendLCDdata(0x01, 0);                   // send 0x01
                Delay1KTCYx(4);                         // wait 2 ms
                SendLCDdata(0x0C, 0);                   // send 0x0C
                SendLCDdata(0x06, 0);                   // send 0x06
        }

        // Display a data RAM-based string (str) at position
        //      Line 1 is at positions 0x80 through 0xA7
        //      Line 2 is at positions 0xC0 through 0xE7

        void DisplayStringRam (char position, char *str)
        {
                char ptr = 0;
                SendLCDdata (position, 0);      // send position
                while (str[ptr] != 0)
                        SendLCDdata (str[ptr++], 1);    // send character
        }

        // Display a program memory based string (str) at position

        void DisplayStringPgm (char position, rom char *str)
        {
                char ptr = 0;
                SendLCDdata(position, 0);       // send position
```

```
        while (str[ptr] != 0)
                SendLCDdata (str[ptr++], 1);    // send character
}
// Display time and date at right of Line 2

void DisplayTimeDate (void)              // once per second
{
        char timedatestring[19];
        int a;
        char Mode;
        char ptr = 0;
        int hours = time / 36000;        // get hours
        char amPM = 'A';
        int minutes = (time % 36000) / 600;   // get minutes
        int seconds = ((time % 36000) % 600) / 10;  // get seconds
        int year1 = year;

        Mode = eeRead (timeMode);        // get timeMode from EEPROM

//      display time

        if (Mode == 1)                   // 12-hour mode
        {
                if (hours >= 12)
                {
                        hours -= 12;
                        amPM = 'P';
                }

                if (hours == 0)
                        hours = 12;
        }

        if ((hours / 10) == 0)
                timedatestring[ptr++] = ' ';
        else
                timedatestring[ptr++] = hours / 10 + 0x30;
        timedatestring[ptr++] = hours % 10 + 0x30;
        timedatestring[ptr++] = ':';
        timedatestring[ptr++] = minutes / 10 + 0x30;
        timedatestring[ptr++] = minutes % 10 + 0x30;
        timedatestring[ptr++] = ':';
        timedatestring[ptr++] = seconds / 10 + 0x30;
        timedatestring[ptr++] = seconds % 10 + 0x30;

        if (Mode == 1)
        {
                timedatestring[ptr++] = ' ';
                timedatestring[ptr++] = amPM;
                timedatestring[ptr++] = 'M';
        }
        timedatestring[ptr] = 0;
        DisplayStringRam (0xe0, timedatestring);

// display date

        ptr = strlenpgm (monthName[month]);
        strcpypgm2ram (timedatestring, monthName[month]);
        if (day / 10 == 0)
                timedatestring[ptr] = ' ';
        else
                timedatestring[ptr++] = day / 10 + 0x30;
        timedatestring[ptr++] = day % 10 + 0x30;
        timedatestring[ptr++] = ',';
        timedatestring[ptr++] = ' ';
        for (a = 1000; a > 0; a /= 10)
        {
                timedatestring[ptr++] = year1 / a + 0x30;
                year1 -= year1 / a * a;
        }
```

```
            timedatestring[ptr] = 0;
            DisplayStringRam (0xcc, timedatestring);
}
// Increment the RTC clock
//    do not incrment date
void DoClock (void)                    // Timer 1 interrupt handler
{
        PIR1bits.TMR1IF = 0;           // clear Timer 1 request
        WriteTimer1 (-25000);          // reload count
        time++;                                    // increment time
        if (time == 864000)            // if new day (864000&sim;1/10 sec)
                time = 0;                  // to 0:00:00:0
        if ((time % 10) == 0)
                DisplayTimeDate();
}
// Read a key from the keypad
unsigned char GetKey (void)
{
        int a;
        unsigned char keyCode;
        PORTC = 0x00;   // select all key columns
        do              // wait for release
        {
                while ((PORTB & 0x0F) != 0x0F)
                        ClrWdt();              // wag Spot's tail
                Delay1KTCYx(30);
        }while ((PORTB & 0x0F) != 0x0F);
        do              // wait for press
        {
                while ((PORTB & 0x0F) == 0x0F)
                        ClrWdt();              // wag tail
                Delay1KTCYx(30);
        }while ((PORTB & 0x0F) == 0x0F);
        PORTC = 0xFE;                  // select a leftmost column
        while ((PORTB & 0x0F) == 0x0F)         // while no key is found
        {
                PORTC = (PORTC << 1) | 1;      // get next column
                keyCode += 4;          // add 4 to keycode
        }
        for (a = 1; a != 0; a <<= 1)
        {                                              // find row
                if ((PORTB & a) == 0)
                        break;
                keyCode++;
        }
        return lookupKey[keyCode];             // look up correct key code
}
char GetCode (int ptr)
{
        int a;
        int currentTime;
        int temp = 0;
        int oneHalfBits = oneBits / 2;
        int oneAndOneHalfBits = oneBits + oneBits / 2;
        for (a = 0; a < 6; a++)
        {
                currentTime = cardBuffer[ptr] + cardBuffer[ptr + 1] << 8;
                if (currentTime > oneHalfBits &&
                    currentTime < oneAndOneHalfBits)
                {
                        temp <<= 1;
                        if (cardBuffer[ptr+2] == 1)
                                temp |= 1;
                }
```

```
                 else if (currentTime > oneAndOneHalfBits &&
                         currentTime < (oneAndOneHalfBits + oneBits))
                 {
                         temp <<= 2;
                         if (cardBuffer[ptr+2] == 1)
                              temp |= 3;
                 }
                 else if (currentTime > (oneAndOneHalfBits + oneBits) &&
                         currentTime < (oneAndOneHalfBits + oneBits * 2))
                 {
                         temp <<= 3;
                         if (cardBuffer[ptr+2] == 1)
                              temp |= 7;
                 }
                 else if (currentTime > (oneAndOneHalfBits + oneBits * 2) &&
                         currentTime < (oneAndOneHalfBits + oneBits * 3))
                 {
                         temp <<= 4;
                         if (cardBuffer[ptr+2] == 1)
                              temp |= 0x0F;
                 }
                 ptr += 3;
          }
          for (a = 0; a < 107; a++)
                 if (code128[a] == temp)
                      break;
          return a;
   }

   char GetCodeB (int ptr)
   {
          int a;
          int currentTime;
          int temp = 0;
          int oneHalfBits = oneBits / 2;
          int oneAndOneHalfBits = oneBits + oneBits / 2;
          for (a = 0; a < 6; a++)
          {
                 currentTime = cardBuffer[ptr] + cardBuffer[ptr + 1] << 8;
                 if (currentTime > oneHalfBits &&
                      currentTime < oneAndOneHalfBits)
                 {
                         temp >>= 1;
                         if (cardBuffer[ptr+2] == 1)
                              temp |= 0x8000;
                 }
                 else if (currentTime > oneAndOneHalfBits &&
                         currentTime < (oneAndOneHalfBits + oneBits))
                 {
                         temp >>= 2;
                         if (cardBuffer[ptr+2] == 1)
                              temp |= 0xC000;
                 }
                 else if (currentTime > (oneAndOneHalfBits + oneBits) &&
                         currentTime < (oneAndOneHalfBits + oneBits * 2))
                 {
                         temp >>= 3;
                         if (cardBuffer[ptr+2] == 1)
                              temp |= 0xE000;
                 }
                 else if (currentTime > (oneAndOneHalfBits + oneBits * 2) &&
                         currentTime < (oneAndOneHalfBits + oneBits * 3))
                 {
                         temp >>= 4;
                         if (cardBuffer[ptr+2] == 1)
                              temp |= 0xF000;
                 }
                 ptr += 3;
          }
```

```
        temp >>= 5;
        for (a = 0; a < 107; a++)
             if (code128[a] == temp)
                 break;

        return a;
}

void ProcessCard (void)
{
        int ptr = 0;
        char check;
        char temp;
        int mod103 = 0;
        char count = 0;
        char buffer[10];
        oneBits = cardBuffer[4] << 8 + cardBuffer[3];
        flag = 0;
        if (cardBuffer[2] == 0 && cardBuffer[5] == 1)
        {              // could be good start
             temp = GetCode(ptr);
             if (temp == 103)          // if forward code
             {
                  while (count != eeRead (idLength))
                  {
                       ptr += 18;
                       buffer[count] = GetCode(ptr);
                       mod103 = buffer[count] * (count + 1);
                       buffer[count] += 32;
                       count++;
                  }
                  ptr += 18;
                  if ((mod103 % 103) == GetCode (ptr))         // good check
                  {
                       for (ptr = 0; ptr < eeRead(idLength); ptr++)
                       {
                            dataQueue[dataQueuePtr++] = buffer[ptr];
                            dataQueue[dataQueuePtr++] = time / 36000;
                            dataQueue[dataQueuePtr++] = (time % 36000) /
                                                          600;
                       }
                  }
             }
             else if (temp == 107)     // of backward code
             {
                  ptr += 3;            // skip rest of STOP code
                  check = GetCodeB(ptr);
                  while (count != eeRead(idLength))
                  {
                       ptr += 18;
                       buffer[count] = GetCodeB(ptr);
                       mod103 = buffer[count] * (count + 1);
                       buffer[count] += 32;
                       count++;
                  }
                  ptr += 18;
                  if ((mod103 % 103) == check)                  // good check
                  {
                       for (ptr = 0; ptr < eeRead(idLength); ptr++)
                       {
                            dataQueue[dataQueuePtr++] = buffer[ptr];
                            dataQueue[dataQueuePtr++] = time / 36000;
                            dataQueue[dataQueuePtr++] = (time % 36000) /
                                                          600;
                       }
                  }
             }
        }
}
```

```
void ProcessNetData (char data)
{
      int a, checksum, tempPtr;
      if (state == 0)                 // process all commands
      {
            packetPtr = 0;
            if (data == 0x17)
                  state = 1;
            else if (data == 0x1A)
                  state = 2;
            else if (data == 0x18)
                  state = 3;
            else if (data == 0x1b)
                  state = 4;
      }
      else if (state == 1)            // process command 0x17 (set time & data)
      {
            packetBuffer[packetPtr++] = data;
            if (packetPtr == 8)
            {
                  checksum = 0;
                  for (a = 0; a < 8; a++)
                        checksum ^= packetBuffer[a];
                  if (checksum == 0)                  // good crc
                  {
                        time = packetBuffer[0] * 36000 + packetBuffer[1] *
                              600 + packetBuffer[2] * 10;
                        day = packetBuffer[3];
                        month = packetBuffer[4];
                        year = packetBuffer[5] * 100 + packetBuffer[6];
                  }
                  state = 0;          // look for next command
            }
      }
      else if (state == 2)            // process change ID address
      {
            packetBuffer[packetPtr++] = data;
            if (packetPtr == 5)
            {
                  checksum = 0;
                  for (a = 0; a < 8; a++)
                        checksum ^= packetBuffer[a];
                  if (checksum == 0)                  // good crc
                  {
                        if (eeRead(unitAddressL) == packetBuffer[0] &&
                              eeRead(unitAddressH) == packetBuffer[1])
                        {
                              eeWrite(unitAddressL, packetBuffer[2]);
                              eeWrite(unitAddressH, packetBuffer[3]);
                        }
                  }
                  state = 0;
            }
      }
      else if (state == 3)            // process read request
      {
            packetBuffer[packetPtr++] = data;
            if (packetPtr == 3)
            {
                  checksum = 0;
                  for (a = 0; a < 8; a++)
                        checksum ^= packetBuffer[a];
                  if (checksum == 0)                  // good crc
                  {
                        PORTAbits.RA6 = 1;            // set transmit
                        if (eeRead(unitAddressL) == packetBuffer[0] &&
                              eeRead(unitAddressH) == packetBuffer[1])
```

```
                                {
                                    while (SaveOutQueue(0x06) == 0x100)     //

                                                                        send ACK
                                        ClrWdt();
                                    while (TXSTAbits.TRMT == 0);    // wait for
                                                            completetion
                                    if (dataQueuePtr != 0)
                                    {
                                        checksum = 0;
                                        tempPtr = 0;
                                        while (tempPtr != dataQueuePtr)
                                        {
                                            checksum ^= dataQueue[tempPtr];
                                            while (SaveOutQueue(
                                            dataQueue[tempPtr]) == 0x100)
                                                ClrWdt();
                                            tempPtr++;
                                        }
                                        while (SaveOutQueue(0xFF) == 0x100)
                                            //send EOM
                                            ClrWdt();
                                        while (SaveOutQueue(checksum) == 0x100
                                        )  //send checksum
                                            ClrWdt();
                                        dataQueuePtr = 0;
                                    }
                                }
                                    SaveOutQueue (0X21);              // send NAK
                                PORTAbits.RA6 = 0;                    // set receive
                                state = 0;
                            }
                }
        }
        else if (state == 4)
        {
            packetBuffer[packetPtr++] = data;
            if (packetPtr == 1)
            {
                eeWrite (idLength, packetBuffer[0]);
                state = 1;
            }
        }
}

void SaveBigQueue (char data)
{
    if (data == 0xFE)
        dataQueueBusy = 1;
    else if (data == 0xFF)
    {
        dataQueue[dataQueuePtr++] = time / 36000;
        dataQueue[dataQueuePtr++] = (time % 36000) / 600;
        dataQueueBusy = 0;
    }
    else
        dataQueue[dataQueuePtr++] = data;
}

void ProcessKey (void)
{
    int a;
    char temp = GetKey();
    if (temp == 0 && idNumberCount < eeRead (idLength) &&
        idNumberCount == 0)
    {
        DisplayStringPgm (0x84, mes4);
        idBuffer[idNumberCount] = temp;
```

```
                        SendLCDdata (0x8C + idNumberCount++, 0);
                        SendLCDdata (temp + 0x30, 1);
        }
        else if (temp >= 1 && temp <= 9 && idNumberCount < eeRead(idLength))
        {
                if (idNumberCount == 0)
                        DisplayStringPgm (0x84, mes4);
                if (alpha == 0)
                {
                        idBuffer[idNumberCount] = temp;
                        SendLCDdata (0x8C + idNumberCount++, 0);
                        SendLCDdata (temp + 0x30, 1);
                }
                else
                {
                        idBuffer[idNumberCount++] = alpha * 10 + temp;
                        SendLCDdata (0x8C + idNumberCount++, 0);
                        SendLCDdata (alpha * 10 + temp + 0x40, 1);

                }
        }
        else if (temp == 10)      // alpha select up
        {
                if (alpha != 0 && alpha != 3)
                        alpha++;
                if (alpha == 1)
                        DisplayStringPgm (0xC0, mes2);
                else if (alpha == 2)
                        DisplayStringPgm (0xC0, mes6);
                else if (alpha == 3)
                        DisplayStringPgm (0xC0, mes7);
        }
        else if (temp == 11)      // alpha select down
        {
                if (alpha != 0 && alpha != 1)
                        alpha--;
                if (alpha == 1)
                        DisplayStringPgm (0xC0, mes2);
                else if (alpha == 2)
                        DisplayStringPgm (0xC0, mes6);
                else if (alpha == 3)
                        DisplayStringPgm (0xC0, mes7);
        }
        else if (temp == 12)      // F1 enter key
        {
                if (idNumberCount - 1 == eeRead(idLength))
                {
                        SaveBigQueue (0xFE);
                        for (a = 0; a < idNumberCount; a++)
                        {
                                if (idBuffer[a] >= 10 && idBuffer[a] <= 18)
                                        idBuffer[a] += 0x31;
                                else if (idBuffer[a] >= 19 && idBuffer[a] <= 27)
                                        idBuffer[a] += 0x27;
                                else if (idBuffer[a] >= 28)
                                        idBuffer[a] += 0x37;
                                SaveBigQueue (idBuffer[a]);
                                SaveBigQueue (0xFF);
                        }
                        DisplayStringPgm (0x84, mes1);
                }
                else
                {
                        DisplayStringPgm (0x82, mes2);
                        idNumberCount = 0;
                }
        }
```

```
        else if (temp == 13) // F5 backspace key
        {
                if (idNumberCount != 0)
                {
                        SendLCDdata (0x8C + idNumberCount--, 0);
                        SendLCDdata (' ' , 1);
                }
        }
        else if (temp == 14)
        {

        }
        else if (temp == 15)       // F7 numeric select key
        {
                alpha = 0;
                DisplayStringPgm (0xC0, mes3);
        }
        else if (temp == 16)       // F2 clear key
        {
                DisplayStringPgm (0x84, mes1);
                idNumberCount = 0;
        }
        else if (temp == 17)
        {

        }
        else if (temp == 18)
        {

        }
        else if (temp == 19)       // F8 alpha select
        {
                alpha = 1;
                DisplayStringPgm (0xC0, mes2);
        }
}

// **************** MAIN PROGRAM ********************

void main  (void)
{
        int temp;

// set up port pins

        OSCCON = 0x72;              // selects an 8-MHz internal clock
        ADCON1 = 0x0F;             // ports are digital
        TRISA = 0x00;              // Port A is output
        PORTA = 0x00;              // net is input to ID card reader
        TRISB = 0xFF;              // Port B is input
        TRISC = 0xA0;              // Port C is output except RC7
        PORTC = 0x00;              // select all columns on the keyboard

// check for default reset jumper

        if (PORTCbits.RC5 == 0)  // on hard intitialization
        {                                        // reset defaults
                eeWrite (timeMode, 1);           // set AM/PM mode
                eeWrite (unitAddressL, 0);       // set unit address to 0x0000
                eeWrite (unitAddressH, 0);       // address set to 0x0000
                eeWrite (passWord0, 0);          // password set to 00411
                eeWrite (passWord1, 0);
                eeWrite (passWord2, 4);
                eeWrite (passWord3, 1);
                eeWrite (passWord4, 1);
                eeWrite (idLength, 10);          // default ID length = 10
        }

// set up clock and clock interrupt

        time = 0;                                // initialize time to midnight
        IPR1bits.TMR1IP = 0;                     // Timer 1 is low priority
```

```
            WriteTimer1 (-25000);                    // every 100 ms
            OpenTimer1 (TIMER_INT_ON &
                    T1_16BIT_RW &
                    T1_SOURCE_INT &
                    T1_PS_1_8);

// set up USART and caches

            IPR1bits.TXBIP = 0;                      // select low priority
            IPR1bits.RCIP = 0;                       // TXBIP is errata
            inPo = inPi = outPo = outPi = 0;         // set up queues
            state = 0;                               // set USART state to 0
            dataQueuePtr = dataQueueBusy = 0;
            OpenUSART (USART_TX_INT_OFF &            // USART operates at 9600 Baud
                    USART_RX_INT_OFF &
                    USART_ASYNCH_MODE &
                    USART_EIGHT_BIT &
                    USART_SINGLE_RX &
                    USART_BRGH_HIGH,
                    51);

// set up interrrupt on change for optical card reader

            flag = alpha = 0;           // show no data
            INTCON2bits.RBIP = 1;       // high priority
            INTCONbits.RBIE = 1;        // enable bit change interrupt
            WriteTimer0 (0);            // initialize Timer 0

// start interrupt system

            RCONbits.IPEN = 1;          // IPEN = 1 to enable priority interrupts
            INTCONbits.GIEH = 1;        // enable high-priority interrupt
            INTCONbits.GIEL = 1;        // enable low-priority interrupt

// the only main system program

            InitLCD();                               // initialize LCD
            DisplayStringPgm (0x84, mes1);           // sign on message
            DisplayStringPgm (0xC0, mes3);
            DisplayTimeDate();          // initialize date & time display

            while (1)                   // main loop (this is it)
            {
                ClrWdt();              // pet spot
                if ((PORTB & 0x0F) != 0x0F)           // if any key
                    ProcessKey();
                if (flag != 0)
                    ProcessCard();                   // if card code
                temp = GetInQueue();
                if (temp != 0x100)                   // if not empty
                    ProcessNetData(temp);            // process net data
            }
        }
```

Example System 2

Example 2 is a continuation of Example 1. Example 2 is the power station discussed in Example 1. In order for the ID card reader system to function, it needs to be located in a network. The network discussed here is an RS-422 network, which requires a special card for the PC that hosts the system. To keep the system as simple as possible, a power station that supplies power to all the ID card readers is included. As a sideline of the power station, is its operation to convert from RS-232C on a PC to RS-422 for the network. All that is needed from the host PC is a serial COM port connection, or a USB adapter to drive the system from a USB port on the PC. In either case, no modification to the PC is required. Figure 8-13 shows the schematic diagram of the power station needed for the ID card reader system.

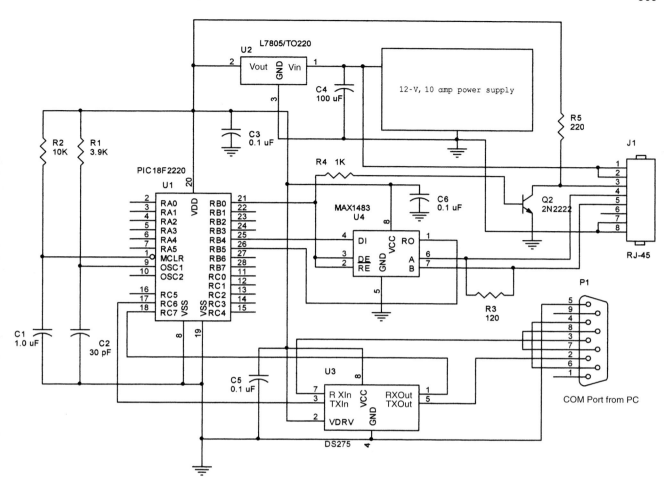

FIGURE 8-13 Power station.

Notice that the power station uses another microcontroller to control the two serial ports: one interfaces to the RS-232C COM port on the host PC, and the other interfaces to the ID card readers on the RS-422 network. The microcontroller also controls turning the communications around on the RS-422 network, which is operated in half duplex mode. The software for the power station is much shorter than the software for the ID card unit because all the power station needs to do is turn the line around or relay information between the PC and the RS-422 network. This data relay is handled by two serial interfaces: the software UART and the hardware USART in the microcontroller. The software UART is by default set to use Port B pins RB4 and RB5 and the hardware USART uses pins RC6 and RC7. Example 8-10 lists the software for the power unit.

EXAMPLE 8-10

```
// Power unit example
//

#include <p18cxxx.h>
#include <sw_uart.h>
```

```c
#include <delays.h>
#include <usart.h>

/* Set configuration bits
 *   - set RC oscillator
 *   - disable watchdog timer
 *   - disable low-voltage programming
 *   - disable brownout reset
 *   - enable master clear
 */

#pragma config OSC = RC
#pragma config WDT = OFF
#pragma config LVP = OFF
#pragma config BOR = OFF
#pragma config MCLRE = ON

// ********* DATA MEMORY VARIABLES ************

#pragma udata

char readQueue[32];
char writeQueue[32];

char readInPtr;
char readOutPtr;
char writeInPtr;
char writeOutPtr;

// **************** FUNCTIONS *****************

void MyHighInt (void);

#pragma interrupt MyHighInt
#pragma code high_vector=0x08

void high_vector (void)
{
      _asm GOTO MyHighInt _endasm
}

#pragma code

int GetWriteQueue (void)
{
      int temp;
      if (writeInPtr == writeOutPtr)
            return 0x100;                   // if empty
      temp = writeQueue[writeOutPtr];        // get data
      writeOutPtr = (writeOutPtr + 1) & 0x1F;
      return temp;
}

int SaveWriteQueue (char data)
{
      if (writeInPtr == ((writeOutPtr + 1) & 0x1F))
            return 0x100;                   // if full
      writeQueue[writeInPtr] = data;
      writeInPtr = (writeInPtr + 1) & 0x1F;
      return 0;
}
int GetReadQueue (void)
{
      int temp;
      if (readInPtr == readOutPtr)
            return 0x100;                   // if empty
      temp = readQueue[readOutPtr];          // get data
      readOutPtr = (readOutPtr + 1) & 0x1F;
      return temp;
}
```

```
int SaveReadQueue (char data)
{
      if (readInPtr == ((readOutPtr + 1) & 0x1F))
            return 0x100;              // if full
      readQueue[readInPtr] = data;
      readInPtr = (readInPtr + 1) & 0x1F;
      return 0;
}

void MyHighInt (void)
{
      int temp;
      if (PIR1bits.RCIF == 1)   // is it a USART receiver?
      {
            PIR1bits.RCIF = 0;         // clear interrupt
            SaveWriteQueue (RCREG);
      }
      else if (PIR1bits.TXIF == 1 && PIE1bits == 1)// is it a USART transmitter?
      {
            PIR1bits.TXIF = 0;         // clear interrupt
            temp = GetReadQueue();     // get queue data
            if (temp == 0x100)
            {
                  while (TXSTAbits.TRMT == 0);
                  PIE1bits.TXIE = 0; // transmitter off
            }
            else
                  TXREG = temp;        // send data
      }
}


//
// ************ TIME DELAYS FOR SOFTWARE UART *****************
//    These delays must be provided to use the software UART in
//    the C18 library
//

void DelayRXHalfBitUART (void)
{
      Delay10TCYx(1);
      Delay1TCY();
      Delay1TCY();
      Delay1TCY();
      Delay1TCY();
      Delay1TCY();
      Delay1TCY();
      Delay1TCY();
      Delay1TCY();
}

void DelayRXBitUART (void)
{
      Delay10TCYx(3);
      Delay1TCY();
      Delay1TCY();
      Delay1TCY();
      Delay1TCY();
      Delay1TCY();
      Delay1TCY();
      Delay1TCY();
      Delay1TCY();
      Delay1TCY();
}

void DelayTXBitUART (void)
{
      Delay10TCYx(4);
      Delay1TCY();
}
```

```
// *************** MAIN PROGRAM ************************

void main (void)
{
        char temp;
        char count = 0xFF;
        ADCON1 = 0x0F;                      // make port pins digital
        TRISC = 0x80;               // program ports
        TRISB = 0x20;
        PORTB = 1;                          // set up transmit to Net

        RCONbits.IPEN = 0;          // only high-priority interrupt

// set up USART and queue

        IPR1bits.RCIP = 0;                  // TXBIP is errata

        readInPtr = readOutPtr = writeInPtr = writeOutPtr = 0;

        OpenUSART (USART_TX_INT_OFF &       //USART operates at 9600 baud
                        USART_RX_INT_OFF &
                        USART_ASYNCH_MODE &
                        USART_EIGHT_BIT &
                        USART_SINGLE_RX &
                        USART_BRGH_HIGH,
                        12);

// Open software UART

        OpenUART();
        while(1)
        {
                if (PORTBbits.RB0 == 1)
                {
                        INTCONbits.GIE = 1;       // enable high-priority interrupt
                        temp = GetWriteQueue();        // get from COM port
                        if (temp != 100)
                        {
                                INTCONbits.GIE = 0;     // disable
                                WriteUART (temp);       // write to Net
                                if (count != 0xFF)
                                {
                                        count--;
                                        if (count == 0)
                                        {
                                                count = 0xFF;
                                                PORTBbits.RB0 = 0;      // turn line around
                                        }
                                }
                                if (temp == 0x18)
                                        count = 3;
                        }
                }
                else
                {
                        INTCONbits.GIE = 0;
                        temp = ReadUART();
                        SaveReadQueue (temp);
                        PIE1bits.TXIE = 1;              // transmitter on
                        INTCONbits.GIE = 1;
                        if (temp == 0xFF || temp == 6 || temp == 0x21)
                                PORTBbits.RB0 = 1;                      // turn line around
                }
        }
}
```

The final piece of this sample system has no software or microcontroller. The expansion module has two RS-422 transceivers controlled by the signal on pin 3 of the network bus. This circuit is illustrated in Figure 8-14. This repeater is a low-cost network-bus-powered device that uses the LM7805 voltage regulator to provide 5 V for the two MAX1483 bus drivers.

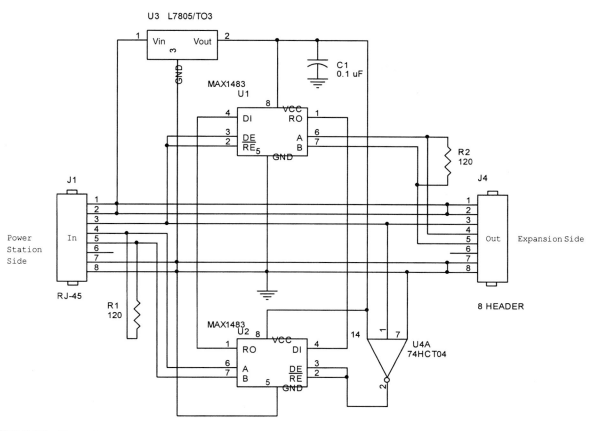

FIGURE 8-14 Repeater.

8-5   SUMMARY

1. The interrupt structure of the microcontroller can include high- and low-priority interrupts or only high-priority interrupts as determined by the IPEN bit in the reset control register (RCON).

2. Interrupt priorities become active only when both types of interrupts occur simultaneously, when the high-priority interrupt takes precedence.

3. Individual interrupts are controlled by three bits: IP, IE, and IF. The IP bit (interrupt priority) selects low (0) or high (1) priority, the IE bit (interrupt enable) selects the enabled (1) or disabled state (0) of the interrupt, and the IF bit (interrupt flag) indicates whether the interrupt is requested (1) or not (0).

4. The interrupt service procedure is installed in a C18 program by (a) defining a prototype for the interrupt, (b) defining a #pragma for interrupt (high) or interrupt low (low), (c) placing a GOTO assembly language instruction in the vector location, and (d) installing the interrupt service procedure at the vectored address for the interrupt.

5. The interrupt service procedure must detect which interrupting device caused the interrupt by testing the IF for the interrupt, then it clear the IF flag bit before ending.

6. The USART is a hardware component in the microcontroller that converts between serial and parallel data in either the synchronous or asynchronous format. As with other programmable devices within the PIC, this device contains interrupts. When the receiver receives a

serial datum, the receiver causes an interrupt and when the transmitter is ready to transmit another datum, the transmitter causes an interrupt. Both the receiver and transmitter each have IP, IE, and IF control bits. The baud is programmable for the USART.

7. A special type of interrupt called interrupt on change detects whether any Port B bit RB4–RB7 changed states. When a state change is detected on these four pins, an interrupt on change occurs. This is extremely useful, in conjunction with a timer, for detecting and measuring the width of an input pulse or pulses because many types of data encode information in their pulse widths.

8. Many systems use multiple interrupts for operation, for example the ID card system discussed in this chapter. The ID card reader system uses an interrupt for a real-time clock, an interrupt on change interrupt for an optical bar code reader, and two interrupts for a USART that communicates through a network. All four of these interrupts are coordinated through the two priority interrupt vectors in the microcontroller.

9. The code 128 bar code is explained in this chapter because it is one of the more common bar codes in use for inventory control and other applications. The process of scanning a code and decoding the scanned information is presented as an example of using the change on interrupt.

10. A small half-duplex network is used for communications between an embedded system and a PC. This interface uses both the RS-232C and the RS-422 (RS-485) protocols for its interface. Also detailed is the conversion between an RS-232C connection and an RS-422 network. RS-422 is chosen here because of the 4,000 feet cabling lengths allowed by the standard.

8-6   QUESTIONS AND PROBLEMS

1. The interrupt vector used by the high-priority interrupt is at memory location _____.
2. The interrupt vector used by the low-priority interrupt is at memory location _____.
3. If priority interrupts are disabled with the IPEN bit, which vector is used for all interrupts?
4. Where is the IPEN bit located?
5. Describe the purposes of the IP, IE, and IF control bits associated with an interrupt.
6. If the high-priority interrupt first services the interrupt on change interrupt followed by the USART receiver interrupt, which interrupt has the highest priority?
7. Describe the significance of clearing the IF bit from within the interrupt service procedure.
8. What is accomplished by the INTCONbits.GIEH = 1; statement in a program?
9. What is accomplished by the IPR1bits.TMR1IP = 0; statement in a program?
10. What is accomplished by the INTCON3bits.INT1IE = 1; statement in a program?
11. Select a statement that places the Timer 2 interrupt input at the low-priority level.
12. Select an instruction that enables the Timer 2 interrupt.
13. What is the RCON register?
14. In which register is the interrupt flag bit (IF) for the interrupt on change interrupt located?
15. Why doesn't the RETFIE instruction appear in an interrupt service procedure?
16. What does an RETFIE instruction accomplish?
17. Why does the program in Example 8-3 use two timers, and what is each of these timers used to accomplish in the program?
18. What is the purpose of the low-priority interrupt service procedure in Example 8-3?
19. Describe how the time is set in the program in Example 8-3.
20. Example 8-3 uses multiplexed displays; how often is a display position switched in this program?

21. Explain how the temperature is read in Example 8-3.
22. What is asynchronous serial data?
23. What is the purpose of the start and stop bits in an asynchronously coded data stream?
24. Most modern asynchronous serial data contains a start bit, _____ data bits, and _____ stop bits.
25. What is the baud rate of a serial transmission?
26. What device is used to convert between asynchronous serial data and parallel data?
27. A USART causes two types of interrupts, what are they?
28. If given the choice, which of the two types of interrupts caused by the USART would be assigned a higher interrupt priority and why?
29. Develop a short sequence of C language statements that select low-priority interrupts for both the transmitter and receiver of the USART and these two interrupts. Do not use the OpenUSART function; instead directly address the control registers.
30. How is the USART programmed for a baud rate of 1200? Find the value programmed into the baud rate register if the clock frequency is 4 MHz.
31. Why are queues often used with I/O devices?
32. How large are the queues used in Example 8-5?
33. Why and when is the transmitter interrupt turned off in the high priority interrupt service procedure in Example 8-5?
34. The receiver interrupt is handled in the high-priority interrupt service procedure in Example 8-5; what is accomplished when it is serviced?
35. Example 8-6 depicts software for Visual C++ on a PC. What is the purpose of the DCB (data control block) in the WriteComPort function?
36. Is it possible to modify the WriteComPort function in Example 8-6 so that it writes more than one byte of data? If so, explain how.
37. Explain how Example 8-7 measures the width of the DCC input pulses.
38. What is a DCC preamble and how long is it?
39. What is a CV for the DCC system?
40. Explain how data EEPROM is accessed in Example 8-7.
41. Which interrupts are used in the program in Example 8-7?
42. Explain how the wait4Bit function operates in the program in Example 8-7.
43. Explain the getByte function in Example 8-7.
44. What is the purpose of Timer 0 in Example 8-7?
45. Explain how the reconfigure pushbutton is used in Example 8-7 and what it accomplishes.
46. Why is the linker script file shown in Example 8-8 and what do the changes made to it accomplish?
47. What is the RS-422 and why is it chosen for the sample system in Figure 8-9?
48. What is the MAX1483?
49. What is a CAT5 cable?
50. What is an RJ-45 connector?
51. What is the code 128 bar code?
52. Is it possible to use the code 128 bar code to place your name on your personal items?
53. Determine and draw the code 128 bar code pattern for the string W4A.
54. Search the Internet and list at least three other bar codes that are used with optical bar code readers.
55. Why is the optical bar code much more common than magnetic codes?
56. What is meant by modulo 103?
57. In the program of Example 8-9 what information is stored in the data EEPROM?
58. Explain where and how the date and time are displayed on the LCD display in Example 8-9.
59. What interrupts are used in Example 8-9, which has the highest priority, and which has the lowest priority?

60. Explain what the main loop in Example 8-9 accomplishes.
61. What is the purpose of the idNumberCount variable in the ProcessKey function in Example 8-9?
62. How does the program in Example 8-9 determine that a command has arrived from the network?
63. What additional commands are suggested to improve the utility of the ID card reader in Figure 8-9?
64. The power station in Figure 8-13 uses a DS275. What is the purpose of this integrated circuit?
65. What is the purpose of the L7805 in Figure 8-13?
66. What is the purpose of the transistor (2N2222) in Figure 8-13?
67. The software in Example 8-10 uses both a USART and a UART; what is the purpose of these two devices?
68. What are the functions of the power station in Figure 8-13?
69. How does the repeater in Figure 8-14 function?

CHAPTER 9

Controlling Systems

Many systems that contain microcontrollers are control systems that control some process or processes. This chapter presents a design methodology for process control systems and several applications. Examples of machines that use process control for operation are the clothes washer, the dishwasher, an assembly line at a factory, traffic light systems, and so forth. These devices perform the same process each time they are activated and there is a method to designing process control systems.

Upon completion of this chapter you will be able to:

1. Explain how to formulate a process for controlling a system.
2. Demonstrate a few simple control systems.
3. Develop a traffic light control system.
4. Develop a system that is similar to the Litter Maid cat litter box.
5. Control the HVAC system in a home.

9-1   FORMULATING THE CONTROL SYSTEM

To make things easier when designing a control system, a simple time/event chart organizes the task so the system can be constructed and software can be written. Suppose that a control system, using a microcontroller, needs to be used in a dishwasher. To limit the amount of effort, this will be a basic system.

   The system has only a few components to control:

1. A fill valve that allows water to fill the dishwasher tub. This typically has a mechanical float valve to stop the water flow when the tub is full.
2. The pump motor squirts water through the dishwasher to rinse and wash the dishes. It also pumps the water out of the dishwasher through a drain hose.
3. A diverter valve that is controlled to select whether the pump motor washes the dishes or pumps water from the dishwasher into the drain.
4. A heating element that dries the dishes, and in more deluxe models heats the water to sterilize the dishes.
5. A soap dispenser solenoid to open a soap dish and dump soap into the wash water.
6. Some dishwashers also have a small fan that blows air across the dishes to speed drying.

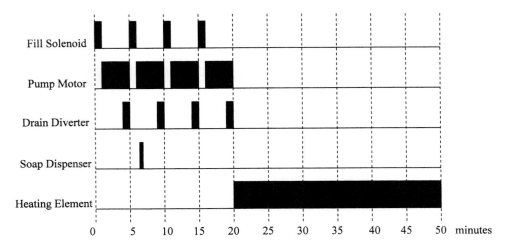

FIGURE 9-1   Process control chart for a simple dishwasher.

Figure 9-1 illustrates a time/event or process chart for designing the software for the dishwasher. The chart lists all of the devices that need to be controlled along with the times at which events occur. This chart is the plan for the system and plots time versus each event. Once this chart is developed, both the software and system are much easier to construct.

From this chart the following information is gleaned:

1. A 1-minute time delay is needed to fill the tub.
2. A 4-minute time delay is needed to run the pump motor.
3. A 30-minute time delay is needed for the heating element.
4. A short time delay is needed to pop open the soap dispenser (200 ms).

This unit has only one pushbutton on to start the wash cycle. The only other switch is a door interlock that senses when the door is opened to suspend the wash cycle. This machine does a prewash, followed by a wash with soap, followed by two rinse cycles, followed by a drying cycle. The pump motor, heating element, drain diverter, and soap dispenser solenoid are all outputs controlled by solid state relays. This system has two inputs, the start pushbutton and the door interlock. The microcontroller needed to manage this system, which has two inputs and five outputs, can be any PIC18 microcontroller, so the least expensive is used, the PIC18F1220. The system schematic appears in Figure 9-2. The interlock pushbutton produces a logic one when the door is closed and a logic zero when it is opened.

The software uses an interrupt for the door interlock switch to interrupt the operation of the dishwasher whenever the door is opened during a cycle. Otherwise the software is fairly straightforward because of the process flow chart in Figure 9-1. Example 9-1 lists the complete program for controlling this machine. The only complicated part is the door interlock switch, which must stop the machine at any time the door is opened. When the door is closed, the dishwasher must continue from the point of interruption. Here an interrupt is used to sense the door open condition. At any time that the door is opened, the INT0 interrupts input on RB0 receives a negative edge. This interrupts the interrupt service procedure at the MyHighInt function. In this function, the current state of the machine is saved, future INT0 interrupts are enabled, and the function waits for the door to close. When the door is closed (notice the debouncing on the door interlock switch), the state of the machine is restored so the operation can continue at exactly the same point as when the door was opened.

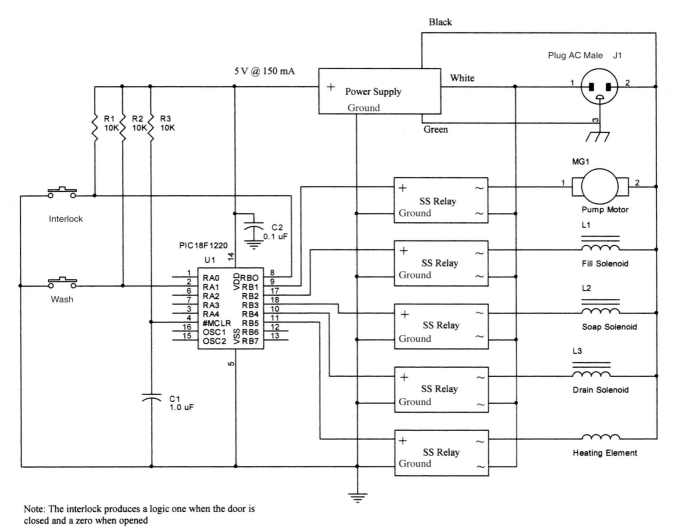

Note: The interlock produces a logic one when the door is
closed and a zero when opened

FIGURE 9-2 PIC18F1220-controlled dishwasher.

EXAMPLE 9-1

```
/*
 * Dishwasher program
 */

#include <p18cxxx.h>
#include <delays.h>

/* Set configuration bits
 *  - set internal oscillator
 *  - disable watchdog timer
 *  - disable low-voltage programming
 *  - disable brownout reset
 *  - enable master clear
 */

#pragma config OSC = INTIO2
#pragma config WDT = OFF
#pragma config LVP = OFF
#pragma config BOR = OFF
#pragma config MCLRE = ON
```

```c
void MyHighInt (void);                // prototypes for interrupts

#pragma interrupt MyHighInt           // MyHighInt is an interrupt
#pragma code high_vector=0x08         // high_vector is the vector at 0x08

void high_vector (void)
{
      _asm GOTO MyHighInt _endasm
}

#pragma code

// Interrupt service procedure
//     occurs anytime door is opened
//

void MyHighInt (void)
{
      char a;
      a = PORTB;                      // save the current state
      PORTB = 0;                      // stop everything
      do
      {
            while (PORTBbits.RB0 == 0);   // while door is open
            Delay100TCYx(10);         // 16 ms debounce
      }while (PORTBbits.RB0 == 0);
      INTCONbits.INT0IF = 0;          // clear INT0IF flag
      PORTB = a;                      // door definitely closed
}

//
// The system clock is 250 KHz or 4 us
//          an instruction cycle is 16 us
//

void Wait200ms (void)
{
      Delay100TCYx(125);             // 16 * 125 * 100 = 200 ms
}

void WaitMinute (char howmany)
{
      int a, b;                      // 300 * .2 sec = 1 minute
      for (a = 0; a < howmany; a++)
            for (b = 0; b < 300; b++)
                  Wait200ms();
}

void DoCycles (void)
{
      char a;
      for (a = 0; a < 4; a++)         // repeat 4 times
      {
            PORTBbits.RB2 = 1;        // fill valve on
            WaitMinute(1);           // wait 1 minute
            PORTBbits.RB2 = 0;        // fill valve off
            PORTBbits.RB1 = 1;        // pump motor on
            if (a == 1)              // wash with soap
            {
                  PORTBbits.RB3 = 1;
                  Wait200ms();
                  PORTBbits.RB3 = 0;
            }
            WaitMinute(3);           // wait 3 minutes
            PORTBbits.RB4 = 1;        // drain solenoid on
            WaitMinute(1);           // wait 1 minute
            PORTB = 0;               // all off
      }
      PORTBbits.RB5 = 1;             // heater on
      WaitMinute(30);               // dry for 30 minutes
      PORTB = 0;                    // all off
}
```

```
void main (void)
{
        OSCCON = 0x23;                // 250-KHz internal clock
        ADCON1 = 0x0F;                // all inputs are digital
        TRISA = 2;                    // Port A bit 1 = input
        TRISB = 1;                    // Port B bit 0 = input
        PORTB = 0;                    // turn off system

        RCONbits.IPEN = 0;            // only high-priority interrupt on
        INTCON2bits.INTEDG0 = 0;      // make INT0 negative edge triggered
        INTCONbits.INT0IE = 1;        // enable INT0
        INTCONbits.GIE = 1;           // enable interrupts

        while (1)
        {
                while (PORTAbits.RA1 == 1);   // Wait until the wash cycle
                DoCycles();
        }
}
```

Other cycles can be added to the system without much additional software. For example, a heavy-duty wash cycle for pots can be added by adding a second soap dispenser to RB6 and a locking pushbutton switch to RA0. The switch could be labeled Normal/Pots. In the pots position, a second wash cycle with the second soap dispenser would scrub pots with a little more soap. Many additional features could be added to the dishwasher for a minimal cost and more money could be charged when the product is sold. Each new feature requires an additional process chart to illustrate its operation and make software development easier.

9-2 SAMPLE SYSTEMS

This section develops a few sequential control systems using the technique presented in Section 9-1.

Example 1

Another sample system is a traffic control system. The system developed here is for a single intersection, but has inputs so it can be expanded into a network of intersections that communicate with each other and rules that can change the basic cycles of each intersection under certain conditions.

Figure 9-3 shows the basic intersection process chart. Directions are indicated as NS for north/south and EW for east/west. The times indicate when a particular direction is activated and

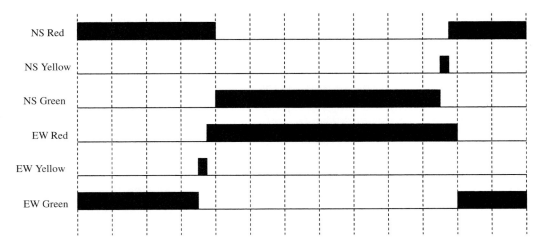

FIGURE 9-3 Process chart for a traffic light and one cycle.

are programmable. The chart follows the standard practice of red in both directions for a short time to prevent collisions. Times are not assigned to the chart because they are programmable.

The system uses the PIC18F4220 microcontroller, a telephone-style keypad, a 2 × 16 LCD display for programming purposes, inputs for traffic sensors and walk pushbuttons for pedestrians, outputs to control the six lamps, and a serial interface for cases where more than a single intersection is interconnected. The interconnection requires some interface standard that can handle fairly long distances. The RS-422 or RS-485 interface handles 4000 feet, which is nearly a mile, so either standard is suitable. The circuitry for the system appears in two schematics: Figure 9-4 shows the main circuit board and Figure 9-5 (which can be on the same board) shows the driver, walk pushbutton, and trip plate board. The system uses the same 4-pin connectors for all connections to the lights, sensors, and walk pushbuttons. This system is designed to control the new LED-style traffic lights and walk lights. The newer lamps require from 13 watts to 25 watts, and the drivers provide enough current for two sets of lights. Red and green lights require 13 watts and yellow lights require 25 watts.

The software for the system is illustrated in Example 9-2. The system has most of the necessary features, but only the commands to synchronize the traffic lights connected to the network. The program seems fairly long yet it compiles into just 2605 words of program memory, so almost 2,000 locations remain for additional tasks.

EXAMPLE 9-2

```
/*
 * Traffic light controller
 */

#include <p18cxxx.h>
#include <delays.h>
#include <timers.h>
#include <usart.h>

/* Set configuration bits
 *   - set internal oscillator
 *   - disable watchdog timer
 *   - disable low-voltage programming
 *   - disable brownout reset
 *   - enable master clear
 */

#pragma config OSC = INTIO2
#pragma config WDT = OFF
#pragma config LVP = OFF
#pragma config BOR = OFF
#pragma config MCLRE = ON

// ********* DATA MEMORY VARIABLES ************

void MyHighInt (void);            // prototypes for interrupt
void MyLowInt (void);             // service procedures

#pragma interrupt MyHighInt       // MyHighInt is an interrupt
#pragma code high_vector=0x08     // high_vector is the vector at 0x08

void high_vector (void)
{
      _asm GOTO MyHighInt _endasm
}

#pragma interruptlow MyLowInt     // MyLowInt is an interrupt
#pragma code low_vector=0x18      // low vector is at 0x0018

void low_vector (void)            // low-prioity vector
{
      _asm GOTO MyLowInt _endasm       // goto low software
}
```

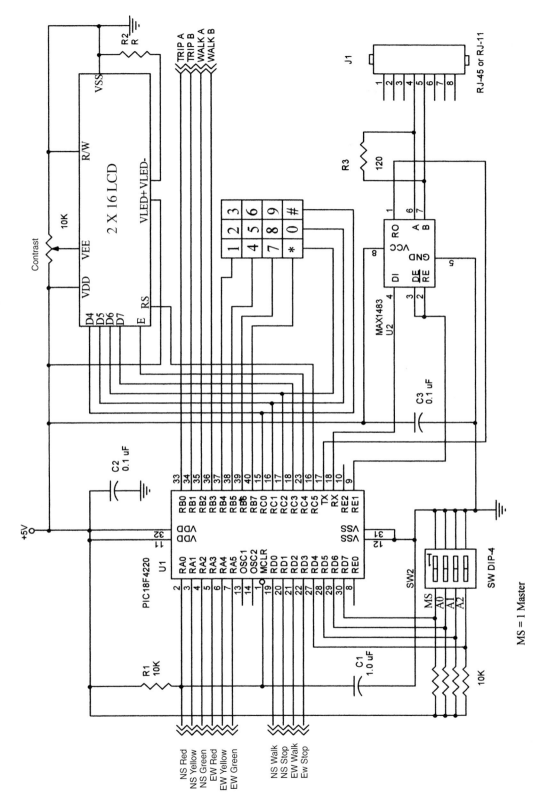

FIGURE 9-4 Traffic light controller main board.

MS = 1 Master

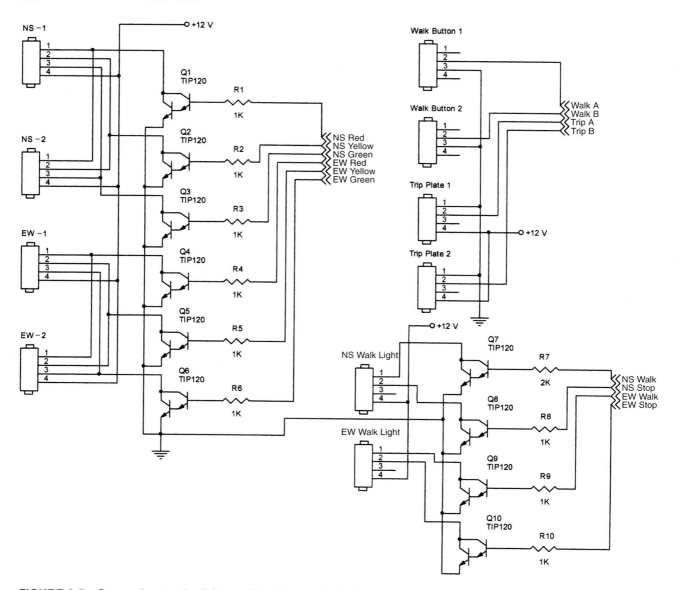

FIGURE 9-5 Connections to the lights, walk buttons, and trip plates.

```
//
// data EEPROM variable
//

#define BlinkStartHours 0
#define BlinkStartMinutes 1
#define BlinkStopHours 2
#define BlinkStopMinutes 3
#define EWYellowTime 4
#define NSGreenTime 5
#define NSYellowTime 6
#define EWGreenTime 7

//
// Program memory data
//
```

```
rom near char lookupKey[] =
{
      1, 4, 7, 10,          // left column
      2, 5, 8, 0,           // middle column
      3, 6, 9, 11           // right column
};

rom near char str1[] = "Ready to control";
rom near char str2[] = "  the traffic!  ";
rom near char str3[] = "                ";
rom near char str4[] = " Enter the time ";
rom near char str5[] = "Set blink start ";
rom near char str6[] = "Set blink stop  ";
rom near char str7[] = " Set EW yellow  ";
rom near char str8[] = "  Set EW green  ";
rom near char str9[] = " Set NS yellow  ";
rom near char str10[] = "  Set NS green  ";
rom near char str11[] = "The current time";
rom near char str12[] = "   Blink start  ";
rom near char str13[] = "   BlinK stop   ";
rom near char str14[] = "    EW yellow   ";
rom near char str15[] = "    EW green    ";
rom near char str16[] = "    NS yellow   ";
rom near char str17[] = "    NS green    ";
rom near char str18[] = "Enter a command ";
rom near char str19[] = "     1 -- 8     ";

//
// data memory variables
//

#pragma udata

char transmitQueue[16];
char receiverQueue[16];
char inTrans;
char outTrans;
char inRecv;
char outRecv;
char seconds;
char minutes;
char hours;
char NSGreen;
char EWGreen;
char state;
char buffer[4];
int ptr;

#pragma code

// Interrupt service procedure
//     occurs each second
//

// Reads a data EEPROM location from address

char eeRead (char address)
{
      EECON1bits.EEPGD = 0;
      EEADR = address;
      EECON1bits.RD = 1;
      return EEDATA;
}

// Write a data EEPROM location at address with data

void eeWrite (char address, char data)
{
      INTCONbits.GIEH = 0;
      INTCONbits.GIEL = 0;
      EECON1bits.EEPGD = 0;
```

```
            EECON1bits.WREN = 1;
            EEADR = address;
            EEDATA = data;
            EECON2 = 0x55;
            EECON2 = 0xAA;
            EECON1bits.WR = 1;
            while (PIR2bits.EEIF == 0);
            PIR2bits.EEIF = 0;
            EECON1bits.WREN = 0;
            INTCONbits.GIEH = 1;
            INTCONbits.GIEL = 1;
    }

    void sendLCDdata (char data, char rs)
    {
            PORTC = data >> 4;        // send left nibble
            PORTCbits.RC5 = rs;       // control RS
            PORTCbits.RC4 = 1;        // pulse E
            PORTCbits.RC4 = 0;
            Delay1TCY();         // delay 48 us
            Delay1TCY();
            Delay1TCY();
            PORTC = data & 0x0F;      // send right nibble
            PORTCbits.RC5 = rs;       // control RS
            PORTCbits.RC4 = 1;        // pulse E
            PORTCbits.RC4 = 0;
            Delay1TCY();         // delay 48 us
            Delay1TCY();
            Delay1TCY();
            PORTC = 0;
    }

    void initLCD (void)
    {
            int a;
            Delay1KTCYx(2);           // wait 32 ms
            for (a = 0; a < 3; a++)
            {
                    sendLCDdata (0x20, 0);   // send 0x20
                    Delay1KTCYx (1);         // wait 16 ms
            }

            sendLCDdata (0x28, 0);   // send 0x28
            sendLCDdata (0x01, 0);   // send 0x01
            Delay1KTCYx (1);         // wait 16 ms
            sendLCDdata (0x0C, 0);   // send 0x0C
            sendLCDdata (0x06, 0);   // send 0x06
    }

    // Display a program-memory-based string (str) at position
    //     Line 1 is at positions 0x80 through 0xA7
    //     Line 2 is ay positions 0xC0 through 0xE7

    // Display a program-memory-based string (str) at position

    void DisplayStringPgm (char position, rom char *str)
    {
            char ptr = 0;
            sendLCDdata(position, 0);        // send position
            while (str[ptr] != 0)
                    sendLCDdata(str[ptr++], 1);   // send character
    }

    void getNumb (char count, char temp)
    {
            if (ptr < count)
            {
                    if (temp >= 0 && temp <= 9)
                    {
                            sendLCDdata (0xC6 + ptr, 0);
```

```
                    sendLCDdata (temp + 0x30, 1);
                    buffer[ptr++] = temp;
            }
            else if (temp == 10)   // backspace (*)
            {
                    if (ptr != 0)
                        ptr--;
                    sendLCDdata (0xC6 + ptr, 0);
                    sendLCDdata (' ', 1);
            }
        }
    }
}

void Disp3 (char temp)
{
        sendLCDdata (temp / 100 + 0x30, 1);
        temp %= 100;
        sendLCDdata (temp / 10 + 0x30, 1);
        sendLCDdata (temp % 10 + 0x30, 1);
}

void Disp4 (char first, char second)
{
        sendLCDdata (first / 10 + 0x30, 1);
        sendLCDdata (first % 10 + 0x30, 1);
        sendLCDdata (':', 1);
        sendLCDdata (second / 10 + 0x30, 1);
        sendLCDdata (second % 10 + 0x30, 1);
}

int outTransQueue (void)
{
        int temp;
        if (inTrans == outTrans)
                return 0x100;           // if empty
        temp = transmitQueue[outTrans];   // get data
        outTrans = (outTrans + 1) & 0x0F;
        return temp;
}

int inTransQueue (char data)
{
        if (outTrans == ((inTrans + 1) & 0x0F))
                return 0x100;           // if full
        transmitQueue[inTrans] = data;
        inTrans = (inTrans + 1) & 0x0F;
        PIE1bits.TXIE = 1;              // transmitter on
        return 0;
}

int outRecvQueue (void)
{
        int temp;
        if (inRecv == outRecv)
                return 0x100;           // if empty
        temp = receiverQueue[outRecv];   // get data
        outRecv = (outRecv + 1) & 0x0F;
        return temp;
}

int inRecvQueue (char data)
{
        if (outRecv == ((inRecv + 1) & 0x0F))
                return 0x100;           // if full
        receiverQueue[inRecv] = data;
        inRecv = (inRecv + 1) & 0x0F;
        return 0;
}
```

```
void MyHighInt (void)
{
      char a, temp;
      if (INTCONbits.INT0IF == 1)
      {
            INTCONbits.INT0IF = 0;      // clear INT0IF flag
            if (NSGreen > 10)
                  NSGreen = 10;
      }
      else if (INTCON3bits.INT1IF == 1)
      {
            INTCON3bits.INT1IF = 0;
            if (EWGreen > 10)
                  EWGreen = 10;
      }
      else if (INTCONbits.RBIF == 1)
      {
            Delay1KTCYx(1);
            temp = 0;
            if (PORTB & 0xF0 != 0xF0)               //good key
            {
                  PORTC = 0xEF;                      // select a leftmost column
                  while ((PORTB & 0xF0) == 0xF0)   // no key is found
                  {
                        PORTC = (PORTC << 1) | 1;        // get next column
                        temp += 4;                        // add rows to keycode
                  }
                  for (a = 0x10; a != 0; a <<= 1)
                  {                                       // find row
                        if ((PORTB & a) == 0)
                              break;
                        temp++;
                  }
                  temp = lookupKey[temp];
                  switch (state)
                  {
                  case 0:
                        {
                              if (temp == 10)     // any command
                              DisplayStringPgm (0x80, str18);
                              DisplayStringPgm (0xC0, str19);
                              state = 1;
                              break;
                        }
                  case 1:
                        {
                              if (temp >= 1 && temp <= 8)
                              {
                                    ptr = 0;
                                    DisplayStringPgm (0x80, str4 +
                                          (state - 1)* 17);
                                    DisplayStringPgm (0xC0, str3);
                                    state = temp + 1;
                              }
                              else
                              {
                                    DisplayStringPgm (0x80, str1);
                                    DisplayStringPgm (0xC0, str2);
                                    state = 0;
                              }
                              break;
                        }
                  case 2:                    // "Set Current Time" (*1 command)
                        {
                              getNumb (4, temp);
                              if (temp == 11)      // # is enter
                              {
                                    hours = buffer[0] * 10 + buffer[1];
                                    minutes = buffer[2] * 10 + buffer[3];
```

```
                                    DisplayStringPgm (0x80, str1);
                                    DisplayStringPgm (0xC0, str2);
                                    state = 0;
                            }
                            break;
                    }
        case 3:                     // "Set Blink Start" (*2 command)
                    {
                    getNumb (4, temp);
                    if (temp == 11)
                    {
                            eeWrite (BlinkStartHours, buffer[0] *
                                    10 + buffer[1]);
                            eeWrite(BlinkStartMinutes, buffer[2] *
                                    10 + buffer[3]);
                            DisplayStringPgm (0x80, str1);
                            DisplayStringPgm (0xC0, str2);
                            state = 0;
                    }
                    break;
                    }
        case 4:                     // "Set Blink Stop" (*3 command)
                    {
                    getNumb (4, temp);
                    if (temp == 11)
                    {
                            eeWrite (BlinkStopHours, buffer[0] *
                                    10 + buffer[1]);
                            eeWrite (BlinkStopMinutes, buffer[2] *
                                    10 + buffer[3]);
                            DisplayStringPgm (0x80, str1);
                            DisplayStringPgm (0xC0, str2);
                            state = 0;
                    }
                    break;
                    }
        case 5:                     // "Set EW Yellow" (*4 command)
                    {
                    getNumb (3, temp);
                    if (temp == 11)
                    {
                            if (buffer[0] * 100 + buffer[1] *
                                    10 + buffer[2] > 255)
                            {
                                    buffer[0] = 2;
                                    buffer[1] = buffer[2] = 5;
                            }
                            eeWrite (EWYellowTime, buffer[0] *
                                    100 + buffer[1] *
                                    10 + buffer[1]);
                            DisplayStringPgm (0x80, str1);
                            DisplayStringPgm (0xC0, str2);
                            state = 0;
                    }
                    break;
                    }
        case 6:                     // "Set EW Green" (*5 command)
                    {
                    getNumb (3, temp);
                    if (temp == 11)
                    {
                            if (buffer[0] * 100 + buffer[1] *
                                    10 + buffer[2] > 255)
                            {
                                    buffer[0] = 2;
                                    buffer[1] = buffer[2] = 5;
                            }
```

```
                              eeWrite (EWGreenTime, buffer[0] *
                                     100 + buffer[1] *
                                     10 + buffer[1]);
                              DisplayStringPgm (0x80, str1);
                              DisplayStringPgm (0xC0, str2);
                              state = 0;
                          }
                          break;
              }
case 7:               // "Set NS Yellow" (*6 command)
          {
              getNumb (3, temp);
              if (temp == 11)
              {
                  if (buffer[0] * 100 + buffer[1]
                          * 10 + buffer[2] > 255)
                  {
                          buffer[0] = 2;
                          buffer[1] = buffer[2] = 5;
                  }
                  eeWrite (NSYellowTime, buffer[0] *
                         100 + buffer[1] *
                         10 + buffer[1]);
                  DisplayStringPgm (0x80, str1);
                  DisplayStringPgm (0xC0, str2);
                  state = 0;
              }
              break;
          }
case 8:               // "Set NS Green" (*7 command)
          {
              getNumb (3, temp);
              if (temp == 11)
              {
                  if (buffer[0] * 100 + buffer[1] *
                          10 + buffer[2] > 255)
                  {
                          buffer[0] = 2;
                          buffer[1] = buffer[2] = 5;
                  }
                  eeWrite (NSGreenTime, buffer[0] *
                         100 + buffer[1] *
                         10 + buffer[1]);
                  DisplayStringPgm (0x80, str1);
                  DisplayStringPgm (0xC0, str2);
                  state = 0;
              }
              break;
          }
case 9:               // "Display all" (*8 command)
          {
              DisplayStringPgm (0x80, str11 + ptr * 17);
              DisplayStringPgm (0xC0, str3);
              sendLCDdata (0xC5, 0);
              if (ptr == 0)
                  Disp4(hours, minutes);
              else if (ptr == 1)
                  Disp4(eeRead(BlinkStartHours),
                          eeRead(BlinkStartMinutes));
              else if (ptr == 2)
                  Disp4 (eeRead(BlinkStopHours),
                          eeRead(BlinkStopMinutes));
              else if (ptr == 3)
                  Disp3 (eeRead(EWYellowTime));
              else if (ptr == 4)
                  Disp3 (eeRead(EWGreenTime));
              else if (ptr == 5)
                  Disp3 (eeRead(NSYellowTime));
```

```
                                         else if (ptr == 6)
                                                 Disp3 (eeRead(NSGreenTime));
                                         ptr++;
                                         if (ptr == 7)
                                                 state = 10;
                                         break;
                            }
                     case 10:
                             {
                                     DisplayStringPgm (0x80, str1);
                                     DisplayStringPgm (0xC0, str2);
                                     state = 0;

                                     break;
                             }
                     }
              }
              PORTC = 0;
              temp = PORTB;                  // must read PORTB to clear change
              INTCONbits.RBIF = 0;           // clear interrupt
       }
}

void MyLowInt (void)
{
       int temp;
       if (PIR1bits.TMR1IF == 1)        // do clock
       {                                            // as 24-hour clock
              PIR1bits.TMR1IF = 0;
              WriteTimer1 (-15625);
              seconds++;
              if (seconds == 60)
              {
                     seconds = 0;
                     minutes++;
                     if (minutes == 60)
                     {
                            minutes = 0;
                            hours++;
                            if (hours == 24)
                                   hours = 0;
                     }
              }
       }
       else if (PIR1bits.RCIF == 1)
       {
              PIR1bits.RCIF = 0;        // clear interrupt
              inRecvQueue (RCREG);      // store received data in queue
       }
       else if (PIR1bits.TXIF == 1)
       {
              PIR1bits.TXIF = 0;        // clear interrupt
              temp = outTransQueue();
              if (temp == 0x100)
                     PIE1bits.TXIE = 0; // transmitter off
              else
                     TXREG = temp;      // send data
       }
}

char HalfSecond (char count)
{
       int a, temp;
       for (a = 0; a < count; a++)
       {
              temp = outRecvQueue();
              if (temp != 0x100 &&
                     PORTDbits.RD7 == 0  &&    // if a slave
                     ((PORTD & 0xF0) >> 4) == (temp & 7) &&
```

```
                                    temp & 0xF7 == 0)
                                        return 1;
                            Delay1KTCYx (31);
                            Delay10TCYx (25);
                    }
            return 0;
    }

    void DoLights (void)
    {
            if ((hours >= eeRead(BlinkStartHours) &&    // blink
                    hours <= eeRead(BlinkStopHours)) && (
                    minutes >= eeRead(BlinkStartMinutes) &&
                    minutes <= eeRead(BlinkStopMinutes)))
            {
                            PORTAbits.RA1 = 1;          // NS yellow
                            PORTAbits.RA5 = 1;          // EW green
                            PORTDbits.RD1 = 1;          // NS stop
                            PORTDbits.RD3 = 1;          // EW stop
                            HalfSecond(1);                  // wait 1/2 second
                            PORTAbits.RA1 = 0;
                            PORTAbits.RA5 = 0;
                            PORTDbits.RD1 = 0;
                            PORTDbits.RD3 = 0;
                            HalfSecond(1);
            }
            else                                        // normal cycle
            {
                    do                                  // do synchronize
                    {
                    do
                    {
                    do
                    {
                    do
                    {
                    do
                    {
                    do
                    {       NSGreen = eeRead (NSGreenTime);
                            EWGreen = eeRead (EWGreenTime);
                            PORTAbits.RA4 = 1;          // EW yellow
                            PORTAbits.RA5 = 0;
                    }
                    while (HalfSecond (eeRead(EWYellowTime)) == 1);
                            PORTAbits.RA4 = 0;
                            PORTAbits.RA3 = 1;          // EW red
                            PORTAbits.RA0 = 0;
                            PORTAbits.RA2 = 1;          // NS green
                            PORTDbits.RD2 = 1;          // EW walk
                            PORTDbits.RD1 = 1;          // NS stop
                    }
                    while (HalfSecond (eeRead(NSGreen/2)) == 1);
                            PORTDbits.RD2 = 0;
                            PORTDbits.RD3 = 1;
                    }
                    while (HalfSecond (eeRead(NSGreen/2)) == 1);
                            PORTAbits.RA1 = 1;          // NS yellow
                            PORTAbits.RA2 = 0;
                    }
                    while (HalfSecond (eeRead(NSYellowTime)) == 1);
                            PORTAbits.RA1 = 0;
                            PORTAbits.RA0 = 1;          // NS red
                            PORTAbits.RA5 = 1;          // EW green
                            PORTAbits.RA3 = 0;
                            PORTDbits.RD3 = 1;          // EW stop
                            PORTDbits.RD0 = 1;          // NS walk
                    }
```

```
                    while (HalfSecond (eeRead(EWGreen/2)) == 1);
                            PORTDbits.RD0 = 0;
                            PORTDbits.RD1 = 1;
                    }
                    while (HalfSecond (eeRead(EWGreen/2)) == 1);
                    }
            }
    }

void main (void)
{
        int a;
        OSCCON = 0x22;                  // 250-KHz internal clock
        ADCON1 = 0x0F;                  // all inputs are digital
        TRISA = 0;                      // Port A all outputs
        TRISB = 0xFF;           // Port B all inputs
        TRISC = 0x80;           // Port C all outputs except RC7
        TRISD = 0xF0;           // Port D 0-3 outputs, 4-7 inputs
        TRISE = 0;                      // Port E all outputs

        if (PORTDbits.RD7 == 0)
                PORTEbits.RE1 = 0;      // slave
        else
                PORTEbits.RE1 = 1;      // master

        PORTA = 0;                      // all lights off
        PORTC = 0;
        PORTD = 0;
        state = 0;

        initLCD();
        DisplayStringPgm (0x80, str1);
        DisplayStringPgm (0xC0, str2);

        INTCON2bits.RBPU = 0;           // Port B pull-ups on

        OpenTimer1 (TIMER_INT_ON &
                T1_8BIT_RW &
                T1_SOURCE_INT &
                T1_PS_1_4);

        WriteTimer1 (-15625);           // every second

        IPR1bits.TMR1IP = 0;            // Timer 1 low priority
        PIE1bits.TMR1IE = 1;            // Timer 1 interrupt on

        INTCON3bits.INT1IP = 1;         // INT1 is high priority
        INTCON3bits.INT1IE = 1;         // enable INT1

        INTCON2bits.INTEDG0 = 0;        // make INT0 negative edge-triggered
        INTCONbits.INT0IE = 1;          // enable INT0

        INTCON2bits.RBIP = 0;           // low priority
        INTCONbits.RBIE = 1;            // enable bit change interrupt

        IPR1bits.TXIP = 0;              // select low priority
        IPR1bits.RCIP = 0;              //     for USART

        OpenUSART (USART_TX_INT_ON &               // USART operates at 1200 baud
                        USART_RX_INT_ON &
                        USART_ASYNCH_MODE &
                        USART_EIGHT_BIT &
                        USART_CONT_RX &
                        USART_BRGH_HIGH,
                        3);

        inRecv = outRecv = inTrans = outTrans = 0;

        RCONbits.IPEN = 1;              // both priority interrupts on
        INTCONbits.GIEH = 1;            // enable interrupts
        INTCONbits.GIEL = 1;
```

```
// Synchronize all slaves
//
for (a = 0; a < 8; a++)
{
        while (inTransQueue(a) == 0x100);   // send address
                        // all slaves are sent 0000 0aaa to synchronixe
                        // them so they all cycle in sync
                        // the 0000 0 is the sync command
}
while (1)
{
        DoLights();
}
}
```

Example 2

This system is the popular litter box similar to one sold by Litter Maid. Because no one enjoys emptying the cat litter box, all the user does is place clumping litter in the box about once a week, and the machine automatically scoops any cat waste into a disposable plastic box that is emptied by the owner once per week. No smell, no fuss.

The system is fairly simple. It contains sensor switches that indicate when a moveable comb is at each end of the box, a motor to move the comb, a beeper for problems, and a pair of photo-sensors to determine if the cat is in the litter box. If the cat steps into the litter box to do its business, the comb scoops the refuse into the plastic collection box after the cat has left the box for at least 10 minutes. (We certainly do not want to traumatize the cat!) This is a simple system; a system schematic appears in Figure 9-6. A small reed relay (should last for about

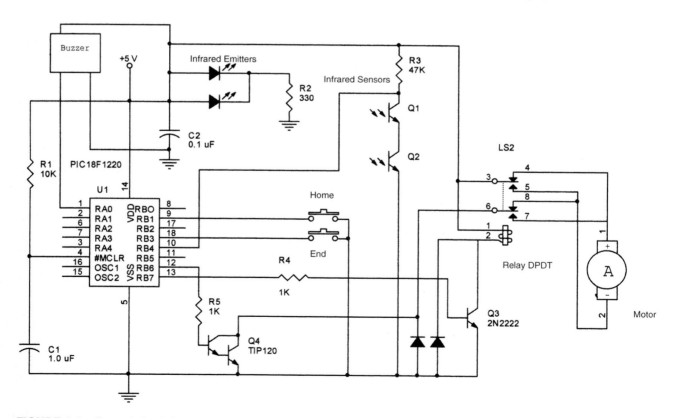

FIGURE 9-6 Control circuit for a litter box.

40,000 cycles) is used to switch directions of the motor because the cost is considerably lower than using a bidirectional motor. The direction can also be changed mechanically. The photo-sensors are also placed in series so that when the cat blocks one or both sensors, the microcontroller receives a logic one on RB4. This series connection performs the same function as an AND gate. The buzzer in the schematic is a Mallory alarm, which is compatible without any additional components.

The program for this machine is shorter than other programs in this chapter. Example 9-3 illustrates the program for controlling the cat litter box. One feature is that if the scoop gets hung up on a large piece of fecal matter, it rocks back and forth trying to push it to the end where the collection box is located. If after 30 attempts at rocking it does not make it to the end, it sets off the alarm and homes the comb where it waits for the cat to do its business again. This is one of the few systems that does not require an interrupt for its operation.

EXAMPLE 9-3

```c
/*
 * Cat litter box
 */

#include <p18cxxx.h>
#include <delays.h>

/* Set configuration bits
 *   - set RC oscillator
 *   - disable watchdog timer
 *   - disable low-voltage programming
 *   - disable brownout reset
 *   - enable master clear
 */

#pragma config OSC = INTIO2
#pragma config WDT = ON
#pragma config WDTPS = 256      // one minute
#pragma config LVP = OFF
#pragma config BOR = OFF
#pragma config MCLRE = ON

#pragma code
void main (void)
{
        unsigned int count, count1;
        OSCCON = 0x22;              // 250-KHz internal clock
        ADCON1 = 0x7F;              // all inputs are digital
        TRISA = 0;
        TRISB = 0x3F;               // program Port B
        PORTB = PORTA = 0;          // motor stop

        while (1)
        {
                while (PORTBbits.RB1 == 1)
                {
                        PORTB = 0x80;       // home comb
                        ClrWdt();
                }
                PORTB = PORTA = 0;          // stop motor & alarm
                while (PORTBbits.RB4 == 0)       // wait for cat
                        ClrWdt();
                while (PORTBbits.RB4 == 1)        // wait for cat to leave
                        ClrWdt();
                count = 0;
                do
                {
                        count++;
                        ClrWdt();
```

```
                Delay1KTCYx(3);           // 12 ms
                if (PORTBbits.RB4 == 0)
                        count = 0;
        }
        while (count != 50000);           // 10 minutes
                count = count1 = 0;
        while (PORTBbits.RB3 == 1)
        {                                 // run comb to End
                count++;
                PORTB = 0xC0;
                Delay1KTCYx(3);
                ClrWdt();
                if (count == 417)         // 5 seconds (must be stuck)
                {
                        PORTB = 0x80;
                        count = 0;
                        count1++;
                        Delay1KTCYx(255); // give it a bump
                }
                if (count1 == 30)         // set off alarm after
                {
                        PORTA = 1;        // 30 bumps
                        break;            // then give up
                }
        }
        PORTB = 0;                        // stop motor
    }
}
```

Example 3

Another example of a simple control system that can be programmed into a microcontroller is the climate control thermostat in a home. The system needs to monitor the room temperature and control the 24-VAC control voltage that operates the furnace and/or air conditioning. Suppose that the system needs some intelligence so it can switch automatically from heating to air conditioning, a feature not found on most home units. To accomplish this requires some type of outdoor temperature sensor. To design this system we need to determine exactly what is being controlled and what is being sensed. The inputs to the system are a control panel with a keypad for programming, two temperature sensors (one for the thermostat and one for the outside temperature), and two small reed relays for turning on the furnace or air conditioner.

The user programming interface consists of a way to enter the time of day and sets of on and off times so the system can be programmed to turn off (or, as an option, lower or raise the temperature) when the user is away for extended periods of time. It is also advantageous to have a weekday and a weekend schedule (one that allows the user to set the days-off schedule because not everyone has Saturday and Sunday off).

The control panel might appear with the programming door open as displayed in Figure 9-7. The number of pushbuttons has been minimized, but there are enough so that programming is intuitive for just about all users. The entire system is built using only three pushbuttons (up, down, and enter) for ease of operation and an LCD display that illuminates when the door covering the pushbuttons is opened. (Some thermostats are located in dark corners and at night are difficult to see for programming). The LCD is a 2 × 24 display panel that is needed for programming and temperature display. During normal operation, the indoor temperature is displayed on line 1 and the outdoor temperature is displayed on line 2. If a user does not thread the outdoor sensor wires to the outside, and does not connect it, then the heating/air conditioning function will not work and no outdoor temperature will be displayed.

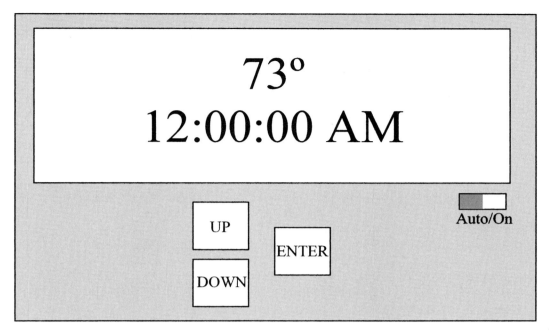

FIGURE 9-7 Control panel for a home HVAC system.

Figure 9-8 illustrates the system's complete schematic diagram. The connections to the heating system are standard in most cases, but some systems have different color codes. The common connection is on one side of a 24-VAC transformer and the other three signal lines carry the voltage back to the heating system to relays that control the furnace, air conditioner, and fan motor. The 24-VAC control voltage is almost always used in heating, ventilation, and air conditioning (HVAC) systems.

The fan motor is controlled by a switch that selects the automatic mode (the fan runs only when heat or air conditioning is needed) or the ON position, which runs the blower motor continuously. The switch in the schematic is shown in the automatic mode.

The software for the system is listed in Example 9-4. Much can be added to this program. It has only a night setback temperature and one setback temperature for the workdays. Additional setback points can be added as needed by adding additional states to the GotKey function. Both the indoor and outdoor sensors must be connected to the system in order for the program to work. The program uses 2841 words of the 4096 available memory locations, so many more features can easily be added without running out of memory space.

EXAMPLE 9-4

```
/*
 *  Home heating/air conditioning thermostat
 */

#include <p18cxxx.h>
#include <delays.h>
#include <timers.h>
#include <string.h>

/* Set configuration bits
 *  - set RC oscillator
 *  - disable watchdog timer
 *  - disable low-voltage programming
 *  - disable brownout reset
```

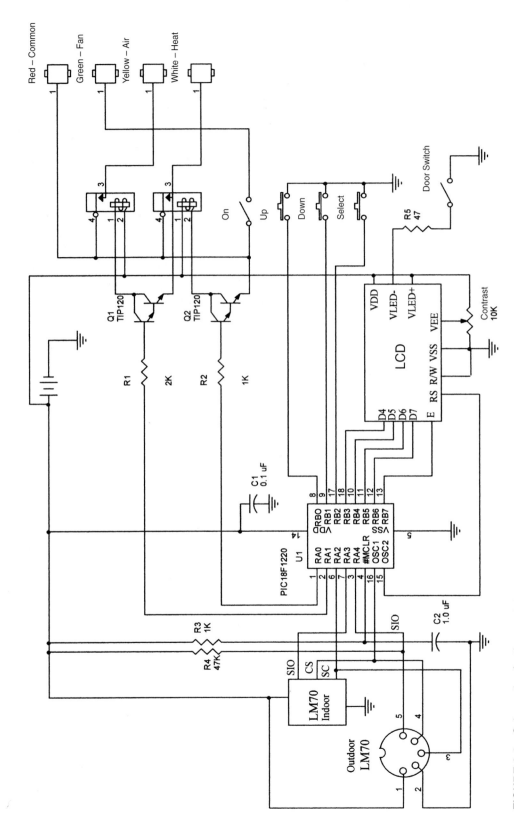

FIGURE 9-8 Schematic of a thermostat.

```
 *   - enable master clear
 */
#pragma config OSC = INTIO2
#pragma config WDT = ON
#pragma config WDTPS = 256        // one second
#pragma config LVP = OFF
#pragma config BOR = OFF
#pragma config MCLRE = ON

//***************** INTERRUPT VECTORS *********************

void MyHighInt (void);           // prototype for interrupt

#pragma interrupt MyHighInt      // MyHighInt is an interrupt
#pragma code high_vector=0x08    // high_vector is the vector at 0x08

void high_vector (void)
{
      _asm GOTO MyHighInt _endasm
}

//***************** DATA MEMORY VARIABLES ********************

short long time;
char hours;
char minutes;
char day;
char displayTimeFlag;
char indoor;
char outdoor;
char setPoint;
char timeOut;
char state;

//******************* PROGRAM MEMORY VARIABLES *****************

char near rom days[][10] = {
      "Sunday     ",
      "Monday     ",
      "Tuesday    ",
      "Wednesday",
      "Thursday   ",
      "Friday     ",
      "Saturday   "
};

char near rom str1[]  = "Indoor                 ";
char near rom str2[]  = "Outdoor                ";
char near rom str3[]  = "    Set time of day    ";
char near rom str4[]  = "                       ";
char near rom str5[]  = "    Set day of week    ";
char near rom str6[]  = " Is a work day (M-F)?  ";
char near rom str7[]  = "   Up = yes, Down = no ";
char near rom str8[]  = "Is                     ";
char near rom str9[]  = " a work day?           ";   // some sample cases
char near rom str10[] = "  Normal temperature   ";
char near rom str11[] = "  Setback temerature   ";
char near rom str12[] = "  Work day ON 1 time   ";
char near rom str13[] = "  Work day OFF 1 time  ";
char near rom str14[] = "    Night OFF time     ";
char near rom str15[] = "   Morning ON time     ";

//**************** DATA EEPROM ********************************

#define temperature 0        // thermostat temp setting
#define workday 1            // work day (M-F) = 1
#define day0 2              // work day = 1, else = 0
#define day1 3
#define day2 4
#define day3 5
```

```
#define day4 6
#define day5 7
#define day6 8
#define normaltemp 9
#define setbacktemp 10
#define workdayOn1h 11
#define workdayOn1m 12
#define workdayOff1h 13
#define workdayOff1m 14
#define nightOffh 15
#define nightOffm 16
#define mornOnh 17
#define mornOnm 18

//********************* FUNCTIONS *****************************
#pragma code

// Read a data EEPROM location from address

char eeRead (char address)
{
     EECON1bits.EEPGD = 0;
     EEADR = address;
     EECON1bits.RD = 1;
     return EEDATA;
}

// Write a data EEPROM location at address with data

void eeWrite (char address, char data)
{
     INTCONbits.GIEH = 0;
     EECON1bits.EEPGD = 0;
     EECON1bits.WREN = 1;
     EEADR = address;
     EEDATA = data;
     EECON2 = 0x55;
     EECON2 = 0xAA;
     EECON1bits.WR = 1;
     while (PIR2bits.EEIF == 0);
     PIR2bits.EEIF = 0;
     EECON1bits.WREN = 0;
     INTCONbits.GIEH = 1;
}

void sendNib (char data, char rs)
{
     PORTB = data;              // send nibble
     PORTAbits.OSC2 = rs;       // control RS
     PORTBbits.RB7 = 1;         // pulse E
     PORTBbits.RB7 = 0;
     Delay1TCY();               // delay 64 us
     Delay1TCY();
}

void sendLCDdata (char data, char rs)
{
     sendNib (data >> 1, rs);
     sendNib (data << 3, rs);
}

void initLCD (void)
{
     int a;
     Delay1KTCYx(1);            // wait 32 ms
     for (a = 0; a < 3; a++)
     {
          sendLCDdata (0x20, 0);   // send 0x20
          Delay100TCYx (2);        // wait 6.4 ms
     }

     sendLCDdata (0x28, 0);    // send 0x28
```

```
        sendLCDdata (0x01, 0);     // send 0x01
        Delay100TCYx (1);          // wait 3.2 ms
        sendLCDdata (0x0C, 0);     // send 0x0C
        sendLCDdata (0x06, 0);     // send 0x06
}

// Display a program-memory-based string (str) at position
//     Line 1 is at positions 0x80 through 0x97
//     Line 2 is at positions 0xC0 through 0xD7

// Display a program-memory-based string (str) at position
void DisplayStringPgm (char position, rom char *str)
{
        char ptr = 0;
        sendLCDdata (position, 0);            // send position
        while (str[ptr] != 0)
                sendLCDdata(str[ptr++], 1);   // send character
}

void GetTemp (void)
{
        char a;
        char b, c;
        PORTAbits.OSC1 = 0;                   // CS = 0
        for (a = 0; a < 9; a++)
        {
                b <<= 1;
                c <<= 1;
                PORTAbits.RA2 = 1;            // SC = 1
                b |= PORTAbits.RA3;
                c |= PORTAbits.RA4;
                PORTAbits.RA2 = 0;            // SC = 0
        }
        for (a = 0; a < 7; a++)
        {
                PORTAbits.RA2 = 1;            // SC = 1
                PORTAbits.RA2 = 0;            // SC = 0
        }
        PORTAbits.OSC1 = 1;                   // CS = 1
        if (c == 0xff)
                c = 0x80;
        indoor = b;
        outdoor = b;
}

void GetTime (short long *temp)
{
        day = *temp / 86400;
        hours = (*temp % 86400) / 3600;
        minutes = ((*temp % 86400) % 3600) / 60;
}

void PutTime (char place)
{
        char temp;
        sendLCDdata (place, 0);
        temp = hours;
        if (hours >= 12)
                hours -= 12;
        if (hours == 0)
                hours = 12;
        if (hours < 10)
                sendLCDdata (' ', 1);
        else
                sendLCDdata (hours / 10 + 0x30, 1);
        sendLCDdata (hours % 10 + 0x30, 1);
        sendLCDdata (':', 1);
        sendLCDdata (minutes / 10 + 0x30, 1);
        sendLCDdata (minutes % 10 + 0x30, 1);
        sendLCDdata (' ', 1);
```

```
        if (temp > 11)
                sendLCDdata ('P', 1);
        else
                sendLCDdata ('A', 1);
}

void PutTemp (char where, char temp)
{
        sendLCDdata (where, 0);
        if (temp < 0)
        {
                temp = -temp;
                sendLCDdata ('-', 1);
        }
        if (temp >= 100)
        {
                sendLCDdata (temp / 100 + 0x30, 1);
                temp -= 100;
        }
        sendLCDdata (temp / 10 + 0x30, 1);
        sendLCDdata (temp % 10 + 0x30, 1);
        sendLCDdata (0xDF, 1);
}

void DisplayTimeDate (void)
{
        GetTemp ();
        GetTime (&time);
        DisplayStringPgm (0x80, str1);
        PutTime (0xD0);
        PutTemp (0x87, indoor);
        if (outdoor != -128)
        {
                DisplayStringPgm (0xC0, str2);
                PutTemp (0xC8, outdoor);
        }
        else
                DisplayStringPgm (0xC0, str4);
}

void DoThermostat (void)
{
        GetTime (&time);
        if (eeRead(nightOffh) == hours &&
                eeRead (nightOffm) == minutes)
                setPoint = eeRead(setbacktemp);
        else if (eeRead(mornOnh) == hours &&
                    eeRead(mornOnm) == minutes)
                setPoint = eeRead (normaltemp);
        else if (eeRead(day0 + day) == 1 &&
                    eeRead(workdayOn1h) == hours &&
                    eeRead(workdayOn1m) == minutes)
                setPoint = eeRead(setbacktemp);
        else if (eeRead(day0 + day) == 1 &&
                    eeRead (workdayOff1h) == hours &&
                    eeRead (workdayOff1m) == minutes)
                setPoint = eeRead (normaltemp);
        else
                setPoint = eeRead (normaltemp);
        if (outdoor < 65)
        {
                if (indoor < setPoint)
                {
                        PORTAbits.RA0 = 1;      // heat on
                        PORTAbits.RA1 = 0;
                }
                else
                {
                        PORTAbits.RA0 = 0;      // heat off
```

```
                                        PORTAbits.RA1 = 0;
                    }
            }
            else
            {
                    if (indoor > setPoint)
                    {
                            PORTAbits.RA0 = 0;
                            PORTAbits.RA1 = 1;        // air on
                    }
                    else
                    {
                            PORTAbits.RA0 = 0;        // air off
                            PORTAbits.RA1 = 0;
                    }
            }
    }
    void MyHighInt (void)
    {
            if (PIR1bits.TMR1IF == 1)                // do clock
            {                                        // as 24-hour clock
                    PIR1bits.TMR1IF = 0;
                    WriteTimer1 (-31250);
                    time++;
                    if (timeOut != 0)
                            timeOut--;
                    if (time == 604800)
                            time = 0;
                    if (displayTimeFlag == 1 && time % 60 == 0)
                    {
                            DisplayTimeDate();
                            DoThermostat();
                    }
            }
    }
    char GetKey (void)
    {
            do
            {
                    while ((PORTB & 7) != 7)
                    {
                            ClrWdt();
                            Delay100TCYx(5);
                            if (timeOut == 0)
                                    return 7;
                    }
            }
            while ((PORTB & 7) != 7);
            do
            {
                    while ((PORTB & 7) == 7)
                    {
                            ClrWdt();
                            Delay100TCYx(5);
                            if (timeOut == 0)
                                    return 7;
                    }
            }
            while ((PORTB & 7) == 7);
            return 0;
    }
    void GotKey (void)
    {
            int a;
            short long b;
            Delay100TCYx(5);                // 16 ms
```

```c
        if ((PORTB & 7) != 3)
                return;                          // ignore all but select
        displayTimeFlag = 0;
        DisplayStringPgm (0x80, str3);
        DisplayStringPgm (0xC0, str4);
        timeOut = 5;
        while (GetKey() == 0)
        {
                timeOut = 5;
                switch (state)
                {
                        case 0:
                        {
                                GetTime (&time);
                                PutTime (0xC8);
                                if (PORTBbits.RB0 == 0)
                                {
                                        time++;
                                        if (time > 604800)
                                                time = 0;
                                }
                                else if (PORTBbits.RB1 == 0)
                                {
                                        time--;
                                        if (time < 0)
                                                time = 604799;
                                }
                                else if (PORTBbits.RB2 == 0)
                                {
                                        state = 1;
                                        DisplayStringPgm (0x80, str5);
                                        DisplayStringPgm (0xC0, str4);
                                        a = 0;
                                }
                                break;
                        }
                        case 1:
                        {
                                a &= 7;
                                DisplayStringPgm (0xC9, days[a]);
                                if (PORTBbits.RB0 == 0 )
                                        a++;
                                else if (PORTBbits.RB1 == 0)
                                        a--;
                                else if (PORTBbits.RB2 == 0)
                                {
                                        time = time % 86400 + a * 86400;
                                        state = 2;
                                        DisplayStringPgm (0x80, str6);
                                        DisplayStringPgm (0xC0, str7);
                                }
                                break;
                        }
                        case 2:
                        {
                                if (PORTBbits.RB0 == 0)
                                {
                                        eeWrite (workday, 1);
                                        state = 4;
                                        DisplayStringPgm (0x80, str10);
                                        DisplayStringPgm (0xC0, str4);
                                        a = indoor;
                                }
                                else if (PORTBbits.RB1 == 0)
                                {
                                        eeWrite (workday, 0);
                                        state = 3;
                                        a = 0;
```

```
                                    DisplayStringPgm (0x80, str8);
                                    DisplayStringPgm (0xC0, str7);
                            }
                            break;
                    }
                    case 3:
                    {
                            DisplayStringPgm (0x83, days[a]);
                            DisplayStringPgm (0x83 + strlenpgm(days[a]),
                                            str9);
                            if (PORTBbits.RB0 == 0)
                                    eeWrite (day0 + a, 1);
                            else if (PORTBbits.RB1 == 0)
                                    eeWrite (day0 + a, 1);
                            else
                                    a--;
                            a++;
                            if (a == 8)
                            {
                                    state = 4;
                                    DisplayStringPgm (0x80, str10);
                                    DisplayStringPgm (0xC0, str4);
                                    a = indoor;
                            }
                            break;
                    }
                    case 4:
                    {
                            PutTemp (0xCA, a);
                            if (PORTBbits.RB0 == 0)
                            {
                                    a++;
                                    if (a >= 100)
                                            a--;
                            }
                            else if (PORTBbits.RB1 == 0)
                            {
                                    a--;
                                    if (a <= 49)
                                            a++;
                            }
                            else if (PORTBbits.RB2 == 0)
                            {
                                    eeWrite (normaltemp, a);
                                    state = 5;
                                    DisplayStringPgm (0x80, str11);
                                    DisplayStringPgm (0xC0, str4);
                            }
                            break;
                    }
                    case 5:
                    {
                            PutTemp (0xCA, a);
                            if (PORTBbits.RB0 == 0)
                            {
                                    a++;
                                    if (a >= 100)
                                            a--;
                            }
                            else if (PORTBbits.RB1 == 0)
                            {
                                    a--;
                                    if (a <= 49)
                                            a++;
                            }
                            else if (PORTBbits.RB2 == 0)
                            {
                                    eeWrite (setbacktemp, a);
```

```
                    state = 6;
                    DisplayStringPgm (0x80, str12);
                    DisplayStringPgm (0xC0, str4);
                    b = 0;
              }
              break;
        }
        case 6:
        {
              GetTime (&b);
              PutTime (0xC8);
              if (PORTBbits.RB0 == 0)
              {
                    time++;
                    if (time > 604800)
                          time = 0;
              }
              else if (PORTBbits.RB1 == 0)
              {
                    time--;
                    if (time < 0)
                          time = 604799;
              }
              else if (PORTBbits.RB2 == 0)
              {
                    state = 7;
                    DisplayStringPgm (0x80, str13);
                    DisplayStringPgm (0xC0, str4);
                    eeWrite (workdayOn1h, hours);
                    eeWrite (workdayOn1m, minutes);
              }
              break;
        }
        case 7:
        {
              GetTime (&b);
              PutTime (0xC8);
              if (PORTBbits.RB0 == 0)
              {
                    time++;
                    if (time > 604800)
                          time = 0;
              }
              else if (PORTBbits.RB1 == 0)
              {
                    time--;
                    if (time < 0)
                          time = 604799;
              }
              else if (PORTBbits.RB2 == 0)
              {
                    state = 8;
                    DisplayStringPgm (0x80, str14);
                    DisplayStringPgm (0xC0, str4);
                    eeWrite (workdayOff1h, hours);
                    eeWrite (workdayOff1m, minutes);
              }
              break;
        }
        case 9:
        {
              GetTime (&b);
              PutTime (0xC8);
              if (PORTBbits.RB0 == 0)
              {
                    time++;
                    if (time > 604800)
                          time = 0;
              }
```

```
                            else if (PORTBbits.RB1 == 0)
                            {
                                    time--;
                                    if (time < 0)
                                            time = 604799;
                            }
                            else if (PORTBbits.RB2 == 0)
                            {
                                    state = 9;
                                    DisplayStringPgm (0x80, str15);
                                    DisplayStringPgm (0xC0, str4);
                                    eeWrite (mornOnh, hours);
                                    eeWrite (mornOnm, minutes);
                                    displayTimeFlag = 1;
                                    DisplayTimeDate ();
                                    return;
                            }
                            break;
                    }
            }
    }
    displayTimeFlag = 1;
    DisplayTimeDate ();
}

//*********************** MAIN STARTUP CODE *************************

void main (void)
{
    OSCCON = 0x02;              // 32 us internal clock
    ADCON1 = 0x0F;             // all inputs are digital
    TRISA = 0x18;              // program Port A
    TRISB = 0x07;              // program Port B
    PORTA = 0x80;
    PORTB = 7;

    initLCD();
    displayTimeFlag = 1;
    DisplayTimeDate ();

    INTCON2bits.RBPU = 0;      // Port B pull-ups on

    OpenTimer1 (TIMER_INT_ON &
            T1_8BIT_RW &
            T1_SOURCE_INT &
            T1_PS_1_1);

    WriteTimer1 (-31250 );     // each second
    INTCONbits.GIEH = 1;       // enable interrupts

    while (1)                  // stays here until a key
    {
            ClrWdt();
            state = 0;
            if ((PORTB & 7) != 7)
                    GotKey();
    }
}
```

9-3   SUMMARY

1. The process chart helps specify the design of a control system by plotting the times that events are to occur.
2. In the sample dishwasher system, an interrupt is used to suspend the operation of the machine.

3. In the traffic light example, a serial RS-422 interface is used to communicate with the traffic light controller so that the operations are synchronized.

4. The traffic light controller uses a variety of input, including: walk pushbuttons, traffic trip plates to control the follow of traffic, and the newer LED traffic lamps as indicators.

5. To keep costs and part counts low, the kitty litter box example uses a relay to control the direction of the motor instead of a more expensive bidirectional motor driver.

6. The HVAC system uses the PIC18 microcontroller to control both heating and air conditioning in a home HVAC system.

9-4   QUESTIONS AND PROBLEMS

1. List the inputs and outputs of the dishwasher highlighted in this chapter.
2. What is the purpose of the interrupt in the dishwasher system?
3. When the door of the dishwasher is opened in the middle of a wash, what happens to the time delays?
4. Modify the process control chart in Figure 9-1 so that the dishwasher has a sanitize setting selected by another switch connected to RA2. This toggle switch is connected to RA2 to select the sanitize cycle when the switch is a logic one and deselects it when a logic zero. To accomplish this, two charts are normally used, one to show the nonsanitize cycle and one to show the sanitize cycle.
5. Modify the software in Example 9-1 so that the switch and sanitize cycle functions properly in the system.
6. Given a washing machine that has two cycles—one for gentle, warm-water washing and one for normal, hot-water washing—develop a process control chart that describes the machine. The control elements are the pump/agitator motor, the fill valve for hot water, the fill valve for cold water, a high-speed gear for the spin cycle controlled by a solenoid, and a drain solenoid for emptying the water from the tub. There is also a door interlock that stops the machine, but only in the spin cycle. The wash cycle should include a wash with spin after washing, and a final rinse and spin.
7. Develop the schematic diagram using a PIC18 of your choice for the washing machine.
8. Write the software to control the washing machine designed in the last two questions.
9. Suppose that a system is needed to control the door locks on an automobile. When the vehicle reaches 15 miles per hour, the doors lock. A new twist is that when the vehicle is stopped and the ignition is turned off, the doors must unlock. If the vehicle is still in motion when the ignition is turned off, the doors will not unlock until a complete stop is achieved. The only control element in the system is the small motor used to unlock the doors. There are two inputs, the ignition signal (logic zero when on) and the speed sensor from the transmission. The speed sensor outputs pulses as the drive shaft or transaxle turns. The ratio is 10 pulses per revolution of the tire. If a 16-inch wheel is used, the sensor outputs 39,610 pulses per mile $\left(\dfrac{5280}{1.333} \times 10 \right)$. The unit produces 11.0028 pulses per second $\left(\dfrac{36,610}{3600} \right)$ for each mile per hour of speed. The 15 miles per hour mark would therefore be 165 pulses per second. A dead stop would be anything below 0.09 mph. Develop the schematic for this system and a process control chart if needed.
10. How many interrupts are needed to develop the software in Question 9?
11. Develop the system program for the problem described in Question 9.
12. How much power is required for a red LED traffic lamp?
13. How long will the red LED traffic lamp last before a replacement is needed?

14. What interface standard is used to connect one traffic light controller to another in the example in this chapter?
15. What are the inputs to the traffic light controller detailed in this chapter?
16. What command from the master to the slaves is illustrated in the traffic light controller in this chapter?
17. Describe what other commands would actually be implemented in a practical multi-intersection traffic control system.
18. Why is a relay used to change the direction of the motor in the cat litter box described in this chapter?
19. How does the software respond when the scoop in the cat litter box becomes stuck behind a big object?
20. Suggest any additional features that might be added to the cat litter box described in this chapter.
21. Sensors are discussed in an earlier chapter. Would it be possible to design a system that scoops up dog feces from the backyard? If so, postulate on how this might be accomplished. If a good system is designed, think of the potential!
22. Detail how the signals in a home HVAC system control the furnace, air conditioner, and circulation fan motor.
23. Because the controls in the system are 24-VAC signals, relays were chosen to control them. Is there an alternate, low-cost, and reliable method for controlling these signals?
24. If no outside air temperature is available to the HVAC controller, is it possible to control both heating and air conditioning without a manual selection? Explain your answer.
25. Is it beneficial to introduce outside air in certain conditions to the HVAC system? If so, explain your answer citing several cases.
26. What role is portrayed by the data EEPROM in the HVAC system described in this chapter?

CHAPTER 10

Advanced Topics

This chapter presents topics that are not used in everyday applications for the microcontroller. An example is additional memory. In earlier chapters you learned that the microcontroller has a limited amount of memory, but not much was said about expanding the memory system. Because there are cases where the onboard I/O is insufficient, this chapter expands the I/O and illustrates how to use the CAN (**controller area network**) and USB (**universal serial bus**) with sample applications.

Upon completion of this chapter you will be able to:

1. Expand the memory in the microcontroller system.
2. Use the UPC code in an inventory control system.
3. Program a microcontroller with a boot block.
4. Develop a boot strap loader program.
5. Expand the I/O connections on the microcontroller.
6. Use the CAN to interconnect microcontrollers in a network.
7. Program and interface the USB to a personal computer.
8. Detail the extended instruction set for the PIC18.

10-1   MEMORY EXPANSION

One problem with the microcontroller is that it contains a finite amount of memory. This is especially true of the SRAM, which often is only 256 or 512 bytes. Some versions have much more internal SRAM, but those devices have no more than 3968 locations. The amount of program memory can also limit applications. The microcontroller may have only 2K or 4K program locations, which is often enough for many systems, but if there is a very large program or a program that has large amount of static data, the memory may need to be expanded. This section explores memory expansion using serial EEPROM, which is the only type of serial memory available for memory expansion.

Adding Serial EEPROM

Serial EEPROMs are available from Microchip for expanding the amount of memory. This is not SRAM and does have a limited life. EEPROM is normally able to handle one million reads or

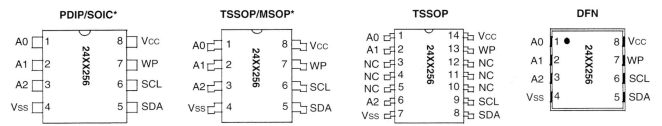

Note: *Pins A0 and A1 are not connected in the MSOP package only.

FIGURE 10-1   Pin-outs of serial EEPROM.

writes, which is sufficient for many applications. To interface a serial EEPROM, the I^{2}C unit or SPI unit is used for the expansion. Software can also be written to use I/O pins, but it is more efficient if the SPI or I^{2}C unit is the interface point because fewer pins are needed on the microcontroller for the interface.

Microchip produces a fairly wide variety of EERPOMs from a tiny 128-bit (16 byte) to a 512K-bit (64K byte) version. Another company, Atmel, produces a flash memory device that stores 4M-bits (512K bytes). Philips Semiconductors also makes a series of serial memory devices.

The serial EEPROM is illustrated in Figure 10-1. These devices are often found on 8-bit integrated circuits so they do not use much board space. The 256K-bit device is illustrated, but all of the devices have a similar pin-out. The A inputs are not address inputs; they are chip select inputs. They are called A inputs because they can be used to select more than one serial EEPROM, hence they select different devices.

Figure 10-2 illustrates two such devices connected to a PIC18F1220 microcontroller using A0 as an address input to select one or the other serial EEPROM. The activity levels of the A inputs are programmed with wires to ground or 5.0 V to select the EEPROM. The three address inputs allow up to eight devices in a system to be connected to the microcontroller. Only two devices can be connected as shown in Figure 10-2 because of the internal drive capability of the serial EEPROM. If additional memory devices are needed, they must be buffered. In this circuit, the U4 memory is selected when A0 = 0 and U5 is selected when A0 = 1, all while A1 and A2 are zeros. The RB6 pin is the serial data connection between the SDA pins on both devices and the microcontroller, and the RB7 pin is the clock signal to the SCL pins on both devices. If only a single device is added to the microcontroller, then the A0 input is connected to ground. In either case, the 10K Ω pull-up resister is needed to operate the memory at the recommended 100-KHz rate.

The serial interface is much like the many others discussed in earlier chapters. When using the PIC18F1220, software must be used to generate the clock for the EEPROM and also to send or receive information from it. The EEPROM usually functions at a 100-KHz rate. Before discussing the data rate, we investigate the operation of the EEPROM. The memory responds to a byte that is sent to it called a **control byte**, as illustrated in Figure 10-3. The control byte has a start bit, the state of the three address inputs, a read/write bit, and an acknowledge bit. The start bit is a negative edge issued to the memory and the clock is a logic one. This is followed by a 1010 and by the state of the address inputs to select a device. Once the command is sent it is followed by the memory address. This sequence writes the address into the memory. A second start pulse is sent to read the data. If data are written, the second write pulse is not sent to the memory, just the data. The ACK signal is received after each byte is sent or read from the memory after a byte is read, but the only time attention is paid to the ACK signal is after a write to memory. The memory requires time to perform the write. After a write, the ACK signal is polled to see if the memory has written the data. If no acknowledge is received, the software must wait until it is

FIGURE 10-2 A pair of
serial EEPROMs interfaced
to the PIC18F1220 micro-
controller.

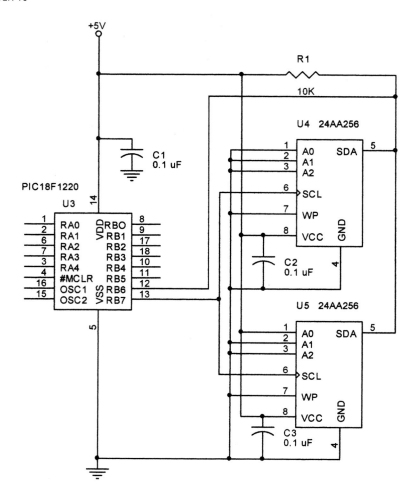

received from the memory. Figure 10-4 illustrates all the transactions that occur between the serial EEPROM and the microcontroller. With the 256K-bit device discussed here, the address is a
15-bit address (32K bytes) with two bytes of addressing information where the A15 bit is unused. The first address byte contains address bits A15–A8 and the second address byte contains
address bits A7–A0.

A page read or write is also possible by reading or writing more than a single byte. The
memory address, which is automatically incremented for each read or write, allows many bytes of
transfer without resending the address. A page by definition for this memory is 64 bytes. Pages
begin with the rightmost 6 bits of the address set to zero. Page 0 is at address 0x0000–0x003F,
Page 1 is at address 0x0040–0x007F, and so forth. If reading more than one byte, the address
wraps around at the end of the page. For example, if address 0x007F is sent to the memory and
2 bytes are read, the content of 0x007F is the first byte and 0x0040 is the second byte.

The software functions required to read or write a byte to the memory appear in Example
10-1. Functions that read or write a block of memory do not appear in the software. In this

FIGURE 10-3 Control byte
of the serial EEPROM.

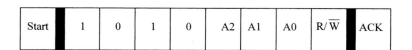

Write Address

S	1	0	1	0	A2	A1	A0	0	ACK	x	A14	A13	A12	A11	A10	A9	A8	ACK	A7	A6	A5	A4	A3	A2	A1	A0	ACK

Followed by for a Read Byte

S	1	0	1	0	A2	A1	A0	1	ACK	D7	D6	D5	D4	D3	D2	D1	D0	ACK	P

or Followed by for a Write Byte

D7	D6	D5	D4	D3	D2	D1	D0	ACK	P

S = Start
P = Stop
ACK = acknowledge

FIGURE 10-4 Signals to the serial EEPROM for a read or a write.

software, #define statements are used so that the port pins used for the memory are easily changed. Here RB7 is used for the SCK signal and RB6 is used for the SDA signal. For the interface to function correctly, the direction in the SDA signal is changed to an input when the memory is sent a logic one signal. This is because the memory circuitry requires a 10K Ω pull-up resister on its SDA connection to properly write a logic one. By switching the direction of the RB6 pin to input for sending a logic one to the SDA pin, the circuit meets the memory device's requirement. When the pin is switched to an output, SDA becomes a zero; when the pin is switched to an input, SDA is an I/O pin.

EXAMPLE 10-1

```
// **************** CONSTANTS ************************

#define SCL PORTBbits.RB7              // RB7 is SCL
#define SDA PORTBbits.RB6              // RB6 is SDA
#define SDA_TRIS TRISBbits.TRISB6

//*************** SERIAL EEPROM FUNCTIONS **************

void SendStart (void)      // send Start
{
      SDA_TRIS = 1;        // SDA = 1
      SCL = 1;             // SCL = 1
      SDA_TRIS = 0;        // SDA = 0
      SCL = 0;             // SCL = 0
}
```

```
void SendStop (void)              // send Stop
{
      SCL = 0;                    // SCL = 0
      SDA_TRIS = 0;               // SDA = 0
      SCL = 1;                    // SCL = 1
      SDA_TRIS = 1;               // SDA = 1
}

char SendSM (char data)           // send a byte
{
      char a, b, c;
      c = 0;
      for (a = 0; a < 8; a++)
      {
            SCL = 0;                    // SCL = 0
            if ((data & 0x80) == 0x80)       // leftmost data bit
                  SDA_TRIS = 1;       // SDA = 1
            else
                  SDA_TRIS = 0;       // SDA = 0
            data <<= 1;                 // shift data left
            SCL = 1;                    // SCL = 1
      }
      SCL = 0;                          // SCL = 0
      SDA_TRIS = 1;
      SCL = 1;                          // SCL = 1
      if (SDA == 1)                     // check SDA for ACK
            c = 1;
      SCL = 0;                          // SCL = 0
      return c;                         // return ACK
}

char ReadSM (void)                // read a byte
{
      char a;
      char b = 0;
      SDA_TRIS = 1;               // set SDA for read
      SCL = 0;                    // SCL = 0
      for (a = 0; a < 8; a++)
      {
            b <<= 1;              // shift left for next bit
            SCL = 1;              // SCL = 1
            if (SDA == 1)
                  b |= 1;         // add in SDA if 1
            SCL = 0;              // SCL = 0
      }
      SDA_TRIS = 0;               // SDA = 0
      return b;                   // return retrieved data
}

void Ack (char control)           // test ACK
{
      char a = 1;
      do
      {
            SendStart();
            a = SendSM (control);    // output control byte
      }
      while (a == 1); #
      SendStop();
}

// Read a byte from memory
//     address is 0x0000 -- 0xFFFF

char ReadByte (int address)
{
      char a = 0;
      SDA = 0;                    // make certain SDA = 0
      SendStart();
```

```
        if ((address & 0x8000) == 0x8000)
            a = 2;
        else
            a = 0;
        SendSM (0xA0 + a);              // command (write address)
        SendSM (address >> 8);
        SendSM (address);
        SendStart ();
        SendSM (0xA1 | a);             // command (read data)
        a = ReadSM ();
        SendStop ();
        return (a);
}

void WriteByte (int address, char data)
{
        char a = 0;
        SDA = 0;                       // make certain SDA = 0
        SendStart ();
        if ((address & 0x8000) == 0x8000)
            a = 2;
        SendSM (0xA0 | a);
        SendSM (address >> 8);
        SendSM (address);
        SendSM (data);
        SendStop ();
        Ack (0xA0 | a);                // wait for ACK
}
```

An Application Using the Extra Memory

Now that we can easily add extra memory to the microcontroller, we need an application that uses it. Suppose that a system is needed to log data and save it for later access. The data that is logged is for an inventory control system in a grocery store. The unit designed here is a calculator-like device that a clerk carries down the aisles. The clerk scans the unit price codes (UPC) on items with an optical reader and then enters the count of the items into the inventory device. The counts and UPC codes are later uploaded to a PC for an inventory update. If society were honest, there would be no need for inventory. Some estimates place shrinkage at 10% in grocery stores, which means that 10% of the merchandise is never paid for at the checkout counter. The system designed here at least allows the grocer to determine what is missing so high-risk items can be moved to better populated areas of the store.

The system needs as input devices, a keypad to enter item counts and an optical wand or reader to scan items. It also needs a link to the PC so that item counts can be transferred to the PC for inventory control. A small screen is used to display the UPC code and the item count when entered by the clerk. A device such as this does not have enough internal EEPROM to store items that would be scanned in an 8-hour workday, so additional memory is needed. With a single 24AA256 serial EEPROM, there are 32K bytes of storage for items and item counts. Is this enough memory? Suppose that a clerk can scan and then count the items on a shelf at the rate of one item per minute. In an 8-hour shift, without any breaks, the clerk can inventory 480 items. A UPC code is either a 6-digit code or a 10-digit code. Suppose all the items are 10 digits. A memory of 4,800 bytes is required to store all the codes. An additional 480 bytes are needed for the item counts if the counts are limited to 255 items. A 32K byte memory is large enough to function without uploading codes and item counts for three work shifts plus a good margin for an extremely speedy worker.

Bar codes were first discussed in Chapter 8 using the bar code (code 128) common in factory inventory systems. The bar code used with grocery store products is different and uses either the UPC-A format or the UPC-E format. The UPC-E version is typically used on smaller items. The UPC-A format is a 10-digit code and the UPC-E format is a 6-digit code. Figure 10-5 illustrates

FIGURE 10-5 Unit pricing code (UPC) barcodes.

UPC-A

UPC-E

examples of both codes. The main difference between code-128 and UPC-A or UPC-E is that the UPC codes allow only numeric data, whereas code-128 code allows alphabetic and numeric data.

In UPC-A and UPC-E, the number is always surrounded by synchronization bits of 101 (bsb, where b is a black bar and s is a white space). In UPC-A, the first code that appears to the right of the leftmost synchronization bits is the number system code (NS) followed by a 5-digit manufacturer's code. To the right of the manufacturer's code is the center synchronization bits that have a pattern of 0101 (sbsb). To the right of the center synchronization pattern is the 5-digit product code. Finally, to the left of the rightmost synchronization bits is the check digit. For a clearer view of this, refer to Figure 10-6. The UPC-E code is different because it contains an implied NS = 0 followed by a 6-digit code that is both the manufacturer's code and the product code. This is not followed by the check digit. This also is illustrated in Figure 10-6. The UPC-A code uses coding OOOOOO followed by RRRRRR for the 12-digit number. (O = odd, R = right, and E = even in Table 10-1). The UPC-E code uses coding EEEOOO from Table 10-1 to encode the 6 digits. In the UPC-E code, the right synchronization field contains a 0101 (sbsb) instead of a 101 (bsb) as in UPC-A.

The conversion here is similar to the code-128 in Chapter 8. The difference is that there are 6 lookup tables, 3 forward and 3 reverse for the UPC-A and UPC-E codes. As in Chapter 8, the interrupt on change function is used to measure the width of each pulse along with a timer.

Figure 10-7 illustrates the complete schematic diagram of this system. The system uses a software UART for interfacing to the PC through the DS-275 driver. The optical wand is interfaced to the system through a software serial interface that uses the interrupt on change feature of the microcontroller. The external memory is a single 32K device that holds the transaction as the system is used to log and count items in the store. At the end of a workday, the content of the EEPROM is uploaded to the PC through the serial interface. Example 10-2 contains the complete program for this system.

FIGURE 10-6 UPC-A and
UPC-E encoding.

UPC-A

UPC-E

TABLE 10-1 UPC-A and UPC-E encodings.

Digit	Odd/Left Code	Even Code	Right Code
0	sssbbsb	sbssbbb	bbbssbs
1	ssbbssb	sbbssbb	bbssbbs
2	ssbssbb	ssbbsbb	bbsbbss
3	sbbbbsb	sbssssb	bssssbs
4	sbsssbb	ssbbbsb	bsbbbss
5	sbbsssb	sbbbssb	bssbbbs
6	sbsbbbb	ssssbsb	bsbssss
7	sbbbsbb	ssbsssb	bsssbss
8	sbbsbbb	sssbssb	bssbsss
9	sssbsbb	ssbsbbb	bbbsbss

EXAMPLE 10-2

```
/*
 * "Inventory control system"
 */

#include <p18cxxx.h>
#include <timers.h>
#include <delays.h>
#include <sw_uart.h>
```

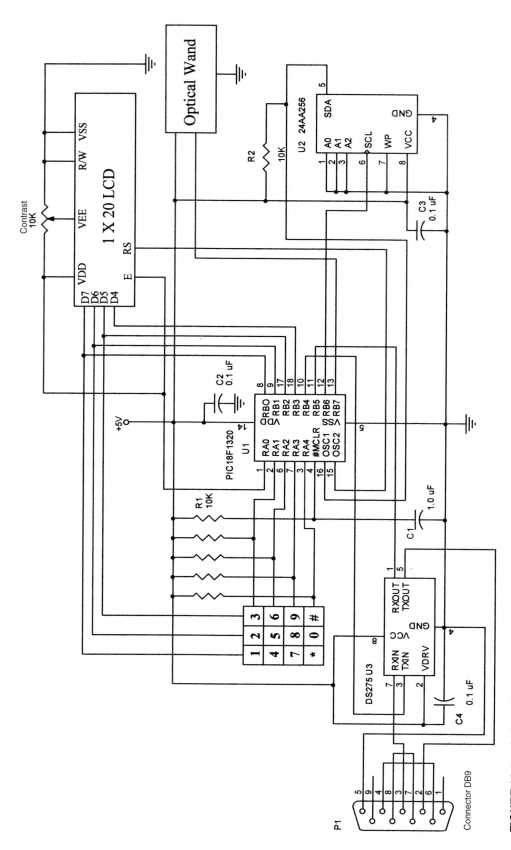

FIGURE 10-7 An inventory control system.

```
/* Set configuration bits
 *   - set internal oscillator
 *   - disable watchdog timer
 *   - disable low-voltage programming
 *   - master clear enabled
 */

#pragma config OSC = INTIO2
#pragma config WDT = OFF
#pragma config LVP = OFF
#pragma config MCLRE = ON

// program memory data

rom near char str1[] = "Uploading to the PC.";
rom near char str2[] = "Ready to swipe code.";
rom near char str3[] = "            Ct=    ";

rom near char odd[] =
{
    0b10010000,      // binary data 0bxxxxxxxx
    0b01010100,
    0b01000101,
    0b00110000,
    0b00001001,
    0b00011000,
    0b00000011,
    0b00100001,
    0b00010010,
    0b10000001
};

rom near char oddbw[] =
{
    0b00000110,
    0b00010101,
    0b01010001,
    0b00001100,
    0b01100000,
    0b00100100,
    0b11000000,
    0b01001000,
    0b10000100,
    0b01000010
};

rom near char even[] =
{
    0b00000110,
    0b00010101,
    0b01010001,
    0b00001100,
    0b01100000,
    0b00100100,
    0b11000000,
    0b01001000,
    0b10000100,
    0b01000010
};

rom near char evenbw[] =
{
    0b10010000,
    0b01010100,
    0b01000101,
    0b00110000,
    0b00001001,
    0b00011000,
    0b00000011,
    0b00100001,
```

```
        0b00010010,
        0b10000001
};
rom near char right[] =
{
        0b10010000,
        0b01010100,
        0b01000101,
        0b00110000,
        0b00001001,
        0b00011000,
        0b00000011,
        0b00100001,
        0b00010010,
        0b10000001
};
rom near char rightbw[] =
{
        0b00000110,
        0b00010101,
        0b01010001,
        0b00001100,
        0b01100000,
        0b00100100,
        0b11000000,
        0b01001000,
        0b10000100,
        0b01000010
};
rom near char lookupKey[] =
{
        1, 4, 7, 10,      // left column
        2, 5, 8, 0,       // middle column
        3, 6, 9, 11       // right column
};
// Define EEPROM addresses

#define addrl 0
#define addrh 1

// Data memory variable

char goodUPC;
char timeOut;
char pulse;
int halfPulseWidth;
char UPC[14];
char UPCptr;
char type;

#pragma interrupt MyHighInt    save=PROD
#pragma code high_vector=0x08          // high_vector is at 0x0008

void high_vector (void)                // high-prioity vector
{
        _asm GOTO MyHighInt _endasm    // goto high software
}

#pragma code

// Read a data EEPROM location from address

char eeRead (char address)
{
        EECON1bits.EEPGD = 0;
        EEADR = address;
        EECON1bits.RD = 1;
        return EEDATA;
}
```

```
// Write a data EEPROM location at address with data

void eeWrite (char address, char data)
{      char temp;
       temp = INTCONbits.GIEH;
       INTCONbits.GIEH = 0;
       EECON1bits.EEPGD = 0;
       EECON1bits.WREN = 1;
       EEADR = address;
       EEDATA = data;
       EECON2 = 0x55;
       EECON2 = 0xAA;
       EECON1bits.WR = 1;
       while (PIR2bits.EEIF == 0);
       PIR2bits.EEIF = 0;
       EECON1bits.WREN = 0;
       if (temp == 1)
              INTCONbits.GIEH = 1;
}

// Send LCD byte data, with RS = rs

#define RS PORTAbits.OSC1
#define E PORTAbits.RA0

void SendLCDdata (char data, char rs)
{
       PORTB = (data >> 4) | 0x80 ;         // send left nibble
       RS = 1;                              // set RS
       E = 1;                               // pulse E
       E = 0;
       Delay10TCYx(4);                      // wait 40 us
       PORTB = (data & 0x0F) | 0x80;        // send right nibble
       RS = 1;                              // set RS
       E = 1;                               // pulse E
       E = 0;
       Delay10TCYx(4);                      // wait 40 us
}

// Initialize LCD

void InitLCD (void)                         // intialize LCD
{
       Delay1KTCYx (20);                    // wait 20 ms
       SendLCDdata (0x20, 0);               // send 0x20
       Delay1KTCYx (6);                     // wait 6 ms
       SendLCDdata (0x20, 0);               // send 0x20
       Delay10TCYx (10);                    // wait 100 us
       SendLCDdata (0x20, 0);               // send 0x20
       SendLCDdata (0x28, 0);               // send 0x28
       SendLCDdata (0x01, 0);               // send 0x01
       Delay1KTCYx (2);                     // wait 2 ms
       SendLCDdata (0x0C, 0);               // send 0x0C
       SendLCDdata (0x06, 0);               // send 0x06
}

// Display a program-memory-based string (str) at position

void DisplayStringPgm (char position, rom char *str)
{
       char ptr = 0;
       SendLCDdata (position, 0);    // send position
       while (str[ptr] != 0)
              SendLCDdata(str[ptr++], 1);    // send character
}

char lookup (rom char* table, char temp)
{
       char a;
```

```c
        for (a = 0; a < 10; a++)
              if (table[a] == temp)
              break;
        return a;
}

void Abort (void)
{
        char temp = UPC[0];
        char a;
        char count = 0;
        CloseTimer0 ();
        WriteTimer0 (0);
        if (UPCptr > 5 && UPCptr < 7)            // regular UPC-E
        {
              if (type = 1)             // backward UPC-E
              {
                    for (count = 0; count < 3; count++)
                          UPC[count] = lookup(oddbw, UPC[count]);
                    for (count = 3; count < 6; count++)
                          UPC[count] = lookup(evenbw, UPC[count]);
                    for (count = 0; count < 3; count++)
                    {
                          temp = UPC[5 - count];
                          UPC[5 - count] = UPC[count];
                          UPC[count] = temp;
                    }

                          goodUPC = 1;
              }
              else                          // forward UPC-E
              {
                    for (count = 0; count < 3; count++)
                          UPC[count] = lookup(even, UPC[count]);
                    for (count = 3; count < 6; count++)
                          UPC[count] = lookup(odd, UPC[count]);
                    goodUPC = 1;
              }
        }
        else if (UPCptr > 7)
        {
              for (a = 0; a < 4; a++)
              {
                    count += temp & 3;
                    temp >>= 2;
              }
              if ((temp & 1) == 0)
              {                                 // backward UPC-A
                    for (count = 0; count < 6; count++)
                          UPC[count] = lookup(rightbw, UPC[count]);
                    for (count = 7; count < 13; count++)
                          UPC[count - 1] = lookup(oddbw, UPC[count]);
                    for (count = 0; count < 6; count++)
                    {
                          temp = UPC[11 - count];
                          UPC[11 - count] = UPC[count];
                          UPC[count] = temp;
                    }
                    goodUPC = 2;
              }
              else
              {                                 // regular UPC-A
                    for (count = 0; count < 6; count++)
                          UPC[count] = lookup(odd, UPC[count]);
                    for (count = 7; count < 13; count++)
                          UPC[count - 1] = lookup(right, UPC[count]);
                    goodUPC = 2;
              }
        }
}
```

```
void MyHighInt (void)
{
      int temp;
      if (PIR1bits.TMR1IF == 1)
      {
            PIR1bits.TMR1IF = 0;
            WriteTimer1(-12500 );
            if (timeOut != 0)
            {
                  timeOut--;
                  if (timeOut == 0)
                        Abort();
            }
      }
      else if (INTCONbits.RBIF == 1 )
      {
            temp = PORTB;                   // must read PORTB to clear change
            INTCONbits.RBIF = 0;            // clear interrupt
            if (ReadTimer0() == 0)          // initial
            {
            OpenTimer0(TIMER_INT_OFF &
                        T0_16BIT &
                        T0_SOURCE_INT &
                        T0_PS_1_64);        // 64 us period
            WriteTimer0 (1);
            timeOut = 3;
            pulse = UPCptr = type = 0;
      }
      else if (pulse == 0)
      {
            halfPulseWidth = ReadTimer0() / 2;
            WriteTimer0 (1);
            pulse++;
            timeOut = 3;
            WriteTimer0 (1);
      }
      else if (pulse < 3)
      {
            halfPulseWidth = (halfPulseWidth + ReadTimer0() / 2 ) / 2;
            WriteTimer0 (1);
            pulse++;
            timeOut = 3;
      }
      else
      {
            temp = ReadTimer0 ();
            WriteTimer0 (1);
            if (temp <= halfPulseWidth * 3)
                  temp = 0;
            else if (temp <= halfPulseWidth * 5)
                  temp = 1;
            else if (temp <= halfPulseWidth * 7)
                  temp = 2;
            else
                  temp = 3;
            UPC[UPCptr] = UPC[UPCptr] << 2 | temp;
            pulse++;
            timeOut = 3;
            if (pulse % 4 == 3)
            {
                  if (pulse == 7)
                  {
                  if ((UPC[UPCptr] & 0x3F) == 0) // UPC-E backward
                        {
                              type = 1;
                              pulse -= 3;
                              UPC[UPCptr] >= 6;
                              UPCptr--;
```

```
                                        }
                                   }
                              UPCptr++;
                         }
                    }
               }
          }

// ***************** CONSTANTS *************************
#define SCL PORTBbits.RB6          // RB6 is SCL
#define SDA PORTAbits.OSC1         // OSC1 is SDA
#define SDA_TRIS TRISAbits.TRISA7

//*************** SERIAL EEPROM FUNCTIONS **************

void SendStart (void)              // send Start
{
     SDA_TRIS = 1;                 // SDA = 1
     SCL = 1;                      // SCL = 1
     SDA_TRIS = 0;                 // SDA = 0
     SCL = 0;                      // SCL = 0
}

void SendStop (void)               // send Stop
{
     SCL = 0;                      // SCL = 0
     SDA_TRIS = 0;                 // SDA = 0
     SCL = 1;                      // SCL = 1
     SDA_TRIS = 1;                 // SDA = 1
}

char SendSM (char data)            // send a byte
{
     char a, b, c;
     c = 0;
     for (a = 0; a < 8; a++)
     {
          SCL = 0;                      // SCL = 0
          if ((data & 0x80) == 0x80)    // leftmost data bit
               SDA_TRIS = 1;            // SDA = 1
          else
               SDA_TRIS = 0;            // SDA = 0
          data <<= 1;                   // shift data left
          SCL = 1;                      // SCL = 1
     }
     SCL = 0;                      // SCL = 0
     SDA_TRIS = 1;
     SCL = 1;                      // SCL = 1
     if (SDA == 1)                 // check SDA for ACK
          c = 1;
     SCL = 0;                      // SCL = 0
     return c;                     // return ACK
}

char ReadSM (void)                 // read a byte
{
     char a;
     char b = 0;
     SDA_TRIS = 1;                 // set SDA for read
     SCL = 0;                      // SCL = 0
     for (a = 0; a < 8; a++)
     {
          b <<= 1;                     // shift left for next bit
          SCL = 1;                     // SCL = 1
          if (SDA == 1)
               b |= 1;                 // add in SDA if 1
          SCL = 0;                     // SCL = 0
     }
     SDA_TRIS = 0;                 // SDA = 0
```

```
        return b;                           // return retrieved data
}

void Ack (char control)                     // test ACK
{
        char a = 1;
        do
        {
            SendStart ();
            a = SendSM ( control);          // output control byte
        }
        while (a == 1);
        SendStop ();
}

// Read a byte from memory
//     address is 0x0000 -- 0xFFFF

char ReadByte (int address)
{
        char a = 0;
        SDA = 0;                            // make certain SDA = 0
        SendStart();
        if ((address & 0x8000) == 0x8000)
            a = 2;
        else
            a = 0;
        SendSM (0xA0 + a);                  // command (write address)
        SendSM (address >> 8);
        SendSM (address);
        SendStart ();
        SendSM (0xA1 | a);                  // command (read data)
        a = ReadSM ();
        SendStop ();
        return (a);
}

void WriteByte (int address, char data)
{
        char a = 0;
        SDA = 0;                            // make certain SDA = 0
        SendStart ();
        if ((address & 0x8000) == 0x8000)
            a = 2;
        SendSM (0xA0 | a);
        SendSM (address >> 8);
        SendSM (address);
        SendSM (data);
        SendStop ();
        Ack (0xA0 | a);         // wait for ACK
}

#define KEYPORT PORTA           // change to match the actual port
#define DELAY 15                // change as needed for time delay

void Switch (char bit)
{
        do                      // wait for release
        {
            while ((KEYPORT & bit) != bit);
            Delay1KTCYx(DELAY);

        }whilen ((KEYPORT & bit) != bit);

        do                      // wait for press
        {
            while ((KEYPORT & bit) == bit);
            Delay1KTCYx(DELAY);

        }while ((KEYPORT & bit) == bit);
}
```

```
unsigned char Key (void)
{
        #define MASK 0x1E          // set mask
        #define ROWS 4             // set number of rows

        int a;
        unsigned char keyCode;

        keyCode = 0;               // clear Port B and keyCode
        PORTB = PORTB & 0xF8;

        Switch (MASK);             // debounce and wait for any key

        PORTB = PORTB & 0xFE;      // select a leftmost column

        while ((PORTA & MASK) == MASK)             // while no key is found
        {
                PORTB = (PORTB << 1) | 1;   // get next column
                keyCode += ROWS;            // add rows to keycode
        }
        for (a = 1; a != 0; a <<= 1)
        {                                          // find row
                if ((PORTA & a) == 0)
                        break;
                keyCode++;
        }
        return lookupKey[keyCode];         // look up correct key code
}

int GetCount (void)
{
        char number[3];
        int retval = 0;
        char count = 0;
        char temp = Key();
        while (temp != 11)
        {
                if (temp == 10 && count != 0)
                {
                        count--;
                        SendLCDdata (0x90 | count + 1, 0);
                        SendLCDdata (' ' , 1);
                        SendLCDdata (0x90 | count + 1, 0);
                }
                else
                {
                        number[count] = temp;
                        if (count != 2)
                                count++;
                }
                temp = Key();
        }
        for (temp = 0; temp < count; temp++)
                retval = retval * 10 + number[temp];
        return retval;
}

void SendPC (void)
{
        int addr;
        int addr1 = 0;
        addr = eeRead (addrl);
        addr = (int) eeRead(addrh) << 8;
        while (ReadUART() != 1);
        WriteUART(1);
        while (addr != addr1)
                WriteUART(ReadByte(addr1++));
        eeWrite (addrl, 0);
        eeWrite (addrh, 0);
}
```

```c
void GetUPC (void)
{
        int addr;
        int count;
        DisplayStringPgm (0x80, str3);
        SendLCDdata (0x80, 0);
        for (count = 0; count < 6; count++)
                SendLCDdata( UPC[count] + 0x30, 1);
        if (goodUPC == 2)
        {
                SendLCDdata ('-', 1);
                for (count = 6; count < 12; count++)
                        SendLCDdata (UPC[count] + 0x30, 1);
        }
        count = GetCount ();
        if (count != 0)
        {
                addr = eeRead(addrl);
                addr = (int) eeRead(addrh) << 8;
                if (goodUPC = 1)
                {
                        WriteByte (addr++, 1);
                        for (count = 0; count < 6; count++)
                                WriteByte (addr++, UPC[count]);
                }
                else
                {
                        WriteByte (addr++, 2);
                        for (count = 0; count < 12; count++)
                                WriteByte(addr++, UPC[count]);
                }
                WriteByte (addr++, count);
                WriteByte (addr++, count >> 8);
                eeWrite (addrl, addr);
                eeWrite (addrh, addr >> 8);
        }
        goodUPC = 0;
        DisplayStringPgm (0x80, str2);
}

void DelayTXBitUART (void)
{
        Delay10TCYx(9);
        Delay1TCY();
        Delay1TCY();
        Delay1TCY();
}

void DelayRXHalfBitUART (void)
{
        Delay10TCYx(4);
        Delay1TCY();
        Delay1TCY();
        Delay1TCY();
        Delay1TCY();
}

void DelayRXBitUART (void)
{
        Delay10TCYx(9);
        Delay1TCY();
}

void main (void)
{
        OSCCON = 0x62;              // internal 4-MHz clock
        ADCON1 = 0x0F;             // I/O is digital
        TRISA = 0x7E;
        TRISB = 0xA0;
```

```
PORTA = 0x80;
PORTB = 0xA0;

InitLCD();
goodUPC = timeOut = 0;

WriteTimer0 (0);

OpenTimer1 (TIMER_INT_ON &     // every 8 us
            T1_16BIT_RW &
            T1_SOURCE_INT &
            T1_PS_1_8);

WriteTimer1 (-12500);

OpenUART();

INTCONbits.RBIE = 1;
PIE1bits.TMR1IE = 1;
INTCONbits.GIEH = 1;

if (eeRead(addrl) == 0xFF && eeRead(addrh) == 0xFF)
{
      eeWrite (addrl, 0);
      eeWrite (addrh, 0);
}
if ((PORTA & 6) == 0 &&
      eeRead--(addrl) != 0 && eeRead(addrh) != 0 )
{
      DisplayStringPgm (0x80, str1);
      SendPC();
}
DisplayStringPgm (0x80, str2);

while (1)
{
      if (goodUPC != 0)
            GetUPC();
}
}
```

10-2   BOOT BLOCK

The PIC18 family members contain a boot block, which allows the microcontroller to reprogram itself in a system. This is very useful in systems that require software upgrades to be applied after the system is installed in an application. This is sometimes called **flashing** the system. It is also called **self-programming**. If you have purchased a main-board for your computer you are probably familiar with flashing the BIOS.

BootStrap Loader

The **boot block** is an area in the memory of 512 bytes that starts at location 0x0000 and ends at location 0x1FF. (Some members have a larger boot block area so the ending address is different). The boot block holds code that is protected so it cannot be overwritten when the device is programmed. The code in the boot block that loads new software to the remainder of the memory is often called a *bootstrap loader*. The bootstrap loader program loads a new operating system into the device when required. Hence, the device is self-programming. For the boot loader to function, it must have a serial connection to the outside world so a new program can be downloaded to the microcontroller when required. A substantial portion of the bootstrap loader is software that controls this serial interface. Figure 10-8 illustrates the boot block in the system program memory. The microcontroller chosen for a project determines the total memory size, but in all cases at least the first 512 bytes of the memory is reserved as the boot block.

FIGURE 10-8 A memory map illustrating the placement of the boot block.

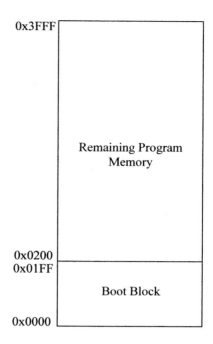

0x3FFF

Remaining Program Memory

0x0200
0x01FF

Boot Block

0x0000

Note: the upper memory address varies on family members

The boot block is protected memory, which means that the boot block is not overwritten when the device is programmed. Programming the internal program memory is similar to programming the internal data EEPROM as discussed in Chapter 6. The main difference between the data EEPROM and the flash program memory is that the flash program memory is protected in blocks, with the first block called the boot block. The data EEPROM has no such protection mechanism. The sizes and number of other protected blocks is determined by the microcontroller family member. For example, the PIC18F1220 has a boot block of 512 bytes and two other protected blocks located at addresses 0x0200–0x07FF and 0x0800–0x0FFF. The two blocks beyond the boot block are called block 0 and block 1, respectively. Another example is the PIC18F4550, which has a boot block of 2K bytes (0x0000–0x07FF) and four other blocks at locations 0x0800–0x1FFF (block 0), 0x2000–0x3FFF (block 1), 0x4000–0x5FFF (block 2), and 0x6000–0x7FFF (block 3). Refer to the data sheet of the microcontroller selected for a project for the exact locations and sizes of the blocks of program memory. These blocks are protected through bits located in the configuration registers including the boot block. Example 10-3 shows the shell of a C language program that functions as a boot block space. This program uses the last location in the data EEPROM to determine if a normal reset should ensue or a bootstrap loading of the program memory beginning at location 0x0200. The #pragma code MainCode=0x0200 statement sets the address of the program and interrupt vectors. This template is usable for any PIC18 family members provided the address of the MainCode block is changed to match its memory map. For example, the change for the PIC18F4550 is #pragma code MainCode=0x0800.

EXAMPLE 10-3

```
/*
 * This is a sample boot block
 */

#include <p18cxxx.h>
```

```
/* Set configuration bits
 *   - set internal oscillator
 *   - disable watchdog timer
 *   - disable low-voltage programming
 *   - master clear enabled
 */

#pragma config OSC = HS
#pragma config WDT = OFF
#pragma config LVP = OFF
#pragma config MCLRE = ON

// Define EEPROM addresses

#define bootControl 0xFF

void MyHighInt (void);                          // prototypes
void MyLowInt (void);

#pragma interrupt MyHighInt              save=PROD
#pragma code high_vector=0x08            // high_vector is at 0x0008

void high_vector (void)                  // high-prioity vector
{
      _asm GOTO MyHighInt _endasm              // goto high software
}

#pragma interruptlow MyLowInt            // MyLowInt is an intettupt
#pragma code low_vector=0x18             // low_vector is the vector at 0x18

void low_vector (void)
{
      _asm GOTO MyLowInt _endasm
}

#pragma code BootBlock

void NormalReset (void);

void main (void)
{
      EECON1bits.EEPGD = 0;              // check boot control byte
      EEADR = bootControl;
      EECON1bits.RD = 1;
      if (EEDATA != 0xFF)                // do normal reset
      {
            _asm GOTO NormalReset _endasm
      }

            // bootstrap software goes here

      _asm GOTO NormalReset _endasm
}

#pragma code Main=0x0200

// do not modify anything from MyHighInt to NormalReset

void HighInt (void);                     // prototypes
void LowInt (void);

void MyHighInt (void)                    // new fixed high-interrupt vector
{
      _asm GOTO HighInt _endasm
}

void MyLowInt (void)                     // new fixed low-interrupt vector
{
      _asm GOTO LowInt _endasm
}

void NormalReset (void)                  // new fixed reset vector
{

}
```

```
void HighInt (void)
{

}

void LowInt (void)
{

}
```

Writing to the FLASH Program Memory

In order to write to the FLASH memory, virtually the same software is used that was detailed in Chapter 6 for writing information to the data EEPROM. The main differences are that the blocks in the program memory must be unlocked in order to write to them using the CP (code protect) bit, the WRT (write protect) bit, and the EBTR (external block table read) bit. These code protection bits are located in the configuration register 0x300008. This location is written just as any other program memory location is written. The write protection bits are located in configuration register 0x30000A and the EBTR bits are location at 0x30000C. This is for the PIC18F1220. Other versions of the microcontroller use these locations and some use more. Refer to the data sheet for the microcontroller used in a project.

The software listed in Example 10-4 illustrates the boot block software for downloading or uploading the contents of the program memory. This software uses a different version of the startup software that is provided with the C18 compiler. This version uses less memory so that the boot block software fits in 512 bytes. (The software requires 508 bytes.) The software illustrated in this example responds to only three commands through a serial interface. Because not all versions of the PIC18 contain a hardware USART, this program uses the software UART functions located in the sw_uart.h header file. The 0x01 command allows program memory to be read for uploading code to a PC, the 0x02 command allows program memory to be written or downloaded from a PC, and the 0x03 command enables a normal boot. The only way back into the bootstrap loader is to write a 0xFF to data EEPROM location 0xFF, which is not illustrated in this example. The inline assembler, which is part of the C18 compiler, uses a slightly different command format for the table instructions from the normal assembler detailed in earlier chapters and also in the documentation available from Microchip. See the program listing for details of the change. Inline assembly code is required to fit the bootstrap loader program into the 512 bytes of available memory.

EXAMPLE 10-4

```
/*
 * Boot loader for the PIC18F1220
 */

// In order for this code to fit into the boot block
// the linker file was changed to use the c018.o initialization
// file instead of the C018i.0 file
//
// The effect is that no initialized data memory can exist in the
// program, which is not a very important feature for most systems.

#include <p18cxxx.h>
#include <sw_uart.h>
#include <delays.h>

/* Set configuration bits
 *  - set internal oscillator
 *  - disable watchdog timer
 *  - disable low-voltage programming
 *  - master clear enabled
 */

#pragma config OSC = HS               // external 4-MHz oscillator
#pragma config WDT = OFF
```

```
#pragma config LVP = OFF
#pragma config MCLRE = ON

// Define EEPROM addresses

#define bootControl 0xFF

// Data memory variable

void MyHighInt (void);                    // prototypes
void MyLowInt (void);

#pragma interrupt MyHighInt          save=PROD
#pragma code high_vector=0x08        // high_vector is at 0x0008

void high_vector (void)              // high-prioity vector
{
     _asm GOTO MyHighInt _endasm             // goto high software
}

#pragma interruptlow MyLowInt save=PROD     // MyLowInt is an interrupt
#pragma code low_vector=0x18         // low_vector is the vector at 0x18

void low_vector (void)
{
     _asm GOTO MyLowInt _endasm
}

#pragma code BootBlock

void NormalReset (void);             // prototype

void DelayTXBitUART (void)           // 93 us for UART
{
     Delay10TCYx(9);
     Delay1TCY();
     Delay1TCY();
     Delay1TCY();
}

void DelayRXHalfBitUART (void)       // 44 us for UART
{
     Delay10TCYx(4);
     Delay1TCY();
     Delay1TCY();
     Delay1TCY();
     Delay1TCY();
}

void DelayRXBitUART (void)           // 91 us for UART
{
     Delay10TCYx(9);
     Delay1TCY();
}

void main (void)
{
     char temp;
     char command;
     char length;
     char addrl;
     char addrh;
     char addru;

     EECON1bits.EEPGD = 0;           // check boot control byte
     EEADR = bootControl;
     EECON1bits.RD = 1;
     if (EEDATA != 0xFF)             // do normal reset
     {
          _asm GOTO NormalReset _endasm
     }
     OpenUART();
     do
```

```c
        {
                command = ReadUART();
                length = ReadUART();
                addrl = ReadUART();
                addrh = ReadUART();
                addru = ReadUART();
                _asm
                        MOVFF addrl,TBLPTRL
                        MOVFF addrh,TBLPTRH
                        MOVLW 0
                        MOVFF WREG,TBLPTRU
                _endasm

                // read program memory
                // <01> <len> <addrl> <addrh> <addru>
                // <data byte(s)>

                if (command == 1)
                {
                        while (length != 0)
                        {
                                _asm                            // read program memory
                                        TBLRDPOSTINC
                                        MOVFF TABLAT,temp
                                _endasm
                                WriteUART(temp);
                                length--;
                        }
                }

                // write program memory
                // <02> <len> <addrl> <addrh> <addru>
                // <data byte(s)>

                else if (command == 2)
                {
                        while (length != 0)
                        {
                                temp = ReadUART();
                                _asm
                                        MOVFF temp,TABLAT
                                        TBLWTPOSTINC
                                _endasm
                                EECON1bits.EEPGD = 1;           // select program EEPROM
                                EECON1bits.WREN = 1;            // unprotect writing
                                EECON2 = 0x55;                  // erase the current byte
                                EECON2 = 0xAA;
                                EECON1bits.WR = 1;              // select a write
                        }
                }

                // normal boot control
                // <03> <data> <dummy> <dummy> <dummy>

                else if (command == 3)
                {
                        EECON1bits.EEPGD = 0;           // select data EEPROM
                        EECON1bits.WREN = 1;            // unprotect writing
                        EEADR = bootControl;            // set up the EEPROM address
                        EEDATA = 0;                     // set up the EEPROM data
                        EECON2 = 0x55;                  // erase the current byte
                        EECON2 = 0xAA;
                        EECON1bits.WR = 1;              // select a write operation
                        while (PIR2bits.EEIF == 0);         // wait until finished
                }
        }
    while (command != 0);
}

// Any additional code is placed beyond the boot block software! In the code
// The following code block is named Main.
```

```
#pragma code Main=0x200

void HighInt (void);
void LowInt (void);

void MyHighInt (void)
{
      _asm GOTO HighInt _endasm
}

void MyLowInt (void)
{
      _asm GOTO LowInt _endasm
}

void NormalReset (void)
{
      _asm GOTO MAIN _endsasm
}

// place no code before this point!

void MAIN (void)                    // new Main function
{

}

void HighInt (void)                 // high-priority interrupt
{

}

void LowInt (void)                  // low-priority interrupt
{

}
```

When flashing the program memory, the configuration registers are changed using the write command (0x02) just as for any other location in the program memory. For example, to allow programming of the configuration register at address 0x300008 to unprotect the code blocks in the PIC18F1220, use the sequence of <0x02> <0x01> <0x08> <0x00> <0x30> <0x03>. The command is 0x02, the length of the data is 0x01, the address in little endian format is 0x300008, and the single data byte is a 0x03. The 0x03 stored or written to configuration register locations 0x300008 protects the program memory code blocks. Configuration registers at locations 0x30000A and 0x30000C must also be changed before a new program is downloaded to the microcontroller for flashing. When the download program is developed for the PC, the maximum number of bytes that can be written with a single write command is 256 (a length of 0x00). There is no provision for error checking. The best policy is to write a block of 256 bytes, read it back, and check to see if it was written correctly before proceeding to the next block of 256 bytes. If there is a mismatch, just rewrite the block and check again. Alternately, a checksum can be used to determine if the code is incorrect. In either case, the outcome is exactly the same; both techniques detect an error.

10-3   EXPANDING THE I/O

Although in many cases the I/O is adequate on one of the versions of the microcontroller, there are cases where additional I/O is required so the available I/O must be expanded. This section illustrates a few techniques for expanding the I/O structure of the microcontroller.

Additional Output Connections

Additional digital output connections are obtained by using a shift register to expand the number of output bits. The circuit in Figure 10-9 uses a shift register to provide an 8-bit output

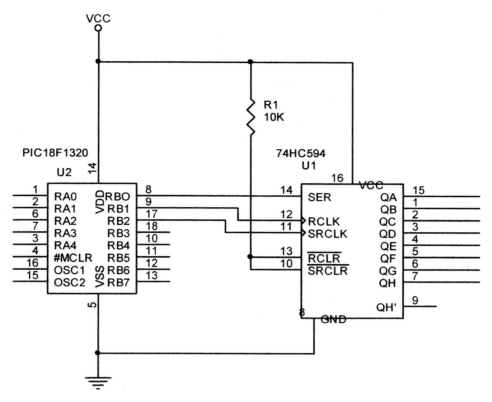

FIGURE 10-9 An 8-bit expansion port.

port to the PIC18F1320 microcontroller. Three I/O pins on the microcontroller are used to operate this interface, so the net gain is only five additional output connections. The software to control this simple interface is also fairly short as illustrated in Example 10-5. This example provides only the functions to initialize the interface and to output a byte to this additional output port.

EXAMPLE 10-5

```
void InitPort (void)
{
     TRISB = TRISB & 0xF8;          // program RB0, RB1, and RB2
     PORTB = PORTB & 0xF8;          // clear RB0, EB1, RB2
}

void OutPort (char data)
{
     char a;
     for (a = 0x80; a != 0; a >> 1)      // shift bit position
     {
          if ((a & data) == a)           // send SER
               PORTBbits.RB0 = 1;
          else
               PORTBbits.RB0 = 0;
          PORTBbits.RB2 = 1;             // clock SRCLK
          PORTBbits.RB2 = 0;
     }
     PORTBbits.RB1 = 1;                   // clock RCLK
     PORTBbits.RB1 = 0;
}
```

The 74HC594 has three inputs. The one called SER is the serial data input connected to RB0 in this circuit. The SRCLK input is the serial input clock and the RCLK input is the register clock. Inside the device there is a shift register and a holding register. In order to obtain glitch-free operation, the data are shifted into the shift register. Afterwards, the holding register is clocked to transfer the data from the shift register to the output pins, which are connected to the holding register. The InitPort function initializes Port B and the OutPort function is used to send data to the port. Note that the data is shifted in with the most significant bit (QH) first.

The shift register is expandable as illustrated in Figure 10-10. The QH output is connected to the next circuit's SER input and the clocks are tied together to produce a 16-bit output port. If a two-wire interface is needed, then a device manufactured by Microchip and others is available that interfaces to the I²C interface on the microcontroller. This does save one connection, but the software is much longer than illustrated in Example 10-5.

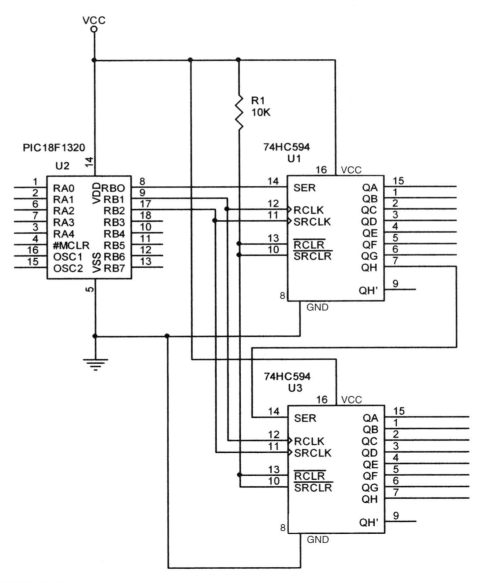

FIGURE 10-10 A 16-bit output port.

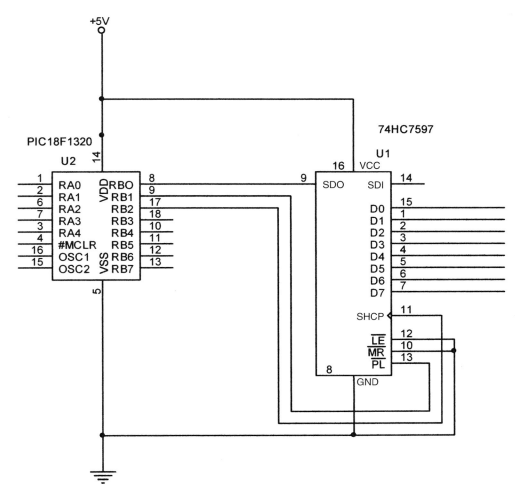

FIGURE 10-11   Adding 8 input pins.

Additional Input Connections

The same technique used for adding additional output pins is used to add additional input pins. The only difference is that instead of using the serial shift register connected to a latch as in the 74HC594, another shift register device is used that is wired internally with a latch feeding a shift register so serial data can be generated from input bits. Figure 10-11 illustrates a connection that adds 8 bits of additional input data bits to the microcontroller. In this example, circuit pin RB0 is the serial data input to the microcontroller, pin RB1 is the shift clock input to the 74HC7595, and RB2 is the parallel load input. The function to initialize and read a byte from the input pins is illustrated in Example 10-6. As with the output port illustrated earlier, the software is similar.

EXAMPLE 10-6

```
void InitPort (void)
{
      TRISB = TRISB & 0xF8;              // program RB0, RB1, and RB2
      TRISBbits.TRISB0 = 1;
      PORTB = PORTB & 0xF8;
      PORTB = PORTB | 1;
}
```

```
char InPort (void)
{
      char a, b;
      PORTBbits.RB1 = 0;                  // load shift regiser
      PORTBbits.RB1 = 1;
      for (a = 0; a <8; a++)              // get all 8 bits
      {
            b <<= 1;
            b |= PORTBbits.RB0;
            PORTBbits.RB2 = 1;
            PORTBbits.RB2 = 0;
      }
      return (b);
}
```

Using a 2-Wire Interface

Although the prior methods of expanding the I/O are acceptable, they do require three connections to the microcontroller. A 2-wire interface is required in some cases, and although there is more software, this does save one connection on the microcontroller.

Figure 10-12 illustrates a pair of PCF8574s interfaced to the PIC18F1320 microcontroller to expand the number of I/O connections. The PCF8574s from Texas Instruments use a 2-wire I^2C interface. Unlike the other port expansion devices, this is a bidirectional device that is either

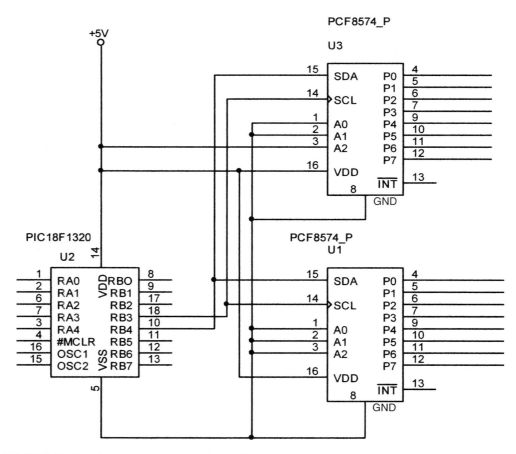

FIGURE 10-12 A 2-wire expansion of the I/O ports.

an input or an output interface. The device features an 8-bit quasi-bidirectional I/O port (P0–P7) pins, including latched outputs with high-current drive capability for directly driving LEDs. The maximum logic zero output current is 25 mA on a port pin and the maximum logic one output current is 300 μA. This is enough current to drive most LEDs. Each quasi-bidirectional I/O pin is used as an input or output without the use of a TRIS register to program the direction. The U3 device is connected to respond to address 100 and U1 is connected to respond to address 000. The software for these devices uses the sw_i2c.h header file in the libraries provided with the C18 compiler to operate the two I/O ports. In the two functions provided in Example 10-7 the address is either 0 or 4 for U1 and U3, respectively. Pins RB3 (clock) and RB4 (data) must be used unless the sw_ic2.h file is recompiled for other pins.

EXAMPLE 10-7

```
// remember #include <sw_i2c.h>

char ReadPort (char address)
{
        char temp;
        SWStartI2C ();
        SWWriteI2C (0xA0 | address);    // control byte
        SWAckI2C();
        temp = SWReadI2C();             // read from U1 or U3
        SWAckI2C();
        SWStopI2C();
        return temp;
}

void WritePort (char address, char data )
{
        SWStartI2C();
        SWWriteI2C (0xA0 | address);    // control byte
        SWAckI2C();
        SWWriteI2C (data);         // write to from U1 or U3
        SWAckI2C();
        SWStopI2C();
}
```

Digital-to-Analog Converters

There is no digital-to-analog converter (DAC) in the PIC microcontroller and at times this function is needed. A peripheral is available from Microchip that converts a digital value into an analog voltage that is inexpensive and easy to interface and control from the microcontroller.

Microchip manufacturers several DAC circuits with serial inputs that are available in 8-pin PDIP integrated circuit packages. The TC1320 is one of them that converts a 8-bit digital number into an analog voltage. Figure 10-13 illustrates the pin-out of the TC1320. In addition to the VDD and VSS connections, a Vref input sets the full-scale output voltage found on the Vout pin, the SCL input for a serial clock, and the SDA connection for the serial data. A DAC-OUT pin provides the output directly from the DAC, but normally the Vout connection is used for the output signal. The DAC_OUT connection is used if an external operational amplifier is used in a

FIGURE 10-13   TC1320 DAC.

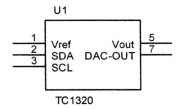

TC1320

system. This is fairly rare because the internal amplifier and the amplified output available at Vout normally provides enough drive for most applications.

The output voltage from the TC1320 is shown in the first equation in Example 10-8. If a 5 V reference is used as the input connection to Vref, the output step voltage is 0.00196 V and the full-scale output is 5 V. Other Vref values can be used to adjust the step voltage and the full-scale output. The second equation in Example 10-8 is used to calculate the value of the digital data applied to the converter to generate a particular output voltage.

EXAMPLE 10-8

$$\text{Vout} = \text{Vref}\left(\frac{\text{DATA}}{256}\right)$$

$$\text{DATA} = \frac{\text{Vout} \times 256}{\text{Vref}}$$

The software interface to the TC1320 normally uses the SPI (serial peripheral interface) using either the software function provided with the C18 libraries or a hardware SPI provided by the internal MSSP interface. The circuit in Figure 10-14 uses the PIC18F1220, which does not have the MSSP unit, so a software SPI feature is added using sw_spi.h located in the C18 libraries.

The operation of the DAC requires that it is placed into normal mode by sending it a 0x90, 0x01 followed by a 0x00. The slave address is factory programmed as 1001 000w and the right bit (w) is a zero for a write operation. If a different slave address is needed, Microchip can provide up to eight different slave addresses at an additional cost. The 0x90, 0x01, followed by a 0x00 programs the DAC to operate in the normal mode. To conserve power, when not converting, the standby mode is entered by sending 0x90, 0x01 followed by a 0x01. To convert a digital value into an analog voltage, a 0x80, 0x00 is followed by the data to be converted. For example, if the Vref is 5 V and the Vout pin is set to 3 V, the sequence of 0x90, 0x00 is followed by a 0x9A. The 0x9A is calculated from the second equation in Example 10-8.

The software to control the interface of Figure 10-14 appears in Example 10-9. The circuit shows a 3-VDC motor connected to the output of the DAC. The speed of the motor is controlled

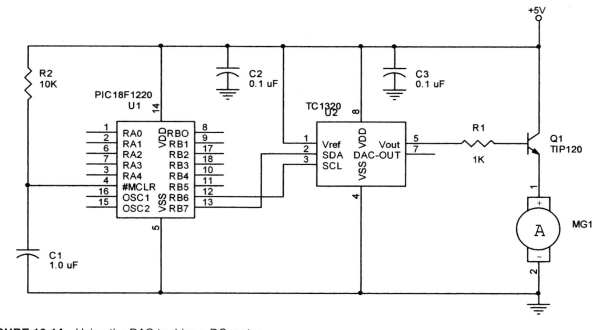

FIGURE 10-14   Using the DAC to drive a DC motor.

by the DAC and the function Speed provided in the software. At times a motor can be controlled in this fashion because it may not start reliably using the pulse width modulator circuit presented in Chapter 7. The TIP120 is a Darlington pair (base to emitter drop of 1.4 V) so the DAC output must range between 1.4 V to 4.4 V to drive the motor from 0 V to 3 VDC. The TIP120 is used as an emitter follower amplifier with a voltage gain of 1.0 V. The output of the DAC is listed at a maximum current of 3 mA. The minimum gain of the TIP120 is 1000. This means that the largest motor that this circuit is able to drive is a 3A motor. The Speed function has uses motor speeds 0 through 152 where 0 is stopped and 152 is full speed. These 153 speed steps should provide enough control over the speed of the motor for most applications.

EXAMPLE 10-9

```c
/*
 * DAC motor control example
 */

#include <p18cxxx.h>
#include <sw_spi.h>

/* Set configuration bits
 *  - set internal oscillator
 *  - disable watchdog timer
 *  - disable low-voltage programming
 *  - disable brownout reset
 *  - enable master clear
 */

#pragma config OSC = INTIO2
#pragma config WDT = OFF
#pragma config LVP = OFF
#pragma config BOR = OFF
#pragma config MCLRE = ON

// function to control the speed of the motor
// through the DAC. Speeds are 0 through 152

void Speed (unsigned char speed)
{
      if (speed >= 0 && speed <= 152)
      {
            speed += 72;          // bias the speed by 1.4 volts
            WriteSWSPI (0x90);         // send speed
            WriteSWSPI (0x00);
            WriteSWSPI (speed);
      }
}

#pragma code

// main program

void main (void)
{
      ADCON1 = 0x0F;          // digital I/O pins
      TRISB = 0;              // Port B is output
      OSCCON = 0x73;          // 8-MHz internal clock

      OpenSWSPI();    // configure SPI port pins

      WriteSWSPI (0x90);         // set to noraml mode
      WriteSWSPI (0x01);
      WriteSWSPI (0x00);

      while (1)
      {
            // do other stuff
      }
}
```

10-4   CAN

The CAN is the **controller area network** used to interface microcontrollers together in a system where microcontrollers must communicate with each other. This interfacing unit called CAN or ECAN (enhanced CAN) is not available on all PIC18F family members. The CAN was originally developed for the automotive industry, but has many applications in many fields where multiple microcontrollers must communicate with each other.

CAN Interconnect

The CAN is like any other area network where data is sent and received between computers or any device that contains the interface. The CAN's main difference is that it is designed to be used with the microcontroller for communication between microcontrollers in a local area in a noisy environment, which is similar to the networks used with the PC except for the amount of noise. The CAN is designed to function in the noisy environments encountered in automobiles and similar environments. The CAN consists of a protocol engine, messaging unit, and control unit, and is fully integrated in various members of the PIC18 family of microcontrollers. As with all networks, the CAN is a serial network that requires two pins on the microcontroller to function. The C18 compiler supports its operation in C language so that software development is efficient.

The RB2 pin is used to transmit data (CANTX) and the RB3 pin is used to receive data (CANRX). These two pins are the only control connections, and are connected to an external bus driver that drives the CAN. Activity on the CAN occurs in either direction between controllers. A microcontroller can be a listener or a talker/listener, which allows for many microcontrollers to be interconnected through the CAN. Figure 10-15 illustrates the typical interconnection of microcontrollers on the CAN. The CAN bus length is 1000 meters without the use of special drivers at a data rate of 40 Kbps. If bus lengths are kept below 40 meters, the bus speed is 1 Mbps. This speed is adequate for most applications that require a network.

The bus nodes are interconnected with a twisted pair of wire (CAT5) as with most networks. The nodes can be connected and disconnected with power applied without any damage to the system. Up to 20 nodes can be used on the CAN without any additional buffering. The CAN bus must be terminated at each end of the bus with a 120 Ω resister. The terminating resister prevents reflections on the bus and in most applications is usually connected across the bus pair at

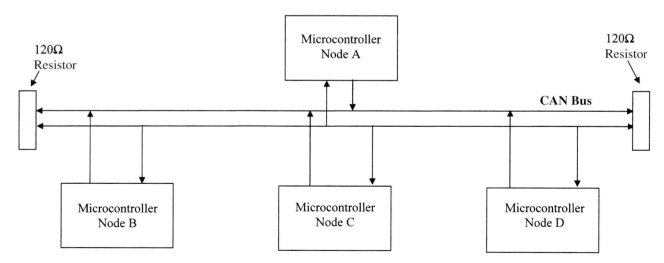

FIGURE 10-15   Microcontrollers connected together on the CAN.

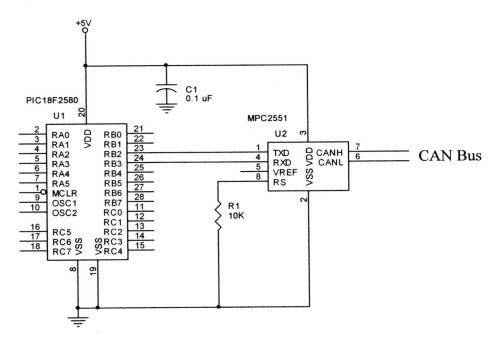

FIGURE 10-16 CAN node.

each node. Figure 10-16 illustrates the typical node with a PIC18F2580 microcontroller and a MPC2551 CAN bus interface device from Microchip. The value for RS is chosen to reduce EMI. The larger its value, the longer the slope of the CAN bus signal and lower the EMI.

CAN Application

A hypothetical application for the CAN is an elevator system. The system needs some type of indicator at each floor to show where the elevator is located, and buttons so that a passenger can summon an elevator to go up or down to the next floor. The elevator car needs a control panel for the passenger to enter the desired floor and a control system to move the elevator car up and down the shaft. The system developed in Figure 10-17 functions for up to a 10-story building, but is easily modified for any number of floors. The greatest change in the system is that buffers are needed because the number of connections to the CAN is limited to 20. Each floor requires it own node on the network, which limits the size of the building without additional buffers.

Each floor needs a controller that is exactly the same as any other floor except for the floor number. The controller has two pushbuttons: one labeled UP and one labeled DOWN, and internal DIO switches to set the floor number. The only exceptions are that the top floor does not have the UP pushbutton and the basement does not have the DOWN pushbutton. It also has large arrow-shaped LED indicators that indicate the direction of travel for the elevator. Above the entrance to the elevator is a column of large LEDs next to numbers that indicate the current floor position of the elevator. Figure 10-17 illustrates the components found with the microcontroller at each floor.

Figure 10-18 illustrates the microcontroller and the components required for the controller used at each floor for the elevator system. The board has four integrated circuits: the PIC18F2580 microcontroller, a CAN interface, and two LED drivers. A set of four DIP switches are used to program the floor number into the controller. A larger building requires a larger DIP switch. Internal weak pull-ups are used on Port B for the two pushbuttons for UP and DOWN and also for the four DIP switches for setting the floor number. The board is small enough to place behind the pushbuttons or the enunciator board containing the floor indicators.

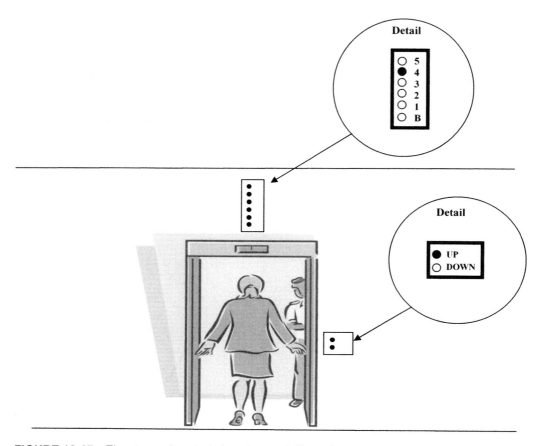

FIGURE 10-17   Elevator and controls found at each floor of a 5-story building.

The software for this system uses the ECAN library provided on the Microchip Website. By copying the ECAN.h, ECAN.def, and ECAN.c files to the project directory and adding the #include <ECAN.h> statement to the program file, the ECAN unit within the microcontroller can be programmed and used in a system. The program for the elevator floor controller appears in Example 10-10. This system transmits whenever the UP or DOWN pushbutton is pressed to inform the elevator car controller, discussed next, that a passenger is waiting at a given floor. The software also listens to the CAN bus for floor announcements by the elevator car controller so the proper floor can be indicated on the enunciator above the door.

EXAMPLE 10-10

```
/*
 * Elevator floor controller
 */

#include <p18cxxx.h>
#include <delays.h>
#include <ECAN.h>

/* Set configuration bits
 *   - set HS oscillator
 *   - disable watchdog timer
 *   - disable low-voltage programming
 *   - disable brownout reset
 *   - enable master clear
 */
```

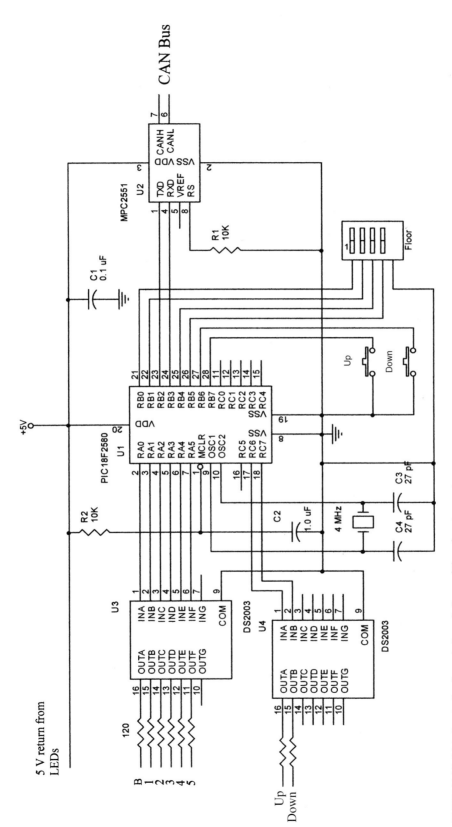

FIGURE 10-18 Controller found at each floor of the elevator system.

```
#pragma config OSC = HS
#pragma config WDT = OFF
#pragma config LVP = OFF
#pragma config BOR = OFF
#pragma config MCLRE = ON

unsigned char floor[1];
char fired;
unsigned long id;
unsigned long iD;
unsigned char data[1];
char dataLen;

#pragma code

// main program

void checkButton (void)
{
    if (PORTB & 0xC0 != 0xC0)
    {
        if (fired == 0)
        {
            Delay1KTCYx(15);              // debounce
            if (PORTB & 0xC0 != 0xC0)
            {
                fired = 1;
                if (PORTBbits.RB7 == 0)
                {
                    PORTCbits.RC6 = 1;         // up
                    floor[0] |= 0x80;
                }
                else                          // down
                {
                    PORTCbits.RC7 = 1;
                    floor[0] &= 0x7F;
                }
                while (!ECANSendMessage(iD,
                    floor,
                    1,
                    ECAN_TX_STD_FRAME));
            }
        }
        else
            fired = 0;
    }
}

void main (void)
{
    ECAN_RX_MSG_FLAGS flags;
    ADCON1 = 0x0f;                     // all port pins digital
    TRISA = 0;                         // program ports
    TRISB = 0xFB;
    TRISC = 0;
    fired = PORTA = PORTC = iD = 0;        // LEDs off
    floor[0] = (PORTB & 3) | ((PORTB & 0x30) << 2);

    ECANInitialize();   // initialize CAN

    while (1)
    {
        checkButton();
        if (ECANReceiveMessage (&id, data, &dataLen, &flags))
        {
            if (id == 1)                    // if from car controller
            {
                PORTA = 1;           // change floor display
                PORTA <<= data[0];
```

```
                         if (data[0] == (floor[0] & 0x7F))
                              PORTC = PORTC & 0x3F;    // UP & DOWN off
                    }
               }
          }
}
```

FIGURE 10-19 Elevator car control panel.

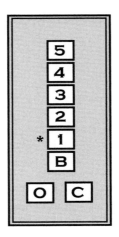

The elevator car also needs a controller that allows a passenger to select a floor, level the elevator car when it stops at a floor, moves the car from one floor to another, and sends the current floor number to all the floor controllers connected to the CAN so they can display the current position of the elevator car. The elevator car controller also announces the floor number and any other information through a speaker in the elevator car. This gives the system a human touch. The DAC is used to power a speaker from a human voice that is sampled at 2 KHz and stored in the memory to provide these vocal messages. Figure 10-19 illustrates the control panel found in the elevator car. This panel is used by passengers to select the floor and contains pushbuttons that are illuminated with LEDs when pressed. What does not appear in this illustration is the speaker that announces floors and the up/down arrow in the car that displays the direction of movement. The elevator-leveling mechanism and motor control system is also not illustrated here.

Figure 10-20 illustrates the schematic for the car controller. The elevator motor is controlled through 24 VAC control voltage signals to the electric motor on the roof. These signals are activated by solid state relays on the elevator car control board. There are two signals: one runs the motor so the car moves up, and the other runs the motor so the car moves down. Leveling is accomplished by timing marks that are located in the elevator shaft and a photo-detector. These are sensed by the controller so that the elevator is stopped in the proper position for a floor.

The DAC is used to generate speech through a speaker that is powered by a TIP120. The TIP120 can produce enough current to drive a very large speaker so a volume control is included to reduce the current through the speaker and the amplitude of the audio messages. Ten audio messages are stored in an array in the program memory. These messages do not appear here and must be generated by the user with another PIC (not shown here) and the ADC to generate the bit streams for the words indicated. These streams must have an amplitude of between 1.4 V and 3.6 V for the system to function correctly.

The elevator car leveling and floor indicator is illustrated in Figure 10-21. These are bar code messages glued to the wall of the elevator shaft and read as the car moves up and down through the shaft. The bar code allows the car to determine the floor number and also position the car so it is level with the floor. The bar code uses binary to encode the floor number with a center mark that allows the system to find the level of the floor.

Example 10-11 lists the software for the elevator car controller. Note that because Port C pins are used in the hardware to control the DAC, the sw_spi.h must be modified so that the

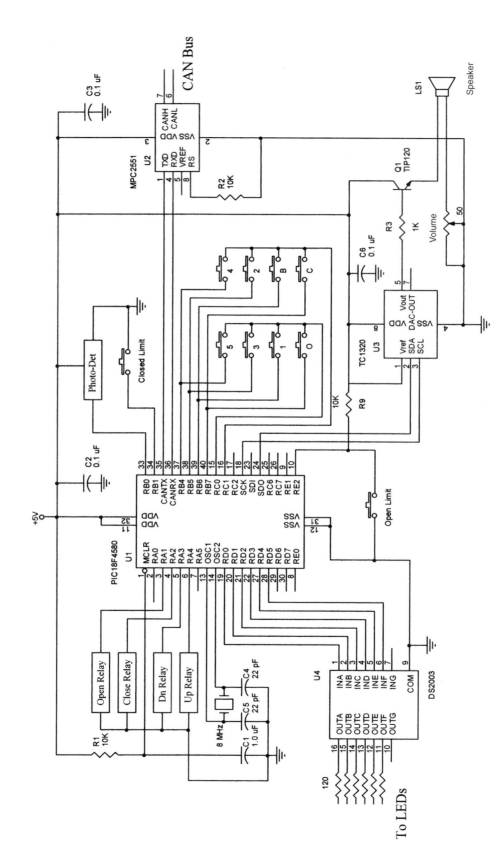

FIGURE 10-20   Schematic of the elevator car controller.

FIGURE 10-21 USB connectors.

Front View

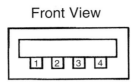

Front View

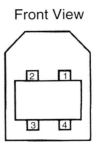

Port C pins are used for the serial clock and serial data connections to the DAC. The ECAN library file must be downloaded from Microchip.com for this software to function.

EXAMPLE 10-11

```
/*
 * Elevator car controller software
 */

#include <p18cxxx.h>
#include <delays.h>
#include <ECAN.h>
#include <spi.h>
#include <timers.h>

/* Set configuration bits
 *  - set HS oscillator
 *  - disable watchdog timer
 *  - disable low-voltage programming
 *  - disable brownout reset
 *  - enable master clear
 */

#pragma config OSC = HSPLL    // 32 MHz
#pragma config WDT = OFF
#pragma config LVP = OFF
#pragma config BOR = OFF
#pragma config MCLRE = ON

// program memory speech
// room for 10 one-second spoken words sampled at 2048 Hz
//
// Audio must be sampled with a minimum level of 1.4 V and a
// maximum level of 3.6 V for the DAC to function correctly.
//          Audio file to be added by the user in the speech array.
//
// 0 = Basement"
// 1 = "First Floor"
// 2 = "Second Floor"
// 3 = "Third Floor"
// 4 = "Fourth Floor"
// 5 = Fifth Floor"
// 6 = "Close the Door"
// 7 = "Going Up"
// 8 = "Going Down"
// 9 = "Initializing"

rom near char speech[10][2048];

rom near char keyt[] =
{
     5, 3, 1, 6, 4, 2, 0, 7
};

void MyHighInt (void);                   // prototypes
```

```
// data memory variables

unsigned char fifo[16];
unsigned char fifoINP;
unsigned char fifoOUTP;
unsigned char SPIbusy;
unsigned char speak;
unsigned int speechCounter;
unsigned int timeDelayCounter;
unsigned char door;
unsigned char callFloor[6];
unsigned char data[1];
unsigned char dataLen;
unsigned currentFloor;
unsigned int barcode[6];
unsigned char barcodePtr;
unsigned char carMoving;        // 1 = up, 2 = down, 0 = stop
char callDirection;                  // 1 = up, 2 = down, 0 = idle
char state;

#pragma interrupt MyHighInt         save=PROD
#pragma code high_vector=0x08       // high_vector is at 0x0008

void high_vector (void)             // high-prioity vector
{
     _asm GOTO MyHighInt _endasm        // goto high software
}

#pragma code

void DoOptical (void);
void DoTimer1 (void);
void DoSPI (void);

void MyHighInt(void)            // new fixed high-interrupt vector
{
     if (INTCONbits.INT0IF = 1)         // if optical reader
          DoOptical();
     else if (PIR1bits.TMR1IF == 1)     // if Timer 1
          DoTimer1();
     else if (PIR1bits.SSPIF == 1)      // if SPI
          DoSPI();
}

void SendFloor (void)
{
     unsigned char floor[1];
     floor[0] = currentFloor;
     while (!ECANSendMessage(1,
                              floor,
                              1,
                              ECAN_TX_STD_FRAME));
}

void WriteFifo (unsigned char data)
{
     if (fifoINP == fifoOUTP && SPIbusy == 0)
     {
          SPIbusy = 1;
          WriteSPI (data);
     }
     else
     {
          fifo[ fifoINP++ ] = data;
          fifoINP &= 15;
     }
}

unsigned int ReadFifo (void)
{
     unsigned int retval = 0x100;
```

```
        if (fifoINP != fifoOUTP)
        {
                retval = fifo[ fifoOUTP++ ];
                fifoOUTP &= 15;
        }
        return retval;
}

void DoSPI (void)
{
        unsigned int data = ReadFifo();
        PIR1bits.SSPIF == 0;
        if (data != 0x100)
                WriteSPI(data);
}

void DoOptical (void)
{
        INTCONbits.INT0IF = 0;
        if (barcodePtr != 0)
                barcode[barcodePtr] = ReadTimer0();
        WriteTimer0 (0);
        barcodePtr++;
}

void DoTimer1 (void)
{
        PIR1bits.TMR1IF == 0;
        WriteTimer1 (-488);
        if (speak != 0)
        {
                WriteFifo (0x90);          // speak!
                WriteFifo (0x00);
                WriteFifo (speech[ speak ][ speechCounter++ ]);
                if (speechCounter == 2047)
                        speak = speechCounter = 0;
        }
        if (timeDelayCounter != 0)     // 1/2048 Hz increments
                timeDelayCounter--;
        if (door == 1 && PORTBbits.RB1 == 0)
        {
                PORTAbits.RA2 = 0;
                door = 0;
        }
        else if (door == 2 && PORTEbits.RE2 == 0)
        {
                PORTAbits.RA1 = 0;
                door = 0;
        }
}

void CloseDoor (void)
{
        while (door != 0);             // wait for door to stop
        PORTAbits.RA2 = 1;             // close door
        door = 1;
}

void OpenDoor (void)
{
        while (door != 0);             // wait for door to stop
        PORTAbits.RA1 = 1;             // open door
        door = 2;
}

void CheckButtons (void)
{
        char but = 0;
        char mask = 0x10;
        PORTC &= 0xFC;
```

```
            if ((PORTC & 0xF0) != 0xF0)
            {
                    Delay1KTCYx(140);
                    if ((PORTC & 0xF0) != 0xF0)
                    {
                            PORTC != 1;
                            if ((PORTC & 0xF0) == 0xF0)
                            {
                                    PORTC != (PORTC & 0xFC) | 2;
                                    but += 4;
                            }
                            while ((PORTC & mask) == mask)
                            {
                                    mask <<= 1;
                                    but++;
                            }
                            but = keyt[but];
                            if (but == 6)
                            {
                                    OpenDoor();
                                    timeDelayCounter = 2048 * 5;    // 5 sec
                            }
                            else if (but == 7)
                                    CloseDoor();
                            else
                                    callFloor[ but ] |= 4;
                    }
            }
    }

// callFloor data:
//   0 = no call
//   1 = call remote up
//   2 = call remote down
//   4 = call local

void CheckCAN (void)
{
        ECAN_RX_MSG_FLAGS flags;
        if (ECANReceiveMessage (0, data, &dataLen, &flags))
        {
                if ((data[0] & 0x80) == 0x80)
                        callFloor[ data[0] & 0x7F ] |= 1;
                else
                        callFloor[ data[0] ] |= 2;
        }
}

void LevelCar (void)
{
        if (carMoving == 1)
        {
                barcodePtr = 0;
                PORTAbits.RA3 = 1;
                while (barcodePtr == 0);
                PORTAbits.RA3 = 0;
                barcodePtr = 0;
                PORTAbits.RA4 = 1;
                while (barcodePtr == 0);
                PORTAbits.RA4 = 0;
                barcodePtr = 0;
                PORTAbits.RA3 = 1;
                while (barcodePtr == 0);
                PORTAbits.RA3 = 0;
        }
        else
        {
                barcodePtr = 0;
                PORTAbits.RA4 = 1;
```

```
                        while (barcodePtr == 0);
                        PORTAbits.RA4 = 0;
                        barcodePtr = 0;
                        PORTAbits.RA3 = 1;
                        while (barcodePtr == 0);
                        PORTAbits.RA3 = 0;
                        barcodePtr = 0;
                        PORTAbits.RA4 = 1;
                        while (barcodePtr == 0);
                        PORTAbits.RA4 = 0;
                }
}

unsigned char ReadBarcode (void)
{
        char a;
        unsigned int b = 0xFFFF;
        for (a = 0; a < 3; a++)
                if (barcode[a] < b)
                        b = barcode[a];
        b *= 3;
        a = 0;
        if (barcode[0] > b)
                a |= 4;
        if (barcode[1] > b)
                a |= 2;
        if (barcode[2] > b)
                a |= 1;
        a--;
        return a;
}

void DoElevator (void)
{
        char a;
        if (state == 0)
        {
                for (a = 0; a < 6; a++)
                        if (callFloor[a] != 0)
                                break;
                if (a != 6)
                {
                        if (currentFloor < a)
                        {
                                PORTAbits.RA4 = 1;        // move car up
                                callDirection = 1;
                                speak = 7;                // say "Going Up"
                        }
                        else
                        {
                                PORTAbits.RA3 = 1;        // move car down
                                callDirection = 2;
                                speak = 8;                // say "Going Down"
                        }
                        state = 1;
                        timeDelayCounter = 4092;
                }
        }
        else if (state == 1)
        {
                barcodePtr = 0;
                state = 2;
        }
        else if (state == 2)
        {
                if (barcodePtr == 4)
                {
                        if (ReadBarcode() == a)
```

```
                        {
                                currentFloor = a;
                                callFloor[a] = 0;
                                PORTAbits.RA4 = 0;
                                PORTAbits.RA3 = 0;
                                LevelCar();
                                OpenDoor();
                                speak = a;              // say floor number
                        }
                        else
                        state = 3;
                }
        }
        else if (state == 3);
        {
                state = 1;
                timeDelayCounter = 4096;
        }
}

void FindFloor (void)
{
        unsigned char flr[1];
        PORTAbits.RA4 = 1;              // move car up
        timeDelayCounter = 4096;       // 2 seconds
        while (timeDelayCounter != 0);     // wait
        PORTAbits.RA4 = 0;             // stop car
        barcodePtr = 0;
        PORTAbits.RA3 = 1;             // move car down
        while (barcodePtr != 4);       // wait for bar code
        flr[0] = ReadBarcode();
        while (!ECANSendMessage(0,
                        flr,
                        1,
                        ECAN_TX_STD_FRAME));
        PORTAbits.RA3 = 0;                  // stop car
        carMoving = 2;
        LevelCar();
        speak = flr[0];                         // say floor number
        currentFloor = flr[0];
        carMoving = 0;
}

void main (void)
{
        char a;
        ADCON1 = 0x0F;                 // select digital I/O
        TRISA = 0;                     // program Port A
        TRISB = 0xFB;                  // program Port B
        TRISC = 0;                     // program Port C
        TRISD = 0;                     // program Port D
        TRISE = 4;                     // program Port E

        PORTA = PORTC = PORTD = SPIbusy = speak = 0;
        timeDelayCounter = speechCounter = door = 0;
        barcodePtr = carMoving = state = callDirection = 0;

        for (a = 0; a < 6; a++)
                callFloor[a] = bar code[a] = 0;

        ECANInitialize();        // initialize CAN

        INTCON2bits.RBPU = 0;          // Port B pullups on

        OpenTimer1 (TIMER_INT_ON &     // 1 us timer
                T1_16BIT_RW &
                T1_SOURCE_INT &
                T1_PS_1_8);

        WriteTimer1 (-488);                    // fire every 488 us
                                               // 1/2048 Hz for audio
```

```
OpenTimer0 (TIMER_INT_OFF &      // 32 us
    T0_16BIT &
    T0_SOURCE_INT &
    T0_PS_1_256);

PIE1bits.SSPIE = 1;              // enable SPI interrupt
RCONbits.IPEN = 0;              // only high-priority interrupt on
INTCON2bits.INTEDG0 = 0;        // make INT0 negative edge triggered
INTCONbits.INT0IE = 1;          // enable INT0
INTCONbits.GIE = 1;             // enable interrupts

OpenSPI (SPI_FOSC_16, MODE_00, SMPEND);   // configure SPI port pins

WriteFifo (0x90);      // set to noraml mode
WriteFifo (0x01);
WriteFifo (0x00);

WriteFifo (0x90);      // speaker off
WriteFifo (0x00);
WriteFifo (0x00);

speak = 9;                      // say "Initilizing"
CloseDoor();
while (door != 0);              // wait for door
FindFloor();
OpenDoor();
timeDelayCounter = 2048 * 5;    // 5 sec

while (1)
{
    CheckButtons();             // see if a local button is pressed
    CheckCAN();                 // see if the CAN has anything
    if (timeDelayCounter == 0)
    {
        CloseDoor();
        DoElevator();           // run the elevator
    }
}
}
```

10-5   USB

The **universal serial bus** (USB) solves a problem with the personal computer system and many embedded applications: high-speed serial data with a bus that provides power to anything connected to it. Other benefits are ease of user connection and access to up to 127 different connections through a four-connection serial cable. This interface is ideal for embedded microcontrollers, keyboards, sound cards, simple video-retrieval devices, and modems. Data transfer speeds are 480 Mbps for full-speed USB 2.0 operation, 11 Mbps for USB 1.1 compliant transfers, and 1.5 Mbps for slow-speed operation. Depending on the version of the PIC, the transfer speed can be USB1.0 through USB 2.0.

Cable lengths are limited to three meters for the full-speed interface and five meters for the low-speed interface. The maximum power available through these cables is rated at 100 mA maximum current at 5 V. If the amount of current exceeds 100 mA, Windows displays a yellow exclamation point next to the device indicating an overload condition.

USB Connector

Figure 10-21, on page 389, illustrates the pin-out of the USB connector. Two types of connectors are specified and both are in use. Each case has four pins on each connector, which contain the signals indicated in Table 10-2. As mentioned, the +5 V and ground signals can be used to power devices connected to the bus as long as the amount of current does not exceed 100 mA per device. The data signals are bi-phase signals. When +data are at 3.3 V, −data are at 0 V, and vice versa.

TABLE 10-2 USB pin configuration.

Pin Number	Signal
1	+ 5.0V
2	– Data
3	+ Data
4	Ground

FIGURE 10-22 Sample interface\ing component.

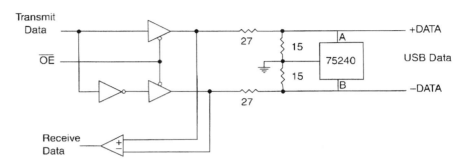

USB Data

The data signals are bi-phase signals that are generated using a circuit such as the one illustrated in Figure 10-22. The line receiver is also illustrated in Figure 10-22. Placed on the transmission pair is a noise-suppression circuit available from Texas Instruments (SN75240). Once the transceiver is in place, interfacing to the USB is complete. The 75773 integrated circuit from Texas Instruments functions as both the differential line driver and receiver for this schematic.

The next phase is learning how the signals interact on the USB. These signals allow data to be sent and received from the host computer system. The USB uses NRZI (non-return to zero, inverted) data encoding for transmitting packets. This encoding method does not change the signal level for the transmission of a logic one, but the signal level is inverted for each change to a logic zero. Figure 10-23 illustrates a digital data stream and the USB signal produced using this encoding method.

The actual data transmitted includes sync bits using a method called *bit stuffing*. If a logic one is transmitted for more than six bits in a row, the bit stuffing technique adds an extra bit (logic zero) after six continuous 1s in a row. Because this lengthens the data stream, it is called bit stuffing. Figure 10-24 shows a bit-stuffed serial data stream and the algorithm used to create it from raw digital serial data. Bit stuffing ensures that the receiver can maintain synchronization for long strings of 1s. The same technique stuffs a one-bit into the stream after six zeros in a row. Data are always transmitted beginning with the least-significant bit first, followed by subsequent bits.

USB Commands

To begin communications, the sync byte (80H) is transmitted first, followed by the packet identification byte (PID). The PID contains eight bits, but only the rightmost four bits contain the

FIGURE 10-23 NRZI data.

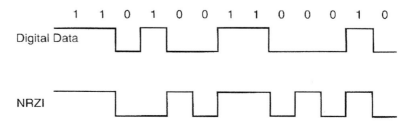

FIGURE 10-24 Data stream and flowchart used to generate USB data.

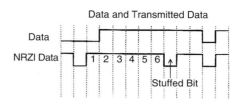

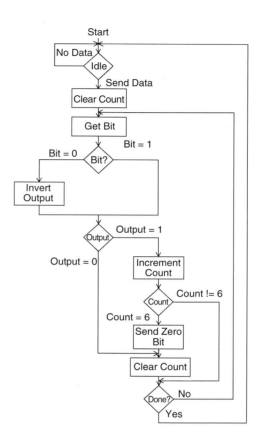

type of packet that follows, if any. The leftmost four bits of the PID are the ones complementing the rightmost four bits. For example, if a command of 1000 is sent, the actual byte sent for the PID is a 0111 1000. Table 10-3 shows the available 4-bit PIDs and their 8-bit codes. Notice that PIDs are used as token indicators, as data indicators, and for handshaking.

Figure 10-25 lists the formats of the data, token, handshaking, and start-of-frame packets found on the USB. In the token packet, the ADDR (address field) contains the 7-bit address of the USB device. As mentioned earlier, up to 127 devices are present on the USB at a time. The ENDP (endpoint) is a 4-bit number used by the USB. Endpoint 0000 is used for initialization, whereas other endpoint numbers are unique to each USB device.

There are two types of CRC (**cyclic redundancy checks**) used on the USB: a 5-bit and a 16-bit CRC (used for data packets). The 5-bit CRC is generated with the $X5 + X2 + 1$ polynomial; the 16-bit CRC is generated with the $X16 + X15 + X2 + 1$ polynomial. When constructing circuitry to generate or detect the CRC, the plus signs represent exclusive-OR circuits. The CRC circuit or program is a serial checking mechanism. When using the 5-bit CRC, a residual of 01100 is received for no error in all five bits of the CRC and the data bits. With the 16-bit CRC, the residual is 1000000000001101 for no error.

TABLE 10-3 PID codes.

PID	Name	Type	Description
E1	OUT	Token	Host → function transaction
D2	ACK	Handshake	Receiver accepts packet
C3	Data0	Data	Data packet (PID even)
A5	SOF	Token	Start of frame
69	IN	Token	Function → host transaction
5A	NAK	Handshake	Receiver does not accept packet
4B	Data1	Data	Data packet (PID odd)
3C	PRE	Special	Host preamble
2D	Setup	Token	Setup command
1E	Stall	Token	Stalled

The USB uses the ACK and NAK tokens to coordinate the transfer of data packets between the host system and the USB device. Once a data packet is transferred from the host to the USB device, the USB device transmits either an ACK (acknowledge) or a NAK (not acknowledge) token back to the host. If the data and CRC are received correctly, the ACK is sent; if not, the NAK is sent. If the host receives a NAK token, it retransmits the data packet until the receiver

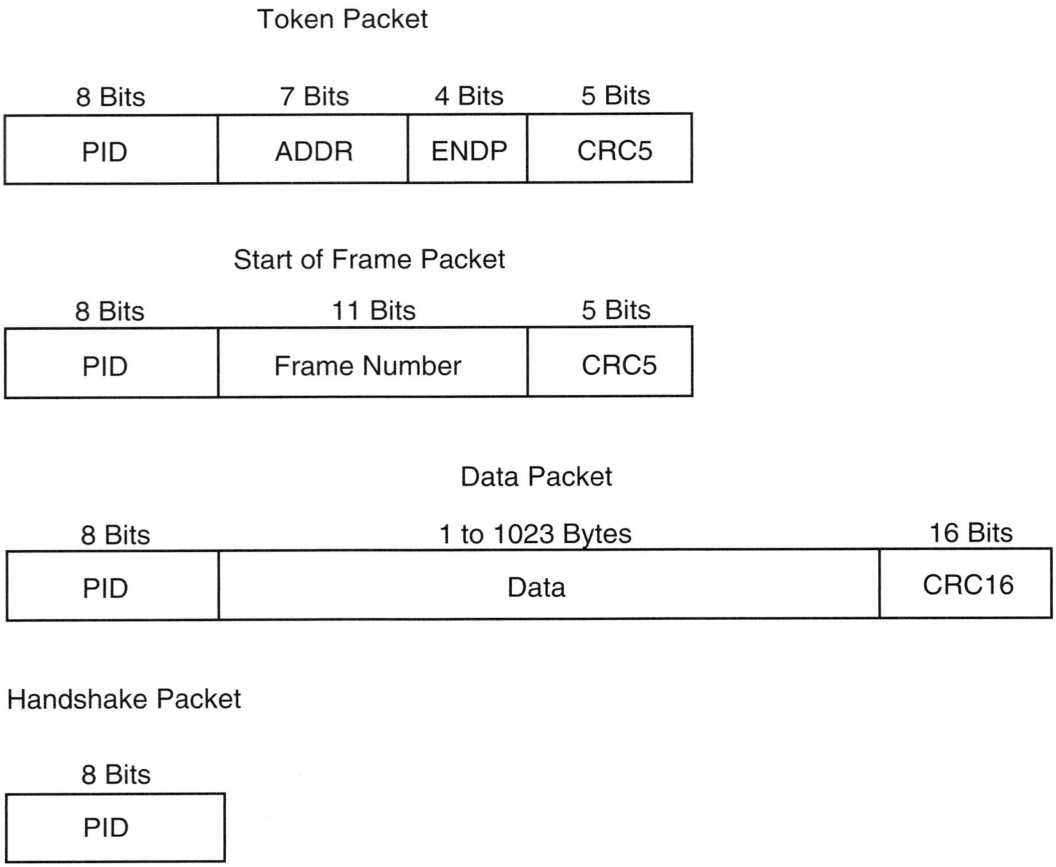

Token Packet

8 Bits	7 Bits	4 Bits	5 Bits
PID	ADDR	ENDP	CRC5

Start of Frame Packet

8 Bits	11 Bits	5 Bits
PID	Frame Number	CRC5

Data Packet

8 Bits	1 to 1023 Bytes	16 Bits
PID	Data	CRC16

Handshake Packet

8 Bits
PID

FIGURE 10-25 Packets on the USB.

finally receives it correctly. This method of data transfer is often called stop and wait flow control. The host must wait for the client to send an ACK or NAK before transferring additional data packets.

PIC18 and the USB

The USB is directly supported by a few of the PIC18 microcontrollers. It allows the microcontroller to be interfaced to a PC and the USB for virtually any application that needs this type of connectivity. A library is provided for the C18 compiler, but must be downloaded from the Microchip website in the High Speed USB demo board section entitled MCHPFSUSB. This section assumes that the file is downloaded and installed so the software illustrated can be used to program the PIC18F4550 microcontroller, which contains a high-speed USB interface. The package contains the library files for the C18 compiler and also the driver for installation on the PC to enable communications to the microcontroller. These functions and the driver are used as the basis of any USB interface between the PC to the microcontroller.

Figure 10-26 illustrates the internal structure of the USB interface in the PIC18F4550 microcontroller. The USB interface on the PIC18 is used in this section to interface to the PC through the usbser.sys, which is a driver that exists in the Windows XP (also available on Windows

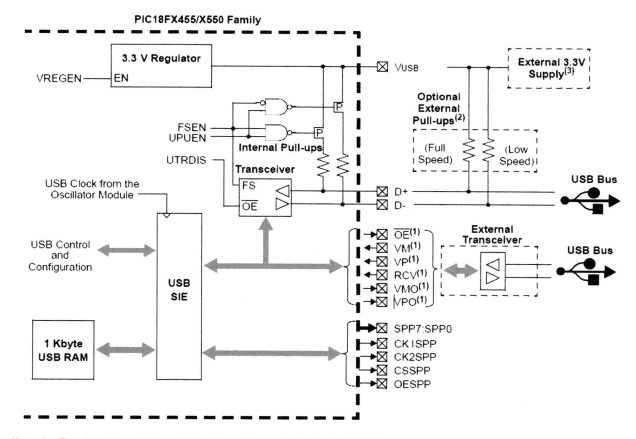

Note 1: This signal is available only if the internal transceiver is disabled (UTRDIS = 1).

    2: The internal pull-up resistors should be disabled (UPUEN = 0) if external pull-up resistors are used.

    3: Do not enable the internal regulator when using an external 3.3 V supply.

FIGURE 10-26 Internal USB structure (Courtesy of Microchip).

FIGURE 10-27   PIC18F4550
connected to the USB.

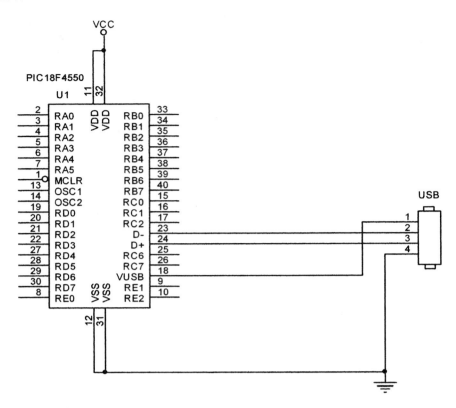

98SE and Windows 2000) operating system. By using this driver a new driver for Windows does not need to be written.

Figure 10-27 illustrates the PIC18F4550 interfaced to a PC through a USB interface connector. Notice that very few components are required for the interface and it is possible to power the PIC18F4550 microcontroller from the USB connector as long as the entire embedded system does not require more than 100 mA. In the circuit illustrated, the internal 3.3 V, VUSB voltage is used to power the USB.

The steps required to illustrate the operation of this simple interface to the USB and how to configure the software for both the microcontroller and the PC follow. To explain this interface the RS-232C emulation for the USB is chosen because it is the easiest to understand and apply to the interface between the PC and the microcontroller. Microchip provides the driver for both the PC and the PIC18 on their website as files for the high-speed USB PicDem board. The software presented here assumes that this software has been downloaded and the driver for the PC is installed. In the IDE, load the workspace (MCHPUSB.mcp) in the C:\MCHPFSUSB\fw\Cdc folder if the package is installed from Microchip. Open the user.c file which contains a place for the user program that uses the USB and its connection to the PC. This software is written for the PIC18F4550 microcontroller. The user.c file, appears in Example 10-12.

EXAMPLE 10-12

```
* FileName:          user.c
* Dependencies:      See INCLUDES section below
* Processor:         PIC18
* Compiler:          C18 2.30.01+
* Company:           Microchip Technology, Inc.
*
* Software License Agreement
*
* The software supplied herewith by Microchip Technology Incorporated
* (the "Company") for its PICmicro® Microcontroller is intended and
* supplied to you, the Company's customer, for use solely and
* exclusively on Microchip PICmicro Microcontroller products. The
* software is owned by the Company and/or its supplier, and is
* protected under applicable copyright laws. All rights are reserved.
* Any use in violation of the foregoing restrictions may subject the
* user to criminal sanctions under applicable laws, as well as to
* civil liability for the breach of the terms and conditions of this
* license.
*
* THIS SOFTWARE IS PROVIDED IN AN "AS IS" CONDITION. NO WARRANTIES,
* WHETHER EXPRESS, IMPLIED, OR STATUTORY, INCLUDING, BUT NOT LIMITED
* TO, IMPLIED WARRANTIES OF MERCHANTABILITY AND FITNESS FOR A
* PARTICULAR PURPOSE APPLY TO THIS SOFTWARE. THE COMPANY SHALL NOT,
* IN ANY CIRCUMSTANCES, BE LIABLE FOR SPECIAL, INCIDENTAL, OR
* CONSEQUENTIAL DAMAGES, FOR ANY REASON WHATSOEVER.
*
* Author               Date        Comment
*~~~~~~~~~~~~~~~~~~~~~~~~~~~~~~~~~~~~~~~~~~~~~~~~~~~~~~~~~~~~~~~~~~~~~~~~~
* Rawin Rojvanit        11/19/04    Original.
********************************************************************/

/********************************************************************
* CDC RS-232 Emulation Tutorial Instructions:
********************************************************************
* Refer to Application Note AN956 for an explanation of the CDC class.
*
* First take a look at Exercise_Example() and study how functions are called.
*
* There are five exercises, each one has a solution in the CDC\user\solutions.
* Scroll down and look for Exercise_01,_02,_03,_04, and _05.
* Instructions on what to do is inside each function.
*
********************************************************************/

/** I N C L U D E S ***********************************************/
#include <p18cxxx.h>
#include <usart.h>
#include "system\typedefs.h"

#include "system\usb\usb.h"

#include "io_cfg.h"              // I/O pin mapping
#include "user\user.h"
#include "user\temperature.h"

/** V A R I A B L E S *********************************************/
#pragma udata
byte old_sw2,old_sw3;

char input_buffer[64];
char output_buffer[64];

rom char welcome[]={"PIC18F4550 Full-Speed USB - CDC RS-232 Emulation
Demo\r\n\r\n"};
rom char ansi_clrscr[]={"\x1b[2J"};               // ANSI Clear Screen Command

/** P R I V A T E   P R O T O T Y P E S ***************************/
void InitializeUSART (void);
void BlinkUSBStatus (void);
BOOL Switch2IsPressed (void);
BOOL Switch3IsPressed (void);
```

```
void Exercise_Example (void);

void Exercise_01 (void);
void Exercise_02 (void);
void Exercise_03 (void);
void Exercise_04 (void);
void Exercise_05 (void);

/** D E C L A R A T I O N S ***************************************************/
#pragma code
void UserInit (void)
{
      mInitAllLEDs();
      mInitAllSwitches();
      old_sw2 = sw2;
      old_sw3 = sw3;

      InitTempSensor();

      InitializeUSART();

}//end UserInit

void InitializeUSART (void)
{
      TRISCbits.TRISC7=1;        // RX
      TRISCbits.TRISC6=0;        // TX
      SPBRG = 0x71;
      SPBRGH = 0x02;             // 0x0271 for 48MHz -> 19200 baud
      TXSTA = 0x24;              // TX enable BRGH=1
      RCSTA = 0x90;              // continuous RX
      BAUDCON = 0x08;            // BRG16 = 1
}//end InitializeUSART

/****************************************************************************
 * Function:        void ProcessIO (void)
 *
 * PreCondition:    None
 *
 * Input:           None
 *
 * Output:          None
 *
 * Side Effects:    None
 *
 * Overview:        This function is a placeholder for other user routines.
 *                  It is a mixture of both USB and non-USB tasks.
 *
 * Note:            None
 ****************************************************************************/
void ProcessIO (void)
{
      BlinkUSBStatus();
      // User Application USB tasks
      if ((usb_device_state < CONFIGURED_STATE)||(UCONbits.SUSPND==1)) return;

      Exercise_Example();

      Exercise_01();
      Exercise_02();
      Exercise_03();
      Exercise_04();
      Exercise_05();

}//end ProcessIO

void Exercise_Example (void)
{
      static byte start_up_state = 0;

      if (start_up_state == 0)
```

```
            {
                if (Switch2IsPressed())
                    start_up_state++;
            }
            else if (start_up_state == 1)
            {
                if (mUSBUSARTIsTxTrfReady())
                {
                    putrsUSBUSART(ansi_clrscr);
                    start_up_state++;
                }
            }
            else if (start_up_state == 2)
            {
                if (mUSBUSARTIsTxTrfReady())
                {
                    putrsUSBUSART("\rMicrochip Technology Inc., 2004\r\n");
                    start_up_state++;
                }
            }
            else if (start_up_state == 3)
            {
                if (mUSBUSARTIsTxTrfReady())
                {
                    putrsUSBUSART(welcome);
                    start_up_state++;
                }
            }
}//end Exercise_Example

void Exercise_01 (void)
{
        /*
         * Write code in this function that sends a literal null-terminated
         * string of text ("Hello World!\r\n") to the PC when switch 2 is
         * pressed.
         *
         * Useful functions:
         *  Switch2IsPressed() returns '1' when switch 2 is pressed.
         *  putrsUSBUSART(...);
         *
         * See examples in Exercise_Example();
         *
         * Remember, you must check if cdc_trf_state is ready for another
         * transfer or not. When it is ready, the value will equal CDC_TX_READY,
         * or use macro: mUSBUSARTIsTxTrfReady()
         */

        /* Insert code here - 3 lines */

        /* End */

}//end Exercise_01

rom char ex02_string[]={"Type in a string here.\r\n"};
void Exercise_02 (void)
{
        /*
         * Write code in this function that sends a null-terminated string
         * of text stored in program memory pointed to by "ex02_string" to
         * the PC when switch 3 is pressed.
         *
         * ex02_string is declared right above this function.
         *
         * Useful functions:
         *  Switch3IsPressed() returns '1' when switch 3 is pressed.
         *  putrsUSBUSART(...);
         *
         * See examples in Exercise_Example();
```

```
      *
      * Remember, you must check if cdc_trf_state is ready for another
      * transfer or not. When it is ready, the value will equal CDC_TX_READY,
      * or use macro: mUSBUSARTIsTxTrfReady()
      */

      /* Insert code here - 3 lines*/

      /* End */

}//end Exercise_02

void Exercise_03(void)
{
      /*
      * Write code in this function that reads data from USB and
      * toggles LED D4 when the data read equals ASCII character '1' (0x31)
      *
      * Useful functions:
      *  byte getsUSBUSART (char *buffer, byte len)   See cdc.c for details
      *  mLED_4_Toggle();
      *
      * Use input_buffer[] to store data read from USB.
      */

      /* Insert code here - 3 lines */

      /* End */

}//end Exercise_03

void Exercise_04 (void)
{
      /*
      * Before starting Exercise_04(), comment out the call to Exercise_01()
      * in ProcessIO(); this function will need to check Switch2IsPressed().
      *
      * Write code in this function that sends the following 4 bytes of
      * data: 0x30,0x31,0x32,0x33 when switch 2 is pressed.
      * Note that these data are not null-terminated and are located in
      * the data memory.
      *
      * Useful functions:
      *  Switch2IsPressed() returns '1' when switch 2 is pressed.
      *  mUSBUSARTTxRam(byte *pData, byte len) See cdc.h for details.
      *
      * Use output_buffer[] to store the four bytes data.
      *
      * Remember, you must check if cdc_trf_state is ready for another
      * transfer or not. When it is ready, the value will equal CDC_TX_READY,
      * or use macro: mUSBUSARTIsTxTrfReady()
      */

      /* Insert code here - 7 lines */

      /* End */

}//end Exercise_04

void Exercise_05 (void)
{
      /*
      * The PICDEM Full-Speed USB Demo Board is equipped with a
      * temperature sensor. See temperature.c & .h for details.
      *
      * All necessary functions to collect temperature data have
      * been called. This function updates the data a few times
      * every second. The program currently sends out the
      * temperature data to the PC via UART.
      *
      * You can check this by hooking up a serial cable and
      * set your serial port to 19,200 baud, 8 bit data, 1 stop,
```

```
      * no parity.
      *
      * The program assumes the CPU frequency = 48 MHz to generate
      * the correct SPBRG value for 19,200 baud transmission.
      *
      * Modify the code to send the ASCII string stored in
      * tempString to the PC via USB instead of UART.
      *
      * The temperature data is stored in tempString array in
      * ASCII format and is null-terminated.
      *
      * Useful function:
      *   putsUSBUSART(...);
      *
      *  Note: It is 'puts' and not 'putrs' to be used here.
      *
      * Remember, you must check if cdc_trf_state is ready for another
      * transfer or not. When it is ready, the value will equal CDC_TX_READY,
      * or use macro: mUSBUSARTIsTxTrfReady()
      */

      static word ex05_count;

      if (ex05_count == 0)
      {
              AcquireTemperature();              // read temperature from sensor
              UpdateCelsiusASCII();              // convert to ASCII, stored in
                                                 // "tempString", see temperature.c

              /* Modify the code below - 3 lines */

                    putsUSART (tempString);
                    ex05_count = 10000;

              /* End */
      }
      else
              ex05_count--;

}//end Exercise_05

/****************************************************************************
 * Function:        void BlinkUSBStatus (void)
 *
 * PreCondition:    None
 *
 * Input:           None
 *
 * Output:          None
 *
 * Side Effects:    None
 *
 * Overview:        BlinkUSBStatus turns on and off LEDs corresponding to
 *                  the USB device state.
 *
 * Note:            mLED macros can be found in io_cfg.h
 *                  usb_device_state is declared in usbmmap.c and is modified
 *                  in usbdrv.c, usbctrltrf.c, and usb9.c
 ****************************************************************************/
void BlinkUSBStatus (void)
{
      static word led_count=0;

      if (led_count == 0)led_count = 10000U;
      led_count--;

      #define mLED_Both_Off()                    {mLED_1_Off();mLED_2_Off();}
      #define mLED_Both_On()                     {mLED_1_On();mLED_2_On();}
      #define mLED_Only_1_On()                   {mLED_1_On();mLED_2_Off();}
      #define mLED_Only_2_On()                   {mLED_1_Off();mLED_2_On();}

      if (UCONbits.SUSPND == 1)
```

```
            {
                if (led_count==0)
                {
                        mLED_1_Toggle();
                        mLED_2 = mLED_1;              // both blink at the same time
                }//end if
            }
            else
            {
                if (usb_device_state == DETACHED_STATE)
                {
                        mLED_Both_Off();
                }
                else if (usb_device_state == ATTACHED_STATE)
                {
                        mLED_Both_On();
                }
                else if (usb_device_state == POWERED_STATE)
                {
                        mLED_Only_1_On();
                }
                else if (usb_device_state == DEFAULT_STATE)
                {
                        mLED_Only_2_On();
                }
                else if (usb_device_state == ADDRESS_STATE)
                {
                        if (led_count == 0)
                        {
                            mLED_1_Toggle();
                            mLED_2_Off();
                        }//end if
                }
                else if (usb_device_state == CONFIGURED_STATE)
                {
                        if (led_count==0)
                        {
                            mLED_1_Toggle();
                            mLED_2 = !mLED_1;        // alternate blink
                        }//end if
                }//end if(...)
            }//end if (UCONbits.SUSPND...)

}//end BlinkUSBStatus

BOOL Switch2IsPressed (void)
{
        if (sw2 != old_sw2)
        {
                old_sw2 = sw2;                       // save new value
                if (sw2 == 0)                        // if pressed
                        return TRUE;                 // was pressed
        }//end if
        return FALSE;                                // was not pressed
}//end Switch2IsPressed

BOOL Switch3IsPressed(void)
{
        if (sw3 != old_sw3)
        {
                old_sw3 = sw3;                       // save new value
                if (sw3 == 0)                        // if pressed
                        return TRUE;                 // was pressed
        }//end if
        return FALSE;                                // was not pressed
}//end Switch3IsPressed

/** EOF user.c *******************************************************************/
```

   To illustrate the use of this shell, first remove all the lines of code except those illustrated in Example 10-13. This removes all of the examples that do not apply to the application developed in the text, and provides a workable shell for developing USB software for interfacing to the personal computer. As can be seen, most of the code is removed to obtain a workable shell for the USB application.

EXAMPLE 10-13

```
/*************** USB Application 2.0 Shell ********************************/
/*     To be used with the files provided by Microchip for the USB
 */

/** I N C L U D E S ****************************************************/
#include <p18cxxx.h>        // add this
#include "system\typedefs.h"
#include "system\usb\usb.h"
#include "user\user.h"

/** V A R I A B L E S *************************************************/
#pragma udata

     // variables go here

/** D E C L A R A T I O N S ******************************************/
#pragma code
void UserInit (void)
{

     // initialization goes here

}//end UserInit

void ProcessIO (void)
{

     // any I/O tasks go here

}//end ProcessIO
```

   Another way to handle the USB is to create a shell that uses the include files in the package and the main function main.c. This technique is used here to develop a simple application, and the shell for main.c appears in Example 10-14. The software in Example 10-14 the include files illustrated in Figure 10-28 are added.

EXAMPLE 10-14

```
/*
 * Example shell that uses USART emulation for the USB
 */

#include <p18cxxx.h>
#include "system\typedefs.h"                   // Required
#include "system\usb\usb.h"                     // Required
#include "io_cfg.h"                             // Required

/* Set configuration bits
 *   - set HS oscillator (4 MHz)
 *   - disable watchdog timer
 *   - disable low-voltage programming
 *   - disable brownout reset
 *   - enable master clear
 */

#pragma config FOSC = HSPLL_HS
#pragma config PLLDIV = 1
#pragma config VREGEN = ON
#pragma config WDT = OFF
#pragma config LVP = OFF
```

FIGURE 10-28 Screen shot of the project files screen (the main program is in example1.c).

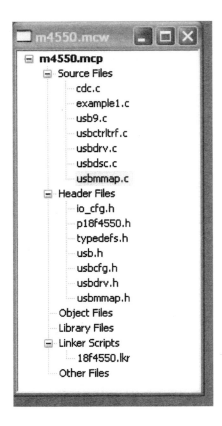

```
#pragma config BOR = OFF
#pragma config MCLRE = ON

void MyHighInt (void);                    // prototypes

#pragma interrupt MyHighInt       save=PROD
#pragma code high_vector=0x08     // high_vector is at 0x0008

void high_vector (void)                   // high-prioity vector
{
      _asm GOTO MyHighInt _endasm          // goto high software
}

#pragma code

void MyHighInt (void)
{

}

// main program

void main (void)
{
          mInitializeUSBDriver();        // see usbdrv.h
}
```

Now that the shell exists for programming the USB, a sample interface is needed to illustrate the function of the USB with the personal computer. Suppose a system is needed to test PC ATX power supplies. This system is designed to test 250-watt power supplies, but can test larger supplies under partial load. The newer supplies have a 24-pin connector so a mating socket is needed to connect to the power supply under test. The operator plugs in the power supply and the system monitors the supply and displays the voltages on the PC. To build this system, the microcontroller is used to monitor the supply voltages and a Visual C application is written to display the voltages on the PC. The visual C application sees the PIC as a serial COM port so no special driver is needed to complete the PC side of the application. The only thing that suffers from this technique is the speed of the transfer, which abides by the transfer rate set for the COM port in the Visual C program. In many applications this presents no problem.

Figure 10-29 illustrates the connection of the PIC18F4550 for this application and Table 10-4 illustrates the pin-out of the ATX power supply connector. The ATX supply has four output voltages: +12 V, −12 V, +5 V and +3.3 V. Each is loaded with an average load and monitored by the ADC inside of the microcontroller. The ADC can accept only a maximum of 5 V, so the voltages are reduced using voltage dividers. In the case of the −12 V supply, an operational amplifier changes the polarity and reduces its voltage. The +12 V signal is reduced by a factor of 4, the −12 V signal is reduced by a factor of 3 and inverted, the +5 V signal is reduced by a factor of 2, and the 3.3 V signal is not reduced. This yields full-scale voltages of 3 V, 4 V, 2.5 V and 3.3 V for the power supply voltages as monitored by the microcontroller—all fall within the acceptable limits of the ADC.

The program that reads the ATX voltages and sends them through the USB to a PC is listed in Example 10-15. This software sends a character string to the PC with the number 1 in ASCII as the first character in the string for the 12 V raw reading from the ADC. The 3.3 V raw reading is sent to the PC beginning with an ASCII 2, the −12 V raw reading is sent beginning with an ASCII 3, and the +5V raw reading is sent beginning with an ASCII 4. The software on the PC must convert the raw reading to the actual voltages. The voltages are read once every half second and sent to the PC in this example.

EXAMPLE 10-15

```
/*
 * ATX power supply monitor using the USB
 */

#include <p18cxxx.h>
#include "system\typedefs.h"
#include "system\usb\usb.h"
#include "io_cfg.h"
#include <timers.h>
#include <adc.h>
#include <stdlib.h>

/* Set configuration bits
 *   - set HS oscillator (4 MHz)
 *   - disable watchdog timer
 *   - disable low-voltage programming
 *   - disable brownout reset
 *   - enable master clear
 */

#pragma config FOSC = HSPLL_HS
#pragma config PLLDIV = 1
#pragma config VREGEN = ON
#pragma config WDT = OFF
#pragma config LVP = OFF
#pragma config BOR = OFF
#pragma config MCLRE = ON

char buffer[4][20];
char state = 4;

void MyHighInt (void);                    // prototypes

#pragma interrupt MyHighInt           save=PROD
#pragma code high_vector=0x08         // high_vector is at 0x0008

void high_vector (void)               // high-prioity vector
{
     _asm GOTO MyHighInt _endasm              // goto high software
}

#pragma code
```

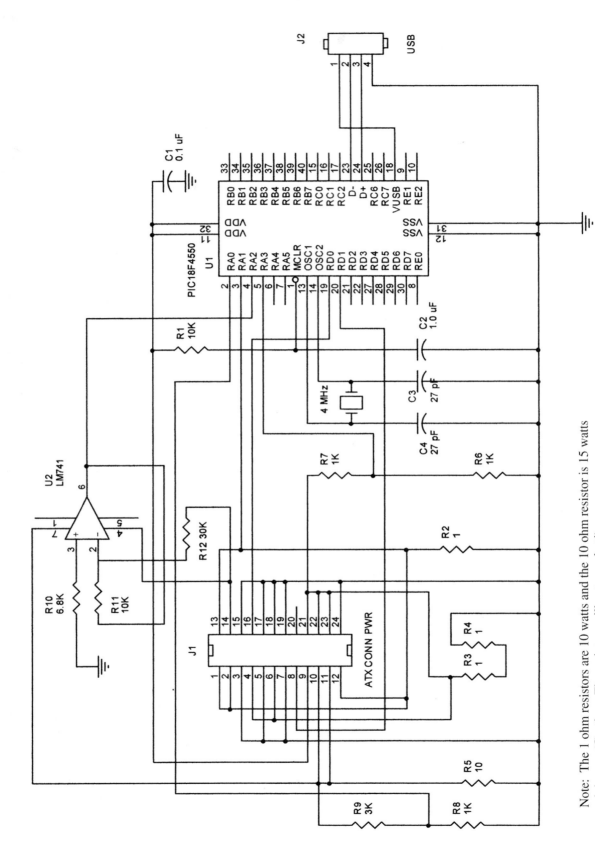

FIGURE 10-29 ATX Power Supply test system.

Note: The 1 ohm resistors are 10 watts and the 10 ohm resistor is 15 watts minimum. (Caution: The resistors will get very hot!)

TABLE 10-4   Pin-out of the ATX power supply.

Pin Number	Voltage	Pin Number	Voltage
1	+3.3 V	13	+3.3 V
2	+3.3 V	14	−12 V
3	Ground	15	Ground
4	+5 V	16	#PS_ON
5	Ground	17	Ground
6	+5 V	18	Ground
7	Ground	19	Ground
8	PWR_OK	20	Not used
9	+5 V (standby)	21	+5 V
10	+12 V	22	+5 V
11	+12 V	23	+5 V
12	+3.3 V	24	Ground

```
void USBTasks (void)
{
     /*
      * Servicing Hardware
      */
     USBCheckBusStatus();                         // must use polling method
     if (UCFGbits.UTEYE!=1)
          USBDriverService();                     // interrupt or polling method

     #if defined(USB_USE_CDC)
     CDCTxService();
     #endif

}// end USBTasks

void GetADC (char channel, char number)
{
     char a;
     SetChanADC (channel);
     ConvertADC();
     for (a = 0; a < 20; a++)
          buffer[number][a] = 0;
     while (BusyADC());
     buffer[number][0] = number + 0x31;
     itoa(ReadADC(), buffer[number] + 1);
}

void MyHighInt (void)
{
     char a;
     if (PIR1bits.TMR1IF == 1)
     {
          PIR1bits.TMR1IF = 0;
          WriteTimer1 (-62500 );
          GetADC (ADC_CH0, 0);
          GetADC (ADC_CH1, 1);
          GetADC (ADC_CH2, 2);
          GetADC (ADC_CH3, 3);
          state = 0;                              // start transmission
     }
}

// main program

void main (void)
{
```

```
            OpenTimer1 (TIMER_INT_ON &              // set timer 1
                        T1_8BIT_RW &
                        T1_SOURCE_INT &
                        T1_PS_1_8);

            WriteTimer1 (-62500 );                  // every half second

            OpenADC (ADC_FOSC_2 &
                     ADC_RIGHT_JUST &
                     ADC_0_TAD,
                     ADC_CH0 &
                     ADC_INT_OFF,
                     0x0B);

        INTCONbits.GIEH = 1;

        mInitializeUSBDriver();                     // see usbdrv.h

        while (1)
        {
            USBTasks();                             // must keep pumping
            if (mUSBUSARTIsTxTrfReady() && state != 4)
            {
                putsUSBUSART(buffer[state]);
                state++;
            }
        }
}
```

The driver for the PC is installed when the PC detects that a USB device is plugged into a USB port. The new hardware message appears on the screen when the system described here is connected. In response to the "add new hardware" dialog box, select the folder: C:\MCHPF-SUSB\Pc\MCHPUSB Driver\Release and install the driver for the PC side of the USB interface. Once the driver is installed, software is written that retrieves the voltages from the ATX power supply test system and displays them on the PC screen. A sample application, written in Microsoft Visual C, appears in Example 10-16 and the dialog box used to display the voltages appears in Figure 10-30. This example uses the HID (human interface device) interface which is part of Windows and does not require any special driver for the PC and Windows. The HID emulates the USB as a COM port and has a maximum transfer rate of 64 KBps, which is adequate for most microcontroller applications. The only software that must be provided to Windows is the inf file, which is located in the package from Microchip and installed as described above.

FIGURE 10-30 Screen shot of the VC program for monitoring the voltages from an ATX power supply.

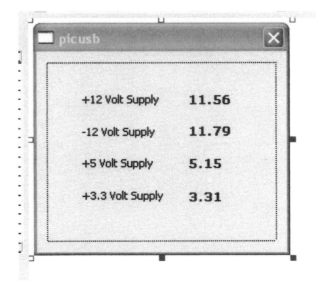

When the inf file is installed by adding new hardware, which appears when the PIC system is connected to the USB port, the PC selects the next available COM port for the USB connection.

The software used in the personal computer appears in Example 10-16, which generates the dialog box in Figure 10-30. The program listing shows only the sections that are added to a standard dialog application. The only exception is that a single line of code—SetTimer(1, 10, 0);—is added to the OnInitDialog function just after the TODO: statement. This starts the software listed in Example 10-16 so it takes a reading after 600 milliseconds from the USB power supply monitor outlined in this section. The software assumes that COM3 is assigned to the device by the hardware enumerator in Windows. For this system to function properly, the microcontroller must be connected to the USB before starting the program in the personal computer.

EXAMPLE 10-16

```
void CpicusbDlg::OnTimer(UINT nIDEvent)// arrives here every 600 msec
{
      char number;
      double actualReading;
      int placeHolder;
      int rawReading;
      char buffer[20];
      if (nIDEvent == 1)
      {
            KillTimer (1);              // kill timer

            number = ReadByte ("COM3");    // read identifier byte
            placeHolder = 0;

            do                        // get ASCII integer
                  buffer[placeHolder] = ReadByte ("COM3");
            while (buffer[placeHolder++] != 0 );
            rawReading = atoi (buffer);

            if (number = '1')              // process +12 V supply
            {
                  actualReading = rawReading * 4 * 0.00488759;
                  _gcvt (actualReading, 5, buffer);
                  Label1.put_Caption (buffer);
            }
            else if (number == '2')/       // process 3.3 V supply
            {
                  actualReading = rawReading * 0.00488759;
                  _gcvt (actualReading, 5, buffer);
                  Label4.put_Caption (buffer);
            }
            else if (number == '3')        // process -12 V supply
            {
                  actualReading = -rawReading * 3 * 0.00488759;
                  _gcvt (actualReading, 5, buffer);
                  Label2.put_Caption (buffer);
            }
            else                           // process 5 V supply
            {
                  actualReading = rawReading * 2 * 0.00488759;
                  _gcvt (actualReading, 5, buffer);
                  Label3.put_Caption (buffer);
            }
            SetTimer (1, 600, 0);          // restart timer
      }
      CDialog::OnTimer (nIDEvent);
}

int CpicusbDlg::ReadByte (CString PortSpecifier)
{
      DCB dcb;
      int retVal;
      BYTE Byte;
      DWORD dwBytesTransferred;
```

```
                    DWORD dwCommModemStatus;

                    HANDLE hPort = CreateFile (
                    PortSpecifier,
                    GENERIC_READ,
                    0,
                    NULL,
                    OPEN_EXISTING,
                    0,
                    NULL
                    );

                    if (!GetCommState(hPort,&dcb))
                          return 0x100;
                    dcb.BaudRate = CBR_9600; //9600 baud
                    dcb.ByteSize = 8; //8 data bits
                    dcb.Parity = NOPARITY; //no parity
                    dcb.StopBits = ONESTOPBIT; //1 stop
                    if (!SetCommState(hPort,&dcb))
                          return 0x100;
                    SetCommMask (hPort, EV_RXCHAR | EV_ERR); //receive character
                                                      event
                    WaitCommEvent (hPort, &dwCommModemStatus, 0); //wait for
                                                           character
                    if (dwCommModemStatus & EV_RXCHAR)
                          ReadFile (hPort, &Byte, 1, &dwBytesTransferred, 0); //read
                    else if (dwCommModemStatus & EV_ERR)
                          retVal = 0x101;
                    retVal = Byte;
                    CloseHandle(hPort);
                    return retVal;
             }
```

10-6   PIC18 EXTENDED INSTRUCTIONS

The extension instructions are used with the C18 compiler, which can be purchased separately.
The free version of the compiler does not use the extension instructions. The extension includes
eight new instructions as illustrated in Table 10-5. These instructions are disabled by default and

TABLE 10-5 PIC18 extended instructions.

Opcode, Operands		Description	Cycles	16-Bit Instruction Word	Status Affected	Notes
ADDFSR	f, k	Add k to FSR	1	1110 1000 ffkk kkkk		
ADDULNK	k	Add k to FSR2 and return	2	1110 1000 11kk kkkk		
CALLW		Call subroutine using WREG	2	0000 0000 0001 0100		
MOVSF	z_s, f_d	Move z_s to f_d	2	1110 1011 0zzz zzzz		
				1111 ffff ffff ffff		
MOVSS	z_s, z_d	Move z_s to z_d	2	1110 1011 1zzz zzzz		
				1111 xxxx xzzz zzzz		
PUSHL	k	Store k at FSR2, decrement FSR2	1	1110 1010 kkkk kkkk		
SUBFSR	f, k	Subtract k from FSR	1	1110 1001 ffkk kkkk		
SUBUNLK	k	Subtract k from FSR2, return	2	1110 1001 11kk kkkk		

f—8-bit data register file location (0x00 through 0xFF)
k—literal value
z_s—7-bit source value of the offset address for indirect addressing
z_d—7-bit destination value of the offset address for indirect addressing

in order to use them they must be enabled by setting the XINST configuration bit. Because the free version of the compiler does not support these instructions, software to use them would be in assembly language form unless you buy the compiler.

As can be seen in Table 10-5, the extended instructions deal mainly with the FSR2 register. The main purpose of the extended instructions is to develop an efficient method of addressing a software data stack. You might recall from Chapter 4 that the examples provided for a data stack used FSR2 as a stack pointer. The software presented with the data stack required two instructions to place a number onto the stack: a literal followed by a move. The extended instructions contain the PUSHL instruction that performs the same task as the two instructions provided in Chapter 4. Example 10-17 illustrates both techniques for placing the number 6 onto a software data stack addressed by FSR2. Notice that if the PUSHL instruction is used in a program it saves one instruction.

EXAMPLE 10-17

```
; code that uses standard instructions to push a 6 onto the data stack

      MOVLW 6                    ;place a 6 onto the stack
      MOVWF POSTDEC2
; code that uses extension instructions to push a 6 onto the data stack

      PUSHL 6
; code to retrieve the 6 from the stack

      MOVF  PREINC2, 0           ;retrieve the 6 from the stack
                                 ;and store it at location 0
```

The extended instruction set also contains two new indexed move instructions. Both use the contents of FSR2 to index memory and both allow a memory-to-memory move operation. The MOVSS instruction moves data from one indexed location to another. The instruction contains both offsets from the location addressed by FSR2. The MOVSF instruction moves the contents a location addressed by FSR2 plus an offset into a data memory location.

The ADDFSR and SUBFSR instructions allow a 6-bit literal to be added to or subtracted from any FSR register. Although this construction has no specific use, it allows small numbers to be added to an FSR register for looking up data in a database or some other similar application.

The ADDUNLK and SUBUNLK instructions are actually return instructions that modify the contents of the FSR2 register by a 6-bit literal. These instructions (especially the SUBUNLK instruction) allows stacked data to be dumped on a return. Example 10-18 illustrates how the SUBUNLK instruction is used to delete data stack information from a function.

EXAMPLE 10-18

```
; function that saves three values on the data stack and then dumps them
; before returning with the SUBUNLK insruction
;
MyFunc:

      PUSHL 6
      PUSHL 8
      PUSHL 3

      ; do things

      SUBUNLK 3                  ; return and unload the stack
```

The final extended instruction is the CALLW instruction, which is similar to a CALL, that replaces the value in PCL with the value in WREG instead of using a function address as in a CALL instruction. This instruction allows various functions or jump addresses to be accessed from a table. Example 10-19 illustrates such an application. In this example the value in WREG selects the function address from a lookup table. The software uses 0 to select function0, 1 to select function1, and so forth. This software has some limitations. The origin must be set to a location that ends at an address containing 0x00 and the value in WREG when Sfunc is called must be between 0 and 0x3F. The maximum number of functions is 64 for this code. This replicates the computed GOSUB function found in the Basic language. It can also simulate the computed GOSUB function by replacing the CALL instruction with GOTO instructions.

EXAMPLE 10-19

```
        ORG 0x1000                 ;must use an origin that ends with 00

; select function could be used for system calls
; WREG must be less than 64

Sfunc:
        RLNCF  WREG          ;multiply WREG times 4
        RLNCF  WREG
        ADDLW  10            ;bias WREG by 8
        CALLW                ;call function

        CALL   Function0        ;for WREG = 0
        CALL   Function1        ;for WREG = 1
        CALL   Function2        ;for WREG = 2

        ;
        ;      additional CALLs here
        ;
```

10-7   SUMMARY

1. Memory is expanded in the PIC18 by adding a serial EEPROM to a few of the port pins. The interface to the EEPROM is handled by software that generates the serial data for the interface.

2. In the rare case where two additional memory devices are not sufficient, another serial interface is created to handle the third and fourth memory device. Memory can be added only singly or in pairs controlled by a single serial interface.

3. The most common codes used on grocery items are the UPC-A or UPC-E code. The UPC-A code encodes a 12-digital numeric product code and the UPC-E code encodes a 6-digit numeric product code.

4. The interrupt on change function along with a timer is used with the UPC code received from the photo wand reader.

5. The boot block is a section of the program memory (at least 512 bytes) that is protected so it is not overwritten when the microcontroller is flashed. The boot block contains a program called a bootstrap loader that loads the memory with a new program before it is flashed.

6. The program memory within the microcontroller is located in EEPROM which is programmed in much the same way as the data EEPROM. The main difference is that a few bits in the configuration registers either enable or block programming. This gives the program memory protection.

7. I/O is expanded in cases where the on-chip I/O is insufficient. Expansion requires a few I/O pins to facilitate a serial interface to an I/O expansion device manufactured by Microchip or one of many other companies. As a rule, it is probably more efficient to purchase a larger microcontroller than expanding the I/O as described in this text. Because some situations require more I/O than is available on any version of the microcontroller, I/O expansion is available.

8. The CAN (controller area network) functions in much the same manner as the Ethernet interface, except the CAN is designed to function in noisy environments.

9. The software layer for controlling the CAN is provided by a Microchip header filed called ECAN.h that provides all common control and data transfer functions for using the CAN.

10. The USB (universal serial bus) is designed to replace the COM and LPT ports on a PC with versions that operate at 1 Mbps, 11 Mbps, and 480 Mbps. The latest version, USB 2.0, operates at 480 Mbps and is sufficient for video streams as well as other functions simultaneously. The USB allows for up to 127 devices on the same bus.

11. The USB is handled by software provided by Microchip and is fairly easy to interface to the PC. The PC requires either a driver program, or it can access the USB as if it were a COM port, which is presented in this text. Because microcontroller applications typically do not need a high-speed pathway to the PC, in most cases this is the best method of interfacing a USB-enabled microcontroller to a PC. This is especially true because no drive needs to be written for the PC to communicate with the microcontroller through this interface.

12. The extended instructions for the PIC18 add additional functionality to the index registers FSR0, FSR1, and FSR2, and in particular to FSR2. Of interest is the PUSHL instruction which enables a data stack to be created with less software overhead.

13. The extended instructions are available only for use in C language if the full version of the C18 compiler is purchased.

10-8 QUESTIONS AND PROBLEMS

1. What type of memory is available for memory expansion of the PIC18 microcontroller?
2. Because the EEPROM is serial, what interface software is used to communicate to the EEPROM from the microcontroller?
3. What sizes of EEPROM are available for memory expansion?
4. The A inputs on serial EEPROM are used for what purpose in the system?
5. How many serial EEPROM can be connected to a single serial interface without buffering?
6. What is the transfer rate of a serial EEPROM interface?
7. The EEPROM contains a control byte. What is its purpose?
8. Describe the function of the serial EEPROM ACK pulse.
9. In Example 10-1, what is the purpose of the ACK function?
10. In Example 10-1, describe how the address selects one memory device or the other.
11. What is the difference between UPC-A and UPC-E code?
12. Draw the UPC-E code obtained for the number 123456.
13. What DS275 driver is used in the interface of Figure 10-7?
14. The main program loop in Example 10-2 is extremely short. What is its purpose in the system?
15. In the program in Example 10-2, how is the UPC code read from the optical wand?
16. In the program in Example 10-2, what operations are used by the GetUPC function?
17. In Example 10-2, how is the information stored in the UPC code buffer sent to the PC?
18. What is the boot block available in many PIC18 microcontrollers?
19. What is the purpose of the bootstrap loader program?
20. Why is it important to be able to flash the program in the microcontroller?
21. Describe how the boot block is a protected memory block in the microcontroller.
22. Describe how Example 10-3 relocates the reset, high-level interrupt, and low-level interrupt vectors.
23. What are the purposes of the CP (code protect) bit, the WRT (write protect) bit, and the EBTR (external block table read) bit.
24. In Example 10-4, explain the difference between the inline assembly language table instruction with the form explained for the assembler program.

25. Briefly explain how I/O is expanded in the PIC18 family of microcontrollers.
26. What is the 74HC594 integrated circuit?
27. How many microcontroller I/O pins are required to interface the 74HC594?
28. What is the two-wire interface for expanding I/O?
29. What is the TC1320?
30. Explain the operation of the software listed in Example 10-9.
31. What is the CAN?
32. What is the maximum length of the CAN bus for 40 Kbps operation?
33. What is the maximum length of the CAN bus for 1 Mbps operation?
34. How many nodes can exist on the CAN bus before buffering is required?
35. What value termination resister is used on the CAN bus and what is its purpose?
36. What library header file is used to control the CAN bus?
37. What are the purposes of the CAN nodes used in the elevator system described in this chapter?
38. Describe the general operation of the floor controller software detailed in Example 10-10.
39. Describe the general operation of the car controller software listed in Example 10-11.
40. What is the USB and what are the speeds of the available versions?
41. The USB supplies power to any device connected to it, but it is limited to _____ mA.
42. The USB is connected through a _____ wire interface cable.
43. What is the simplest interface to the USB as far as developing software for the PC is concerned?
44. Describe the operation of the sample interface program listed in Example 10-15.
45. When using RS-232 emulation with the USB, and the Windows program in the PC provides the driver, which COM port is provided? Explain your answer.
46. The PC side application in Example 10-16 activates every 600 msec when the timer fires. What does the OnTimer function accomplish each time it executes?
47. What is the transfer speed for the RS-232 connection to the USB?
48. How is the inf file installed in the PC for the USB interface?
49. The extended instructions in the PIC18 family are available in the assembler. How can they be used in the C18 compiler?
50. Many of the extended instructions deal with which FSR register?
51. What is the purpose of the PUSHL instruction?
52. What is the purpose of the CALLW instruction?
53. What is the purpose of the SUBUNLK instruction?

PIC18 Family Instruction Set

Opcode, Operands		Description	Cycles	16-Bit Instruction Word				Status Affected	Notes
BYTE-ORIENTED FILE REGISTER INSTRUCTIONS									
ADDWF	f, d, a	Add WREG and f	1	0010	01da	ffff	ffff	C DC Z OV N	1, 2
ADDWFC	f, d, a	Add WREG and f with C	1	0010	00da	ffff	ffff	C DC Z OV N	1, 2
ANDWF	f, d, a	AND WREG with f	1	0001	01da	ffff	ffff	Z N	1, 2
CLRF	f, a	Clear f (0x00)	1	0110	101a	ffff	ffff	Z	2
COMF	f, d, a	Complement f (one's)	1	0001	11da	ffff	ffff		4
CPFSEQ	f, a	Compare f with WREG, skip =	1–3	0110	001a	ffff	ffff		4
CPFSGT	f, a	Compare f with WREG, skip >	1–3	0110	010a	ffff	ffff		4
CPFSLT	f, a	Compare f with WREG, skip <	1–3	0110	000a	ffff	ffff		4
DECF	f, d, a	Decrement f	1	0000	01da	ffff	ffff	C DC Z OV N	1, 2, 3, 4
DECFSZ	f, d, a	Decrement f, skip if 0	1–3	0010	11da	ffff	ffff		1, 2, 3, 4
DCFSNZ	f, d, a	Decrement f, skip if $\neq$ 0	1–3	0100	11da	ffff	ffff		1, 2
INCF	f, d, a	Increment f	1	0010	10da	ffff	ffff	C DC Z O V N	1, 2, 3, 4
INCFSZ	f, d, a	Increment f, skip if 0	1–3	0011	11da	ffff	ffff		4
INFSNZ	f, d, a	Increment f, skip if $\neq$ 0	1–3	0100	10da	ffff	ffff		1, 2
IORWF	f, d, a	Inclusive OR WREG with f	1	0001	00da	ffff	ffff	Z N	1, 2
MOVF	f, d, a	Move f	1	0101	00da	ffff	ffff	Z N	1
MOVFF	f_s, f_d	Move f_s to f_d Source $\rightarrow$ Destination $\rightarrow$	2	1100 1111	ffff ffff	ffff ffff	ffff ffff		
MOVWF	f, a	Move WREG to f	1	0110	111a	ffff	ffff		
MULWF	f, a	Multiply WREG with f	1	0000	001a	ffff	ffff		
NEGF	f, a	Negate f (two's)	1	0110	110a	ffff	ffff	C DC Z O V N	1, 2
RLCF	f, d, a	Rotate f through C left	1	0011	01da	ffff	ffff	C Z N	
RLNCF	f, d, a	Rotate f left	1	0100	01da	ffff	ffff	Z N	1, 2
RRCF	f, d, a	Rotate f through C right	1	0011	00da	ffff	ffff	C Z N	
RRNCF	f, d, a	Rotate f right	1	0100	00da	ffff	ffff	Z N	
SETF	f, a	Set f (0xFF)	1	0110	100a	ffff	ffff		
SUBFWB	f, d, a	Subtract f from WREG with borrow	1	0101	01da	ffff	ffff	C DC Z O V N	1, 2
SUBWF	f, d, a	Subtract f from WREG	1	0101	11da	ffff	ffff	C DC Z O V N	
SUBWFB	f, d, a	Subtract WREG from f with borrow	1	0101	10da	ffff	ffff	C DC Z O V N	1, 2
SWAPF	f, d, a	Swap nibbles in f	1	0011	10da	ffff	ffff		4
TSTFSZ	f, a	Test f, skip if 0	1–3	0110	011a	ffff	ffff		1, 2
XORWF	f, d, a	Exclusive OR WREG with f	1	0001	10da	ffff	ffff	Z N	

Opcode, Operands		Description	Cycles	16-Bit Instruction Word				Status Affected	Notes
BIT-ORIENTED FILE REGISTER INSTRUCTIONS									
BCF	f, d, a	Bit clear f	1	1001	bbba	ffff	ffff		1, 2
BSF	f, d, a	Bit set f	1	1000	bbba	ffff	ffff		1, 2
BTFSC	f, d, a	Bit test f, skip if 0	1–3	1011	bbba	ffff	ffff		3, 4
BTFSS	f, d, a	Bit test f, skip if 1	1–3	1010	bbba	ffff	ffff		3, 4
BTG	f, d, a	Bit toggle f	1	0111	bbba	ffff	ffff		1, 2
CONTROL INSTRUCTIONS									
BC	n	Branch if carry	1–2	1110	0010	nnnn	nnnn		
BN	n	Branch if negative	1–2	1110	0110	nnnn	nnnn		
BNC	n	Branch if no carry	1–2	1110	0011	nnnn	nnnn		
BNN	n	Branch if positive	1–2	1110	0111	nnnn	nnnn		
BNOV	n	Branch if no overflow	1–2	1110	0101	nnnn	nnnn		
BNZ	n	Branch if not zero	1–2	1110	0001	nnnn	nnnn		
BOV	n	Branch if overflow	1–2	1110	0100	nnnn	nnnn		
BRA	n	Branch always	2	1101	0nnn	nnnn	nnnn		
BZ	n	Branch if zero	1–2	1110	0000	nnnn	nnnn		
CALL	n, s	Call subroutine	2	1110	110s	kkkk	kkkk		
				1111	kkkk	kkkk	kkkk		
CLRWDT		Clear watchdog timer	1	0000	0000	0000	0100	$\overline{TO}$ $\overline{PD}$	
DAW		Decimal adjust WREG	1	0000	0000	0000	0111	C DC	
GOTO	n	Go to address	2	1110	1111	kkkk	kkkk		
				1111	kkkk	kkkk	kkkk		
NOP		No operation	1	0000	0000	0000	0000		
NOP		No operation	1	1111	xxxx	xxxx	xxxx		4
POP		Pop top of return stack	1	0000	0000	0000	0110		
PUSH		Push top of return stack	1	0000	0000	0000	0101		
RCALL	n	Relative CALL	2	1101	1nnn	nnnn	nnnn		
RESET		Software RESET	1	0000	0000	1111	1111	All	
RETFIE	s	Return with interrupts enabled	2	0000	0000	0001	000s	GIE/GIEH, PEIE/GIEL	
RETLW	k	Return with literal in WREG	2	0000	1100	kkkk	kkkk		
RETURN	s	Return from subroutine	2	0000	0000	0001	001s		
SLEEP		Enter standby mode	1	0000	0000	0000	0011	$\overline{TO}$ $\overline{PD}$	
LITERAL INSTRUCTIONS									
ADDLW	k	Add k to WREG	1	0000	1111	kkkk	kkkk	C DC Z OV N	
ANDLW	k	AND k with WREG	1	0000	1011	kkkk	kkkk	Z N	
IORLW	k	Inclusive OR k with WREG	1	0000	1001	kkkk	kkkk	Z N	
LSFR	f, k	Move k to FSR f	2	1110	1110	00ff	kkkk		
				1111	0000	kkkk	kkkk		
MOVLB	k	Move k to BSR	1	0000	0001	0000	kkkk		
MOVLW	k	Move k to WREG	1	0000	1110	kkkk	kkkk		
MULLW	k	Multiply WREG by k	1	0000	1101	kkkk	kkkk		
RETLW	k	Return with k in WREG	2	0000	1100	kkkk	kkkk		
SUBLW	k	Subtract WREG from k	1	0000	1000	kkkk	kkkk	C DC Z OV N	
XORLW	k	Exclusive OR k with WREG	1	0000	1010	kkkk	kkkk	Z N	

Opcode, Operands	Description	Cycles	16-Bit Instruction Word	Status Affected	Notes
TABLE INSTRUCTIONS					
TBLRD*	Table read	2	0000 0000 0000 1000		
TBLRD*+	Table read with post-increment	2	0000 0000 0000 1001		
TBLRD*−	Table read with post-decrement	2	0000 0000 0000 1010		
TBLRD+*	Table read with pre-increment	2	0000 0000 0000 1011		
TBLWR*	Table write	2	0000 0000 0000 1100		5
TBLWR*+	Table write with post-increment	2	0000 0000 0000 1101		5
TBLWR*−	Table write with post-decrement	2	0000 0000 0000 1110		5
TBLWR+*	Table write with pre-increment	2	0000 0000 0000 1111		5
EXTENSION INSTRUCTIONS					
ADDFSR f, k	Add k to FSR	1	1110 1000 ffkk kkkk		
ADDULNK k	Add k to FSR2 and return	2	1110 1000 11kk kkkk		
CALLW	Call subroutine using WREG	2	0000 0000 0001 0100		
MOVSF z_s, f_d	Move z_s to f_d	2	1110 1011 0zzz zzzz		
			1111 ffff ffff ffff		
MOVSS z_s, z_d	Move z_s to z_d	2	1110 1011 1zzz zzzz		
			1111 xxxx xzzz zzzz		
PUSHL k	Store k at FSR2, decrement FSR2	1	1110 1010 kkkk kkkk		
SUBFSR f, k	Subtract k from FSR	1	1110 1001 ffkk kkkk		
SUBUNLK k	Subtract k from FSR2, return	2	1110 1001 11kk kkkk		

Notes:
1. When a port register is modified as a function of itself (e.g., MOVF PORTB, 1, 0), the value used will be the value present on the pins themselves.
2. If this instruction is executed on the TMR0 register (and where applicable, d = 1), the prescaler will be cleared if assigned.
3. If the program counter is modified or a conditional test is true, the instruction requires two cycles. The second cycle is executed as a NOP.
4. Some instructions are two-words. The second word will execute as a NOP unless the first word retrieves information embedded in the second word. This ensures that all program memory locations have a valid instruction.
5. If the table write starts the write cycle to internal memory, the write continues until terminated.

Table Keys:
f—8-bit data register file location (0x00 through 0xFF)
d—destination bit (0 = store result in WREG, 1 = store result in f)
a—access bit (0 = access RAM location, 1 = RAM bank specified by BSR)
C—carry (carry from leftmost bit)
DC—digit carry (carry between right and left nibble)
Z—zero (0 = not zero, 1 = zero)
OV—overflow (0 = no overflow, 1 = arithmetic overflow)
N—negative (0 = positive, 1 = negative)
f_s—12-bit data register file source address (0x000 through 0xFFF)
f_d—12-bit data register file destination address (0x000 through 0xFFF)
n—relative address for branch instructions, direct address for CALL/BRA, and return instructions
s—fast CALL select bit (0 = do not update shadow registers, 1 = update shadow registers)
bbb—selects a bit position (0 through 7)
k—literal value (can be 8-, 12-, or 20-bits)
$\overline{TO}$–timeout bit
$\overline{PD}$—power down bit
GIE—global interrupt enable bit
GIEH—global interrupt enable bit (high-priority interrupt)
GIEL—global interrupt enable bit (low-priority interrupt)
PEIE—peripheral interrupt enable bit
z_s—7-bit source value of the offset address for indirect addressing
z_d—7-bit destination value of the offset address for indirect addressing
x—don't care bit

APPENDIX B

Common C Language Library Functions

This appendix contains a condensed listing of the commonly used hardware and software functions provided by the C18, C language library. This is a good reference when writing software for the microcontroller. Each function is listed with one or two examples of its usage. The parameters in some functions will change with different PIC18 versions, so they must be checked in the Microchip documentation.

#include <adc.h> // hardware analog-to-digital converter

Function	Parameters	Purpose	Example
char BusyADC (void)	Returns 1 if busy converting	Tests the ADC to see if it is busy	while (BusyADC());
void CloseADC (void)		Closes the ADC and disables its interrupt	CloseADC();
void ConvertADC (void)	—	Starts a new conversion	ConvertADC();
void OpenADC (char config1, char config2)	**Config1:** **A/D clock source:** ADC_FOSC_2 ADC_FOSC_4 ADC_FOSC_8 ADC_FOSC_16 ADC_FOSC_32 ADC_FOSC_64 ADC_FOSC_RC **A/D result justification:** ADC_RIGHT_JUST ADC_LEFT_JUST **A/D voltage reference source:** ADC_8ANA_0REF ADC_7ANA_1REF ADC_6ANA_2REF ADC_5ANA_1REF ADC_5ANA_0REF ADC_4ANA_2REF	Configures the ADC	Open ADC (ADC_FOSC_16 & ADC_8ANA_0REF, ADC_CH0 & ADC INT_ON);

#include <adc.h> // hardware analog-to-digital converter (*continued*)

Function	Parameters	Purpose	Example
	ADC_4ANA_1REF ADC_3ANA_2REF ADC_3ANA_0REF ADC_2ANA_2REF ADC_2ANA_1REF ADC_1ANA_2REF ADC_1ANA_0REF ADC_0ANA_0REF **Config2:** **Channel:** ADC_CH0 ADC_CH1 ADC_CH2 ADC_CH3 ADC_CH4 ADC_CH5 ADC_CH6 ADC_CH7 **A/D Interrupts:** ADC_INT_ON ADC_INT_OFF		
int ReadADC (void)		Reads the outcome of the conversion	result = ReadADC();
void SetChanADC (char channel)	**Channel:** ADC_CH0 ADC_CH1 ADC_CH2 ADC_CH3 ADC_CH4 ADC_CH5 ADC_CH6 ADC_CH7	Changes the channel	SetChanADC (ADC_CH3);

#include <capture.h> // hardware CCP unit

Function	Parameters	Purpose	Example
void CloseCapturex (void) or **void CloseECapture (void)**	—	Closes a CCP unit x or enhanced unit	CloseCapture1();
void OpenCapturex (char config) or **void OpenECapture (char config)**	**Config:** **Enable CCP Interrupts:** CAPTURE_INT_ON CAPTURE_INT_OFF **Interrupt Trigger:** Cx_EVERY_FALL_EDGE Cx_EVERY_RISE_EDGE Cx_EVERY_4_RISE_EDGE Cx_EVERY_16_RISE_EDGE EC1_EVERY_FALL_EDGE EC1_EVERY_RISE_EDGE EC1_EVERY_4_RISE_EDGE EC1_EVERY_16_RISE_EDGE	Opens a CCP unit x or enhanced unit	OpenCapture1 (CAPTURE_INT_ON & C1_EVERY_RIDE_EDGE);
unsigned int ReadCapturex (void) or **unsigned int ReadECapture (void)**	—	Reads the CCP unit x latch or enhanced unit	value = ReadCapture1();

#include <ic2.h> // hardware I²C unit

Function	Parameters	Purpose	Example
void AckI2C (void) or **void AckI2Cx (void)**	—	Generates a bus acknowledge (ACK) signal	AckI2C(); or AckI2C2();
void CloseI2C (void) or **void CloseI2Cx (void)**	—	Closes the SSP unit	CloseI2C(); or CloseI2C1();

unsigned char DataRdyI2C (void)
or
unsigned char DataRdyI2Cx (void)
	Returns 1 if there is data in the buffer	Tests the buffer for data	if (DataRdyI2C())
			{
			    // Get data

unsigned char getcI2C (void) or **unsigned char getcI2Cx (void)**	Returns a byte from the buffer	Reads the buffer	dat = getcI2C();

#include <ic2.h> // hardware I2C unit

Function	Parameters	Purpose	Example
unsigned char getsI2C (unsigned char *ptr, unsigned char length) or **unsigned char getsI2Cx (unsigned char *ptr, unsigned char length)**	Reads a string of characters from the I2C bus and stores them into the data memory location addressed by ptr; returns 0 if all bytes have been transferred	Reads a string from the I2C bus	getsI2C1(bob, 6);
void IdleI2C (void) or **void IdleI2Cx (void)**	—	Wait until the I2C bus is idle	IdleI2C();
void NotAckI2C (void) or **void NotAckI2Cx (void)**	—	Generates a not acknowledge (NAK) signal	NotAckI2C();
void OpenI2C (unsigned char sync, unsigned char slew) or **void OpenI2Cx (unsigned char sync, unsigned char slew)**	**Sync:** SLAVE_7 SLAVE_10 MASTER **Slew:** SLEW_OFF SLEW_ON	Opens the I2C connection as a master or a slave	OpenI2C (MASTER, SLEW_OFF);
unsigned char putcI2C (unsigned char data) or **unsigned char putcI2Cx (unsigned char data)**	Returns 0 if the write is successful	Writes a byte to the I2C bus	putcI2C(0x31);
unsigned char putsI2C (unsigned char *ptr, unsigned char length) or **unsigned char putsI2Cx (unsigned char *ptr, unsigned char length)**	Returns 0 if the write is successful	Writes a string to the I2C bus	putsI2C(fred, 3);
void RestartI2C (void) or **void RestartI2Cx (void)**	—	Generates a restart condition	RestartI2C();
void StartI2C (void) or **void StartI2Cx (void)**	—	Generates a start condition	StartI2C1();

(continued)

#include <ic2.h> // hardware I2C unit (*continued*)

Function	Parameters	Purpose	Example
void StopI2C (void) or **void StopI2Cx (void)**	—	Generates a stop condition	Stop2C2();
WriteI2C or **WriteI2C**	See putcl2C() or putcl2Cx()	—	—
ReadI2C or **ReadI2Cx**	See getcl2C() or getcl2CCx()	—	—
unsigned char EEAckPolling (unsigned control) or **unsigned char EEAckPollingx (unsigned control)**	The control is 0xA0 ORed with the address of the device; returns a 0 for no errors	Generates an acknowledge polling sequence for Microchip EE memory	pol = EEAckPolling (0xA0);
unsigned char EEByteWrite (unsigned char control, unsigned char address, unsigned char data) or **unsigned char EEByteWritex (unsigned char control, unsigned char address, unsigned char data)**		Writes a byte to the EEPROM address specified in the function	EEByteWrite (0xA0, 0x10, 0x1A);
unsigned char EECurrentAddRead (unsigned char control) or **unsigned char EECurrentAddReadx (unsigned char control)**		Reads a single byte from the EEPROM at the current address	EECurrentAddRead (0xA0);
unsigned char EEPageWrite (unsigned char control, unsigned char address, unsigned char *ptr) or **unsigned char EEPageWritex (unsigned char control, unsigned char address, unsigned char *ptr)**		Writes a page of data memory addressed by ptr to the EEPROM beginning at the current address	EEPageWrite1 (0xA0, 0x10, buffer);

#include <ic2.h> // hardware I2C unit

Function	Parameters	Purpose	Example
unsigned int EERandomRead (unsigned char control, unsigned char address) or **unsigned int EERandomReadx (unsigned char control, unsigned char address)**		Reads a single byte from EEPROM memory requiring a single byte address	EERandomRead (0xA0, 0x20);
unsigned char EESequentialRead (unsigned char control, unsigned char address, unsigned char *ptr, unsigned char length) or **unsigned char EESequentialRead (unsigned char control, unsigned char address, unsigned char *ptr, unsigned char length)**		Reads a string of data from the EEPROM	EESequentialRead (0xA0, 0x30, buffer, 5);

#include <pwm.h> hardware PWM unit

Function	Parameters	Purpose	Example
void ClosePWMx (void)		Closes the PWM channel	ClosePWM1();
void OpenPWMx (char period)	PWM period = (period + 1) x 4 x TOSC x TMR2 prescaler. (Timer 2 must also be opened and programmed before the PWM is opened).	Opens the PWM and sets its period	OpenPWM1 (10);
void SetDCPWNx (int duty_cycle)	Duty cycle is changed for the selected PWM	Changes the PWM duty cycle	SetDCPWM1 (500);
void SetOutputPWMx (unsigned char config, unsigned char mode)	**Config: PWM output:** SINGLE_OUT FULL_OUT_FWD HALF_OUT FULL_OUT_REV **Mode:** PWM_MODE_1 PWM_MODE_2 PWM_MODE_3 PWM_MODE_4	Configures the PWM output	SetOutputPWM1 (SINGLE_OUT, PWM_MODE_1);

#include <spi.h>

Function	Parameters	Purpose	Example
void CloseSPI (void) or **void CLoseSPIx (void)**	—	Closes the SPI unit	CloseSPI();
unsigned char DataRDYSPI (void) or **unsigned char DataRDYSPIx (void)**	Returns a 1 if data is available.	Tests the SPI buffer for available data	if (DataRDYSPI()) { // process data }
unsigned char getcSPI (void) or **unsigned char getcSPIx (void)**	Returns data from the SPI bus	Initiates a SPI read and returns the data	dat = getc (SPI);
void getsSPI (unsigned char *ptr, unsigned char length) or **void getsSPIx (unsigned char *ptr, unsigned char length)**	Reads data from the SPI bus into a buffer addressed by *ptr for the number of bytes indicated by length	Reads a string from the SPI	getsSPI (buffer, 10);
void OpenSPI (unsigned char sync, unsigned char bus, unsigned char smp) or **void OpenSPIx (unsigned char sync, unsigned char bus, unsigned char smp)**	**Sync:** SPI_FOSC_4 SPI_FOSC_16 SPI_FOSC_64 SPI_FOSC_TMR2 SLV_SSON SLV_SSOFF SPI **Bus:** MODE_00 MODE_01 MODE_10 MODE_11 **Smp:** SMPEND SMPMID	Opens the SPI bus and configures it	OpenSPI (SPI_FOSC_4, MODE_00, SMPMID);
unsigned char putcSPI (unsigned char data) or **unsigned char putcSPIx (unsigned char data)**	Return value is 0 for no write collision	Sends a character to the SPI bus	putcSPI (0x2A);
void putsSPI (unsigned char *ptr) or **void putsSPIx (unsigned char *ptr)**	Writes the contents of a buffer addressed by *ptr to the SPI bus	Writes a character string to the SPI	putsSPI (buffer);
unsigned char ReadSPI (void) or **unsigned char ReadSPIx (void)**	See getcSPI or getcSPIx	Reads a byte from the SPI	x = ReadSPI();
unsigned char WriteSPI (unsigned char data) or **unsigned char WriteSPIx (unsigned char data)**	See putcSPI or putcSPIx	Writes a byte to the SPI	WriteSPI(0x10);

#include <timers.h> hardware timer control

Function	Parameters	Purpose	Example
void CloseTimerx (void)	—	Closes the selected timer	CloseTimer20;
void OpenTimer0 (unsigned char config)	**Config:** **Interrupt:** TIMER_INT_ON TIMER_INT_OFF **Width:** T0_8BIT T0_16BIT **Clock Source:** T0_SOURCE_EXT T0_SOURCE_INT **External Clock Trigger:** T0_EDGE_FALL T0_EDGE_RISE **Prescaler Value:** T0_PS_1_1 T0_PS_1_2 T0_PS_1_4 T0_PS_1_8 T0_PS_1_16 T0_PS_1_32 T0_PS_1_64 T0_PS_1_128 T0_PS_1_256	Open Timer 0 and configure	OpenTimer0 (TIMER_INT_ON & T0_16BIT & T0_SOURCE_INT & T0_PS_1_16);
void OpenTimer1 (unsigned char config)	**Config:** **Interrupt:** TIMER_INT_ON TIMER_INT_OFF **Width:** T1_8BIT_RW T1_16BIT_RW **Clock Source:** T1_SOURCE_EXT T1_SOURCE_INT **Prescaler Value:** T1_PS_1_1 T1_PS_1_2 T1_PS_1_4 T1_PS_1_8 **Oscillator:** T1_OSC1EN_ON T1_OSC1EN_OFF **Synchronize Clock Input:** T1_SYNC_EXT_ON T1_SYNC_EXT_OFF	Open Timer 1 and configure	OpenTimer1 (TIMER_INT_ON & T1_16BIT_RW & T1_SOURCE_INT & T1_PD_1_1);

#include <timers.h> hardware timer control (*continued*)

Function	Parameters	Purpose	Example
	Use With **1 or 2 CCPs:** T3_SOURCE_CCP T1_CCP1_T3_CCP2 T1_SOURCE_CCP **For devices with more** **than 2 CCPs:** T34_SOURCE_CCP T12_CCP12_T34_CCP3 45 T12_CCP1_T34_CCP23 45 T12_SOURCE_CCP		
void OpenTimer2 (unsigned char config)	**Config:** **Interrupt:** TIMER_INT_ON TIMER_INT_OFF **Prescaler Value:** T2_PS_1_1 T2_PS_1_4 T2_PS_1_16 **Postscaler Value:** T2_POST_1_1 T2_POST_1_2 : : T2_POST_1_15 T2_POST_1_16 1:16 **Use With** **1 or 2 CCPs:** T3_SOURCE_CCP T1_CCP1_T3_CCP2 T1_SOURCE_CCP **More than 2 CCPs:** T34_SOURCE_CCP T12_CCP12_T34_CCP3 45 T12_CCP1_T34_CCP23 45 T12_SOURCE_CCP	Open Timer 2 and configure	OpenTimer2 (TIMER_INT_ON & T2_PS_1_1 & T2_POST_1_1);
void OpenTimer3 (unsigned char config)	**Config:** **Interrupt:** TIMER_INT_ON TIMER_INT_OFF **Width:** T3_8BIT_RW T3_16BIT_RW **Clock Source:** T3_SOURCE_EXT T3_SOURCE_INT	Open Timer 3 and configure	OpenTimer3 (TIMER_INT_OFF & T3_8BIT_RW & T3_SOURCE_INT & T3_PS_1_1);

#include <timers.h> hardware timer control

Function	Parameters	Purpose	Example
	Prescaler Value: T3_PS_1_1 T3_PS_1_2 T3_PS_1_4 T3_PS_1_8 **Synchronize Clock Input:** T3_SYNC_EXT_ON T3_SYNC_EXT_OFF **Use With** **1 or 2 CCPs:** T3_SOURCE_CCP T1_CCP1_T3_CCP2 T1_SOURCE_CCP **More than 2 CCPs:** T34_SOURCE_CCP T12_CCP12_T34_CCP345 T12_CCP1_T34_CCP2345 T12_SOURCE_CCP		
void OpenTimer4 (unsigned char config)	**Config:** **Interrupt:** TIMER_INT_ON ER_INT_OFF **Prescaler Value:** T4_PS_1_1 T4_PS_1_4 T4_PS_1_16 **Postscaler Value:** T4_POST_1_1 T4_POST_1_2 : : T4_POST_1_15 T4_POST_1_16 1:16	Open Timer 4 and configure	OpenTimer4 (TIMER_INT_ON & T4_PS_1_1 & T4_POST_1_1);
unsigned int ReadTimerx (void);	Returns the value in the selected timer.	Reads the timer count	x = ReadTimer2();
void WriteTimerx (unsigned int)	—	Writes the selected timer	WriteTimer1(-12500);

#include <usart.h> hardware USART

Function	Parameters	Purpose	Example
void **baudUSART (unsigned config)** or void **baudxUSART (unsigned config)**	**Config:** **Clock Idle State:** BAUD_IDLE_CLK_HIGH BAUD_IDLE_CLK_LOW **Baud Rate Generation:** BAUD_16_BIT_RATE BAUD_8_BIT_RATE **RX Pin Monitoring:** BAUD_WAKEUP_ON BAUD_WAKEUP_OFF **Baud Rate Measurement:** BAUD_AUTO_ON BAUD_AUTO_OFF	Sets the baud rate configuration on an EUASRT	baudUSART (BAUD_IDLE_CLK_HIGH & BAUD_16_BIT_RATE & BAUD_WAKEUP_ON & BAUD_AUTO_ON);

char BusyUSART (void)
or
char BusyxUSART (void)
	Returns 1 if transmitter is busy	Tests the USART transmitter	while (BusyUSART());
			or
			while (Busy2USART());

void CloseUSART (void) or **void ClosexUSART (void)**	—	Closes the USART	CloseUSART(); Or Close1USART();
char DataRdyUSART (void) or **char DataRdyxUSART (void)**	Returns 1 if data has been received	Tests the USART receiver	while (!DataRdyUSART());
char getcUSART (void) or **char getcxUSART (void)**	Returns a character received by the USART	Reads a character from the USART	x = getcUSART();
void getsUSART (unsigned char *ptr, unsigned char length) or **void getsxUSART (unsigned char *ptr, unsigned char length)**	Reads a character string from the USART and stores it into the location addressed by *ptr for the number of bytes specified by length.	Reads a character string from the USART	getsUSART (buffer, 4);
void **OpenUSART (unsigned char config, unsigned int brg)** or **void OpenxUSART (unsigned char config, unsigned int brg)**	**Config:** **Interrupts:** USART_TX_INT_ON USART_TX_INT_OFF USART_RX_INT_ON USART_RX_INT_OFF **Mode:** USART_ASYNCH_MODE USART_SYNCH_MODE **Width:** USART_EIGHT_BIT USART_NINE_BIT **Slave/Master:** USART_SYNC_SLAVE USART_SYNC_MASTER	Opens and configures the USART	OpenUSART (USART_TX_INT_ON & USART_RX_INT_ON & USART_ASYNCH_MODE & USART_EIGHT_BIT & USART_SYNC_MASTER & USART_CONT_RX & USART_BRGH_HIGH, 25);

#include <usart.h> hardware USART

Function	Parameters	Purpose	Example
	Reception mode: USART_SINGLE_RX USART_CONT_RX **Baud rate:** USART_BRGH_HIGH USART_BRGH_LOW **BRG:** Determines the Baud rate		
char ReadUSART (void) or **char ReadxUSART (void)**	See getcUSART() or getCxUSART()	Read USART	x = ReadUSART();
void WriteUSART (char data) or **void WritexUSART (char data)**	See putcUSART() or putcxUSART()	Write USART	WriteUSART(0x10);

#include <sw_i2c.h> software I^{2}C interface

By default, the software I^{2}C interface uses RB3 for the clock and RB4 for data. These pin numbers can be changed by modifying the sw_i2c.h file and recompiling the functions.

Function	Parameters	Purpose	Example

char Clock_test (void)	Returns 0 for no error	Used to time clock stretching on a slave	if (Clock_test())
			{
			    // handle error
			}

char SWAckI2C (void)
or
char SWNotAckI2C (void)
	Returns 0 if slave acknowledges	Generates an ACK or a NAK	if (SWAckI2C())
			{
			    // handle error
			}

char SWGetcI2C (void)	Returns data byte from I2C or −1 for an error	Reads the I2C bus	x = SWGetcI2C();
char SWGetsI2C (unsigned char *ptr, unsigned char length)	Reads the I2C bus and stores characters in a buffer addressed by *ptr; the number of bytes read are defined by length. An error is returned as a −1	Reads a string from the I2C bus	SWGetsI2C (buffer, 5);
char SWPutcI2C (unsigned char data)	Sends data to the I2C bus, returns 0 if write is successful	Sends a byte to the I2C bus	SWPutcI2C (0x8A);
char SWPutsI2C (unsigned char *ptr, unsigned length)	Sends the contents of memory addressed by *ptr to the I2C bus. Length is the number of bytes sent; a return 0 is no error	Sends a string to the I2C bus	SWPutsI2C (buffer, 10);

Function	Parameters	Purpose	Example
char SWReadI2C (void)	See SWGetcI2C()	Reads a byte from the I2C	x = SWReadI2C();
void SWRestartI2C (void)	—	Restarts the I2C bus	SWRestartI2C();
void SWStartI2C (void)	—	Starts the I2C bus	SWStartI2C();
void SWStopI2C (void)	—	Stops the I2C bus	SWStopI2C();
char SWWriteI2C (unsigned char data)	See SWPutcI2C()	Writes a byte to the I2C bus	SWWriteI2C (0x22);

#include <sw_spi.h> software SPI interface

By default, the SPI uses RB2 for the $\overline{CS}$ pin, RB3 for the data input pin, RB7 for the data output pin, and RB6 for the clock signal. These pin numbers can be changed by modifying the sw_spi.h file and recompiling the functions.

Function	Parameters	Purpose	Example
void ClearCSSWSPI (void)	—	Clears the chip select pin	ClearCSSWSPI();
void OpenSWSPI (void)	—	Opens the SPI interface	OpenSWSPI();
char putcSWSPI (char data)	Returns the byte read from the SPI	Writes a byte to the SPI	x = putcSWSPI (0x 66);
void SetCSSWSPI (void)	—	Sets the chip select pin	SetCSSWSPI();
char WriteSWSPI (char data)	See putcSWSPI()	Writes a byte to the SPI	x = WriteSWSPI (0x33);

#include <sw_uart.h> software UART interface

By default, the software UART uses RB4 as the transmit data pin and RB5 as the receiver data pin. These pin numbers can be changed by modifying the sw_uart.h file and recompiling the functions. In addition, the following time delays for transmit and receive must be defined in the program as functions: DelayTXBitUART, DelayRXHalfBitUART, and DelayRXBitUART. The DelayTXnitUART function must have a delay time of

$$\left[\left(\frac{Fosc}{Baud \times 2}\right) + 1\right] \times 0.5 - 12,$$ the DelayRXHalfBitUART function must have a delay time of

$$\left[\left(\frac{Fosc}{Baud \times 4}\right) + 1\right] \times 0.5 - 9,$$ and the DelayRXBitUART function must have a delay time of

$$\left[\left(\frac{Fosc}{Baud \times 2}\right) + 1\right] \times 0.5 - 14.$$

Function	Parameters	Purpose	Example
char getcUART (void)	—	Reads a byte from the UART	x = getcUART();
void getsUART (char *ptr, unsigned char length)	—	Reads a string from the UART	getsUART (buffer, 10);

Function	Parameters	Purpose	Example
void OpenUART (void)	—	Opens the UART	OpenUART();
void putcUART (char data)	—	Sends a byte out of the UART	putcUART (0x33);
void putsUART (char *ptr)	Writes a null terminated string to the UART	Writes a string to the UART	putsUART (string1);
char ReadUART (void)	See getcUART()	Reads a byte from the UART	x = ReadUART();
void WriteUART (char data)	See putcUART(data)	Writes a byte to the UART	WriteUART (0x0D);

#include <delays.h> software time delays

Function	Parameters	Purpose	Example
void Delay1TCY (void)	—	Delays 1 instruction cycle time	Delay1TCY();
void Delay10TCYx (unsigned char count)	Count is a number between 0 and 255, where a 0 is count of 256.	Delays 10 x count x the instruction cycle time	Delay10TCYx (10);
void Delay100TCYx (unsigned char count)	Count is a number between 0 and 255, where a 0 is a count of 256	Delays 100 x count x the instruction cycle time	Delay100TCYx (12);
void Delay1KTCYx (unsigned char count)	Count is a number between 0 and 255, where a 0 is a count of 256	Delays 1000 x count x the instruction cycle time	Delay1KTCYx (10);
void Delay10KTCYx (unsigned char count)	Count is a number between 0 and 255, where a 0 is a count of 256	Delays 10,000 x count x the instruction cycle time	Delay10KTCYx (1);

APPENDIX C

Answers to Selected Even-Numbered Questions and Problems

CHAPTER 1

2. Intel Corporation
4. The von Neumann machine is the standard architecture used in computer systems.
6. Complex instruction set computer
8. One or zero
10. 1024 K
12. A nibble is half a byte or 4 bits
14. Address, control, and data
16. The read signal or $\overline{RD}$
18. An EPROM is erased with an ultra-violet light outside of the system whereas an EEPROM is erased in the system electrically.
20. 4K or 4096
22. SRAM or DRAM
24. (a) char sets aside a signed byte of memory, (b) short sets aside a 16-bit signed memory location, (c) int sets aside a 16-bit signed memory location, (d) float sets aside a 24-bit memory location for a floating-point number, and (e) double sets aside a 24-bit memory location for a floating point number.
26. (a) 13.25, (b) 57.125, (c) 43.3125, and (d) $7\frac{1}{16}$
28. (a) 163.1875, (b) 297.75, (c) 172.859375, (d) 4011.1875, and (e) 3000.05078125
30. (a) 0.101_2 0.5_8 $0.A_{16}$, (b) 0.00000001_2 0.002_8 0.01_{16}, (c) 0.10100001_2 0.502_8 $0.A1_{16}$, (d) 0.11_2 0.6_8 $0.C_{16}$, and (e) 0.1111_2 0.74_8 $0.F_{16}$
32. (a) C2, (b) 10FD, (c) BC, (d) 10, and (e) 8BA
34. (a) 0111 1111, (b) 0101 0100, (c) 0101 0001, and (d) 1000 0000
36. (a) DATA "FROG", 0 (b) DATA "Arc", 0 (c) DATA "Water", 0 (d) DATA "Well", 0
38. DATA "What time is it?", 0

40. (a) 0000 0011 1110 1000, (b) 1111 1111 1111 0100, (c) 0000 0011 0010 0000, (d) 1111 0011 0111 0100
42. char Fred1 = -34;
44. The little endian format stores the least significant byte in the lowest numbered byte of memory, and the big endian format stores the most significant byte in the lowest numbered byte of memory.
46. (a) packed: 0000 0001 0000 0010 unpacked: 0000 0001 0000 0000 0000 0010, (b) packed: 0100 0100 unpacked: 0000 0100 0000 0100, (c) packed: 0000 0011 0000 0001 unpacked: 0000 0011 0000 0000 0000 0001, and (d) packed: 0001 0000 0000 0000 unpacked: 0000 0001 0000 0000 0000 0000 0000 0000
48. (a) 89, (b) 9, (c) 32, and (d) 1
50. (a) $+3.5$ (b) -1.0 (c) -12.5

CHAPTER 2

2. The Harvard architecture uses a separate memory space for the program and a separate memory space for the data.
4. Pipelining is where instruction fetches overlap executions, allowing one instruction to be executed per instruction clock, increasing the efficiency of the program.
6. 8
8. 0x000000 (first) and 0x1FFFFF (last)
10. 12
12. The EEPROM is in a special separate area of data memory space addressed by special function registers.
14. 0x000 through 0x2FF
16. 0x007FFF
18. TRIS, LAT, and PORT

20. The working register (W) or (WREG) is the place where most arithmetic and logic instruction place the result of an operation.

22. The WREG register is located in the data RAM at address 0xFE8.

24. The FSR registers are used to indirectly address the data memory.

26. The status register

28. Because a zero selects output and a one selects input, TRISB is loaded with a 0xFE.

30. The 0x33 is sent to Port A by loading the WREG with a 0x33 and then by addressing either the LATA or PORTA register and sending it the 0x33 from WREG.

32. 3

34. The IDE is the integrated development system used to program the PIC microcontroller in either assembly language or C language and to simulate its operations. The IDE can also control the programmer and an external emulator.

36. Emulation and programming.

38. A simulator is a program that functions in the same manner as the microcontroller.

40. (a) Start: is a label that is used to refer to the instruction, GOTO is the instruction of opcode, and Heaven is the location that the GOTO goes to in a program called the operand. (b) ADDLW is the opcode, 0x29 is the operand, and this is followed by a comment. (c) Loopy1: is the label, MOVFF is the opcode, WREG and 0x145 are operands, and this is followed by a comment.

42. The assembly process is where the assembler takes a source file and converts it into an object file.

44. An object file is a file that contains the assembly language source program in its binary machine language version. It is the object of the assembly language process.

46. The .lkr file is the linker file that contains instructions for the linker program.

48. The CODE directive delineates a section of code or instructions.

50. The CODE 0x1000 informs the assembler that the instructions that follow in the program are to be placed at program memory location 0x1000.

52. This statement reserves 2 bytes of memory for the label DATA1.

CHAPTER 3

2. WREG

4. (a) MOVLW 0x34 (b) ADDLW 3 (c) IORLW 6 (d) MOVLB 2 (e) SUBLW 9

6.
```
MOVLW 3
ADDLW 9
IORLW 5
```

8. (a) BCF WREG, 2 (b) BSF 0x11, 3, ACCESS (c) BTG 0x1A, 5, ACCESS (d) BSF STATUS, 2 (e) BCF STATUS, N

10. The BTFSS instruction allows a bit of a data register memory location to be tested. If the bit under test is set (1), the next instruction in the program is skipped.

12. The access bank contains data register memory locations 0x000 through 0x07F and 0xF80 through 0xFFF in many versions of the microcontroller. Note that some versions use a different split.

14. A literal addition adds a constant whereas a byte-oriented addition adds a variable.

16.
```
MOVLW 0x5A
MOVWF 0x10, ACCESS
MOVWF 0x20, ACCESS
MOVEF 0x30, ACCESS
```

18.
```
MUVLW WREG
MOVF  PRODL
```

20. The carry status register bit holds the borrow after a subtraction.

22.
```
MOVF   0x10, ACCESS
ADDWF  0x20, ACCESS
MOVWF  0x20, ACCESS
MOVF   0x11, ACCESS
ADDWFC 0x21, ACCESS
MOVWF  0x21, ACCESS
MOVF   0x12, ACCESS
ADDWFC 0x22, ACCESS
MOVWF  0x22, ACCESS
```

24. Decrement a data register memory location and skip the next instruction if the result is not zero.

26. The POSTINC1 operand uses FSR1 to address memory and then increments the contents of FSR1.

28. The contents of WREG is ones complemented if it is Exclusive-ORed with an 0xFF.

30. The GOTO instruction allows a program to continue at any address in the program memory, whereas the BRA has a limited branch distance of ±1024 from the next instruction in the program.

32. The instruction only allows a branch from +127 to −128 bytes from the next instruction in the program.

34.
```
        LFSR   1,0x000
        MOVLW  WREG
LOOP1:
        ADDWF  POSTINC1
        CPFSEQ FSR1L
        BRA    LOOP1
```

36. When the CALL instruction is executed, the contents of the program counter are placed onto the stack. The RETURN instruction removes the contents of the program counter from the stack. Because the CALL instruction stores the address of the next instruction in the program on the stack, the REUTRN returns to the instruction immediately following the CALL.

38. Example 3-17 uses the DAW instruction to accomplish the conversion by adding a 6 to WREG if the value in the low nibble is greater then 9. Although a 7 must be added to letters, by adding a 6 with DAW the 7 is obtained with an increment of the value in WREG if 0x40 or greater, as it is for the letters after adding a 6.

40.
```
;*********** Prime lookup table ************
;
;
PRIMES    CODE_PACK   0x1F00
PRIME1    DB    1, 2, 3, 4, 5, 7, 11, 13,
                17,19
          DB    23, 29, 31, 37, 41, 43, 47, 53
          DB    59, 61, 67, 71, 73, 79, 83,
                89, 97
          DB    101, 103, 107, 109, 113, 127
          DB    131, 137, 139, 149, 151, 157,
                163
          DB    167, 173, 179, 181, 191, 193,
                197, 199
          DB    211, 223, 227, 229, 233, 239,
                241, 251
          DB    0        ;end of table
;
;************* Function PRIME **************
;
;     Returns Carry = 1 if WREG is prime
;     Returns Carry = 0 if WREG is NOT prime
;
PRIME:    MOVWF   PRODL       ;save number in
                              PRODL
          MOVLW   0           ;address lookup
                              table
          MOVWF   TBLPTRL
          MOVLW   0x1F
          MOVWF   TBLPTRH
          MOVLW   0
          MOVWF   TBLPTRU
PRIME1:
          TBLRD*+             ;get a prime
                              number
          MOVLW 0
          CPFSEQ  TABLAT
          BRA     PRIME2      ;if not end of
                              table
          BCF     STATUS, C   ;clear carry
          MOVFF   PRODL, WREG
          RETURN
PRIME2:
          MOVFF   PRODL, WREG
          CPFSEQ  TABLAT
          BRA     PRIME1      ;if not found
          BSF     STATUS, C   ;set carry if
                              prime
          MOVFF   PRODL, WREG
          RETURN
```

42. The memory required when a macro is invoked is equal to the number required to store the instruction in the macro without any additional bytes for a call or a return.

CHAPTER 4

2. The program stack is an area of memory inside of the microcontroller that holds the return addresses from functions or subroutines.

4. (a) MOVWF POSTDEC1 (b) MOVF PREINC1, 0 (c) MOVFF PRODL, POSTDEC1 and MOVFF PRODH, POSTDEC1

6. PRODL is removed first because the data stack is a first-in, last-out storage device.

8. Software for a queue must detect the full and empty conditions.

10. The memory in the queue is used over and over again by the software so I is considered a cyclic memory.

12. A packed BCD number is stored as two digits per byte, whereas an unpacked BCD number is stored as one digit per byte.

14. BCD numbers are subtracted by using tens complement addition so that the DAW instruction can be used to correct the result.

16. A BCD number is used in a program instead of binary because of the amount of time required to convert between binary and BCD.

18. The multiply instruction is used for 16-bit multiplication with the cross multiply algorithm discussed in the chapter and illustrated in Example 4-9.

20.
```
;************ function MULTAB ************
;
;
MULTAB      MULWF   TABLAT
            MOVFF   PRODL, WREG
            RETURN
```

22. To multiply signed numbers, the numbers are first examined and if negative they are made positive. The result is made negative if the two numbers that are multiples have different sign-bits. Sign changes are accomplished by using the two complement.

24. The best method to divide by 4 is shifting the number to the right two places. This assumes that the number is positive or unsigned.

26. It is by using the algorithm in the text, although the operation will require more time than with smaller divisions.

28. The only change needed to accomplish octal conversion is that the MOVLW .10 instruction must be changed to a MOVLW .8 instruction and the software will generate an octal result.

30. A number is converted from decimal to binary by multiplying the result, which is initially zero, by 10 and then adding the decimal digit to the result. This multiply and add operation is repeated until no additional digits appear in the decimal number.

32. The instruction between BinBCDs and BinBCDs1 require 4.5 μs, the instruction between BinBCDs1 and BinBCDs2 require 4.5 μs plus the time required for Div times the iterations until the number is zero. The Div function requires 54 μs so the time required is

59.5 μs per iteration. The time required for the instructions between BinBCDs2 and the end of the function is 4.5 μs. The total time is 9.0 μs plus 59.5 times the number of iterations. If three iterations are required, then the total time is 187.5 μs.

34. 3.5 μs

36.
```
MyFunc:
        CALL  GetData
        SUBLW 0x59
        BZ    MyFunc1        ;right shift
        ADDLW 0x59
        SUBLW 0xF0
        BZ                   ;release a key
        RETURN
MyFunc1:
        CLRF  0x10, ACCESS
        REUTRN
MyFunc2:
        SUBLW 0x59
        BZ    MyFunct3
        RETURN
MyFunc3:
        SETF  0x10, ACCESS
        RETURN
```

CHAPTER 5

2. Assembly language code is placed between the _asm and _endasm statements in C language. The assembly language used in C language must not contain any semicolons and must include all fields of the instruction.

4. The #pragma config statement is used to configure the microcontroller for operation.

6. The Delay1KTCYx function causes a delay of 1000 instruction cycles. At a clock frequency of 4 MHz, an instruction cycle requires 1.0 μs so the Delay1KTCYx function causes a delay of 1.0 ms. Because 10 ms is required, the Delay1KTCYx(10); instruction is used in the program.

8.
```
while (1)
{
    PORTBbits.RB0 ^=1;  // complement bit
    Delay10KTCYx (100); // wait one second
}
```

10.
```
char bitPattern = 0;
PORTBbits.RB3 = 1;
while (bitPattern != 16)
{
    PORTB = PORTB & 0xF0 | bitPattern;
                    // send out input bits
    if (bitPattern != 0x0F && PORTAbits.RA4
    != 1
        || bitPattern == 0x0F &&
        PORTAbits.RA4 != 0)
        PORTBbits.RB3 = 0;   // bad
    bitPattern++;
}
```

12. The rom near directive specifies program memory data at addresses 0x0000 through 0xFFFF (16-bit pointer), and the rom far directive specifies data at addresses 0x000000 through 0xFFFFFF (24-bit pointer).

14. Each entry is divided into two parts. The rightmost three bits contain the count of the number of dits and dahs and the leftmost five bits contain the dit.dah pattern for the Morse code.

16. while (PORTBbits.RB2 == 1);

18. 0x101

20. The count variable is used to indicate how many 20 μs periods of time for the width of the pulse. If count is 1 or 2, the pulse is a logic one and if count is greater than 2, the pulse is a logic zero.

22. The states[state]() statement references a lookup table of pointers to different functions and calls them using the state variable as a reference.

24.
```
float xc[11];
float cap = 0.000001;
int f;
char count = 0;

for (f = 100; f <= 1000; f += 100)
{
    xc[count++] = 1 / (6.2832 * f * cap);
}
```

CHAPTER 6

2. The PDIP package is the plastic dual in-line package most often used for prototyping.

4. The SSOP package is the shrink small outline package.

6. The VDD is the positive power supply connection that is often 5 V and the VSS connection is the ground connection.

8. 12 mA

10. 8.5 mA

12. 15 ms

14. 0x0000

16. The watchdog timer is used to reset the microcontroller whenever it overflows. Its purpose is to restart the system if the system software hangs for any reason.

18. 4.0 ms

20. The stack reset occurs if the stack overflows or underflows, which does not happen during normal software operation, but can happen if noise affects the operation of the microcontroller.

22. 40 MHz

24. 4

26. The RCIO code is used to operate the microcontroller with an RC oscillator with pin OSC2 used as an I/O pin.

28. By using the #pragma config OSC = HSPLL statement.

32. The dice toss selects random numbers by incrementing a number while the user holds down a pushbutton. Because this counter increments at a high rate, the user

cannot time its outcome by timing how long the push-button is held down.

34. The value is calculated at 475 Ω, but because this value is not a standard resistor value, a 470 Ω resistor is used in the circuit.

36.
```
ADCON1 = 0x7f;    // program for digital pins
TRISB = 0;        // Port B is ouput
```

38. An interrupt is a hardware- or software-initiated function call.

40. These statements direct the compiler to use the high-priority interrupt vector at location 0x0008 for the MyHighInt function.

42. This instruction enables the high-priority interrupt inputs.

44. A 32.768-KHz crystal is connected to the Timer 1 clock input pins and Timer 1 is programmed to divide the count by 32,768, which causes an interrupt exactly once per second.

46. If the interrupt flag (IF) is not cleared for an interrupt, the interrupt will occur immediately upon returning from the interrupt service function.

48. 0.00323 V

50. The ADCON register GO bit is set to a logic one to start a conversion.

52. To read the data EEPROM, the EEPGD bit is cleared, the address is sent to the EEADR register, the RD bit is set, and the data are then read from the EEDATA register.

54. 4 ms

56. The CCP input is programmed to trigger on each positive edge. The software counts out the distance between positive edges at an 8-μs interval so the frequency is determined by using the count.

CHAPTER 7

2. 10 to 20 ms

4. Switch (0x81);

6. Pull-up resistors are needed because an open circuit is not a valid input state to a microcontroller. An open input is affected by noise.

8.
```
//
// key codes for a 4 X 6 style keypad
//   stored as constants in the program memory
//

rom near char lookupKey[] =
{
    1,  4,  7,  10,      // left column
    2,  5,  8,  0,
    3,  6,  9,  11,
    12, 13, 14, 15,
    16, 17, 18, 19
};
```

```
//
// uses function Switch from Example 7-1
//

unsigned char Key (void)
{
    #define MASK 0x0f      // set mask
    #define ROWS 6         // set number of
                              rows

    int a;
    unsigned char
    keyCode;

    PORTB = keyCode = 0; // clear Port B and
                            keyCode

    Switch (MASK);        // debounce and
                             wait for any key

    PORTB = 0xFE;         // select a leftmost
                             column

    while ((PORTA & MASK)    // while no key
    == MASK)                 is found
    {
        PORTB = (PORTB       // get next
        << 1) | 1;           column
        keyCode += ROWS;     // add rows to
                                keycode
    }
    for (a = 1; a != 0; a <<= 1)
    {                        // find row
        if ((PORTA & a) == 0)
            break;
        keyCode++;
    }

    return lookupKey        // look up
    [keyCode];              correct key
                            code

}
```

10. (See Figure C-1)

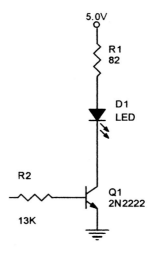

FIGURE C-1

12. (See Figure C-2)

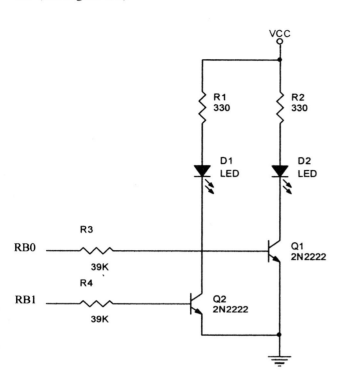

FIGURE C-2

```
// ****************** function Flash
*******************
//
//   uses delays.h
//
void Flash (void)
{
    char a;
    PORTB = 1;
    for (a = 0; a < 20; a++)
    {
        Delay10KTCYx(50);    // 1/2 second
        PORTB ^= 3;
    }
}
```

14. The persistence of vision in the human eye makes them appear to be lit 100% of the time.

16. 6

18. The VEE connection on an LCD display is connected to the wiper of a 10-KΩ potentiometer that is connected between 5 V and ground.

20. The E input pin to the LCD functions as a clock input to strobe data into the LCD from the data input pins.

22. The cursor is moved by sending the position to the display with a logic one in bit position 7. This position is sent with the RS pin equal to a logic zero.

24.
```
// ************ function Cursor ***********
//
//
void Cursor (char position)
{
    PORTB = position >> 4;
    PORTBbits.RB4 = 1;
    PORTBbits.RB4 = 0;
    PORTB = position & 0x0F;
    PORTBbits.RB4 = 1;
    PORTBbits.RB4 = 0;
}
```

26. A 0x01 as follows in function Clear.
```
// ************ function Clear ***********
//
//
void Clear (void)
{
    PORTB = 0;
    PORTBbits.RB4 = 1;
    PORTBbits.RB4 = 0;
    PORTB = 1;
    PORTBbits.RB4 = 1;
    PORTBbits.RB4 = 0;
}
```

28. The main advantage of the VFD is it brightness in all types of lights and the ability to dim it if needed.

30. The stepper motor is called a stepper motor because it moves in discrete steps.

32. A full step is where the stepper motor is moved from one pole piece to another, whereas the half step is moved to a pole piece on one step and the on the next step to a point halfway between pole pieces. Half steps are accomplished by energizing opposing pole piece coils and adjacent pole piece coils.

34. By varying the pulse width, the average amount of current through the motor is varied, changing it speed.

36. The kickback or damper diode shunts the current produced by the collapsing magnet field of a motor, relay, or coil around the transistor to prevent damage.

38. About half speed.

40. (See Figure C-3)

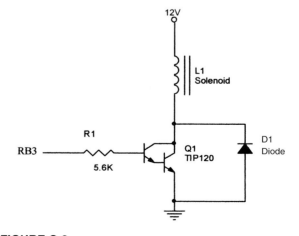

FIGURE C-3

```
// ************* function Fire *************
//
//   uses delays.h
//
void Fire (void)
{
    PORTBbits.RB3 = 1;
    Delay10KTCYx(70);
    PORTBbits.RB3 = 0;
}
```

42. 0000, 0001, 0010, 0011, 0101, 0111, 0110, 0100
44. 125
46. If the rightmost bit of b is a logic 1, then temp is incremented to round it up.
52. 38 KHz

CHAPTER 8

2. 0x0018
4. The IPEN bit is located in the RCON register.
6. The interrupt on change is serviced first so it has the high priority.
8. The high-priority interrupts are enabled.
10. The interrupt INT1 input is enabled.
12. PIE1bits.TMR2IE = 1;
14. INTCON
16. The return from interrupt sets the GIE/GEIH flag bit to re-enable interrupts. If the RETFIE 1 instruction is executed, the GEIL flag bit is set to re-enable the low-priority interrupts.
18. The low-priority interrupt in Example 8-3 handles the real-time clock.
20. Once every 1.024 ms.
22. Asynchronous serial data is data that is sent without a clock. In place of the clock a start bit and stop bit are sent in addition to the data.
24. 8 data bits and 1 stop bit
26. A UART or a USART
28. The receiver interrupts so received data is not lost.
30. One of the equations in Example 8-4 is used to calculate the value placed in the SPBRG register. To set the baud rate at 1200, SPBRG is programmed with 51 for high speed or 207 for low speed.
32. Each queue is 16 bytes.
34. The receiver interrupt flag is cleared and the data from the RCREG is placed in the input queue.
36. The third parameter in
    ```
    bool retVal = WriteFile(hPort,Data,1,
    &byteswritten,NULL);
    ```
    is changed to the length of the string to write more than one byte.
38. The preamble is at least 10 bits of logic ones in a row.
40. The EEPROM is accessed by the EERead and EEWrite functions listed in the software.

42. The wait4Bit function checks the flag bit to see if a bit has arrived from the tracks in a DCC system.
44. Timer 0 is used to measure the pulse width in the DCC system in increments of $\frac{1}{4}$ μs.
46. Whenever an array is larger than 256 bytes, the linker script must be modified to accommodate the larger array.
48. The MAX1483 is a line driver/receiver that supports the RS-422 interface standard for long-haul serial data transmission of up to 4000 feet.
50. The RJ-45 connector is the standard connector used with CAT5 cable and networks.
52. Yes, the code 128 bar code supports alphabetic data.
56. This is a check of the validity of the data in bar code 128.
58. The time is displayed at the left of line 2 and the data is displayed at the right of line 2.
60. The main program loop checks to see if a key is pressed, if the network has data, or if a card has been swiped.
62. It checks the input queue to see if data has arrived from the network.
64. The DS275 is an interface to the serial COM port on the PC. The signal levels on the COM port are ±12 V signals that must be converted to TTL levels for the microcontroller.
66. The transistor sends a signal that turns the line around on the half-duplex network.
68. The power station provides power to the system, it interfaces the system to the COM port on the PC, and it drives the network.

CHAPTER 9

2. The interrupt is used to detect when the door is opened during a wash cycle.
4. (See Figure C-4)
10. Two: one from the transmission speed sensor and one from a time base or clock.
12. 13 watts
14. RS-422 or RS-485
16. Synchronize all slaves
18. A relay is used only because of the system cost. A motor controller that allows both-way control costs more.
22. The controls in a HVAC system are 24 VAC signals that are controlled in the application using relays driven by TIP120 drivers.
24. No. The outside air temperature must be known to determine if air conditioning or heating is needed.
26. The EEPROM is used to store settings for the system.

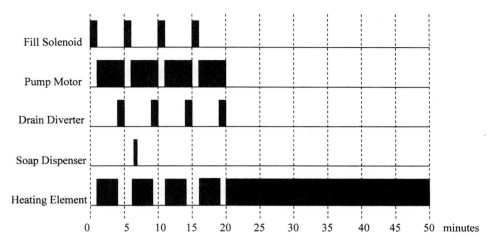

FIGURE C-4 (Diagram for the sanitize cycle; water is heated).

CHAPTER 10

2. The interface to the EEPROM is serial so the SPI or I²C software can be used for the interface.

4. The A inputs are programmed to select the EEPROM.

6. 100 KHz

8. The ACK signal is tested after a write to EEPROM and indicates that a write has been completed.

10. One EEPROM is selected if the address is less than 0x8000 and the other is selected if the address is 0x8000 through 0xFFFF.

14. The main program loop tests to see if a good UPC code has been swiped.

16. The code is displayed on the LCD screen, the user is asked for the item count, and the count is stored in the EEPROM queue for later transmission to the PC through the serial interface.

18. The boot block is a protected area of memory that allows space to store a bootstrap loader program.

20. The ability to flash the microcontroller program memory is so that software updates can be applied to a system in the product without removing the microcontroller.

22. The software uses goto instructions to relocate the reset and interrupt vectors to a location above the protected boot block.

24. The mnemonic opcode is different in the inline assembly version.

26. This integrated circuit is a shift register that contains a parallel latch used as a buffer for glitch-free operation.

28. The two-wire interface uses the I²C interface and saves one port pin.

30. The software sends a biased speed to the motor through the DAC.

32. 1000 meters

34. 20

36. ECAN.h

38. Each floor controller monitors the UP and DOWN pushbuttons and sends press information to the car controller when either is pressed. The floor controller also displays the floor number of the elevator as it is received from the car controller.

40. The USB is the universal serial bus available as versions 1.0 (1Mbps), 1.1 (11 Mbps), and 2.0 (480 Mbps).

42. 4

44. The program reads the voltages from the interface every half second and transmits them to the PC through the USB.

46. The OnTimer function calculates the voltage from the raw data received through the USB and displays it on the active-X labels that correspond to each voltage.

48. The inf file is installed the first time that the microcontroller is connected to the USB on the PC. The installation is automatic except for selecting the location of the inf file on the PC.

50. FSR2

52. The CALLW instruction is used to call a function addressed by the PC with the rightmost 8 bits equal to WREG.

APPENDIX D

A PIC-Based System Using a USB Interface

This appendix provides a schematic and the software for the implementation of a PIC-based system with a built-in bootstrap loader program that connects to a PC through the USB bus. The system can be used in a laboratory environment for the study of the PIC microcontroller or as a low-cost development system. The PIC system costs between $30 and $40, which is much less expensive than anything on the market and provides a valuable experience building a fairly complete small system. Construction is accomplished with wire-wrap (which cost a little more) or by bread-boarding because of the relatively small number of connections needed to construct the system. Because the system uses USB connectivity and the PIC is programmed in place, it is ideal for any modern experimentation. The only time that it must be placed in a PIC programmer is for initial programming to place the boot loader onto the PIC. After this initial programming, the PIC board is reprogrammed with the resident boot block loader program. This allows it to be programmed in place without the PIC programmer.

A PIC18F4550 I/P plastic PDIP version is selected for the system because of its built-in high-speed USB interface which is operated with a speed of 1M bps or 12M bps (USB 2.0 according to Microchip). The system is powered from the computer through the USB cable, which allows up to 500 mA of current before a supplemental power supply is required. The PIC will draw only 200 mA of current if fully loaded so just about anything can be interfaced to the board without additional power. The most current version of the program is provided here and on the Website http://members.ee.net/brey at the bottom of the "Technical Stuff" section. The program contains a small real-time operating system as well as the boot loader program.

Figure D-1 contains the schematic diagram of the system, which is fully debugged and will function if built as shown. All unused I/O pins (21 pins) are connected to a 30-pin SIP (single inline package) socket, which can be snapped to size for 21 pins, purchased at http://www.elexp.com so it is fairly easy to experiment with this board and connect just about anything to it using some wires connected to the SIP pins. All of the parts can be purchased just about anywhere, including http://www.mouser.com, http://www.digikey.com, http://www.allelectronics.com, and http://www. elexp.com. The type A USB connector is mounted on the board because the USB standard states that if a computer is connected to a computer (if you classify a PIC as a computer), both systems will have a type A connector for interconnectivity. Included is an A-to-A USB cable that is 6 feet (2 meter) in length. Either a 4 × 4 matrix keypad or a 3 × 4 matrix telephone-style keypad will function with only a slight software modification.

Because of the simplicity of this circuit, a printed circuit board is not needed. Direct wiring works just as well and costs far less. The operating system provides services for the LCD display, keypad, and internal real-time clock. The real-time delays are in milliseconds (up to 64K) and seconds (up to 64K), so timing is also very easy. Because the time delays are in real time they are

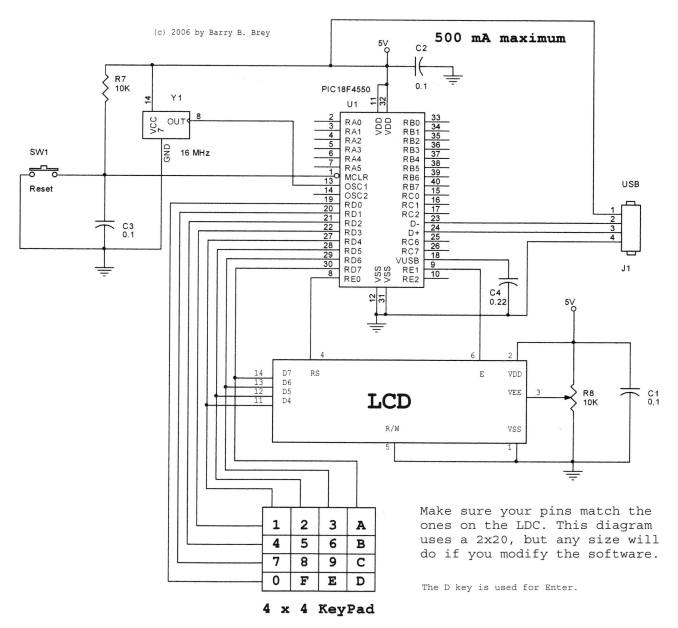

500 mA maximum

Make sure your pins match the ones on the LDC. This diagram uses a 2x20, but any size will do if you modify the software.

The D key is used for Enter.

4 x 4 KeyPad

FIGURE D-1

accurate to within ±1 ms, which is good enough for most control type applications. Since the microcontroller operates with a 16-MHz clock, it executes 4,000,000 instructions per seconds, which provides plenty of headroom for any microcontroller-based application. The operating system does use it, but less than 1% of the time. It also uses about 10% of the available memory (both program and data). The time of day can be displayed in either 24- or 12-hour format and can be switched on an off at anytime via a software switch.

FIGURE D-2

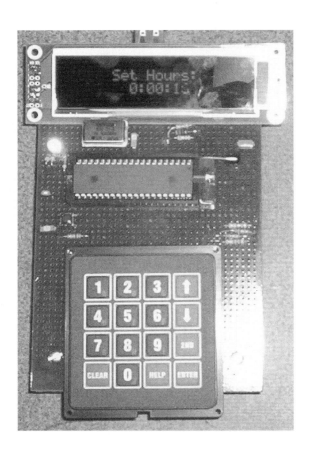

Figure D-2 contains a photograph of my board. I placed a three-color LED on Port Pins RC0, RC1, and RC2 using three 150-ohm current-limiting resistors. The system is way too cool for words!

Parts List:	What I paid:
1 PIC18F4550 I/P PDIP microcontroller	$ 6.72
1 40 pin DIP socket	$ .25
1 14 pin DIP socket (oscillator)	$ .15
1 30 pin SIP socket for I/O connections	$ 1.90
1 2 × 16 LCD display	$ 7.50
1 4 × 4 matrix keypad	$ 5.95
1 10K resistors	$ .06
1 10 K pot	$ .45
1 16 MHz crystal oscillator	$ 1.95
1 no Pushbutton switch	$ .50
3 0.1 uF capacitors	$ .30
1 0.22 uF capacitor	$ .15
1 USB 2 meter cable Type A to Type A	$ 3.50
1 USB connector Type A	$ .56
Total	**$30.21**

Example D-1 illustrates the test software for the system. This is before the USB is activated to allow the system to be tested. It allows the system to function as a simple clock and its purpose is to merely test the hardware. The program uses the blcd.h file which controls the LCD display and provides precision time delays. The header is also available at http://members. eenet/brey and appears in Example D-2.

EXAMPLE D-1

```c
#include <p18f4550.h>
#include <timers.h>
#include <blcd.h>

/*    The initial test program for the EET-387 project
      Copyright (c) 2006 by Barry B. Brey

      Version 1.1 ... 5/26/06

            requires:

                            timers.h  .. timer control
                            blcd.h    .. kcd and time delays

            linker:         18f4550.lkr
*/

// program configuration bits

#pragma config WDT = OFF              // spot, the watchdog turned on
#pragma config WDTPS = 32768          // watchdog count (131 seconds)
#pragma config BOR = ON               // brown-out voltage
#pragma config MCLRE = ON             // master clear on
#pragma config PWRT = OFF             // power up timer on
#pragma config PBADEN = OFF           // Port B digital
#pragma config PLLDIV = 4             // clock for full speed USB
#pragma config FOSC = ECPLLIO_EC      // PIC USB clock is 48 MHz
#pragma config CPUDIV = OSC1_PLL2     // External Oscillator with PLL
#pragma config USBDIV = 2             // high speed USB
#pragma config VREGEN = ON            // turn on USB regulator
#pragma config STVREN = ON            // stack overflow reset
#pragma config LVP = OFF              // low voltage program
#pragma config WRTB = ON              // boot block write protection
void HighInt (void);

#pragma interrupt HighInt
#pragma code _HIGH_INTERRUPT_VECTOR = 8
void _high_ISR (void)
{
      _asm goto HighInt _endasm   // goto high software
}

/*    program memory constants    */

rom near char keyLookUp[] = { 0,  7,  4,  1,   // may need to change
                             15,  8,  5,  2,
                             14,  9,  6,  3,
                             13, 12, 11, 10 };

// these are for a 2 x 20 display

rom near char mes1a[] = " The PIC 4550 Demo  ";
rom near char mes1b[] = "  Test Program      ";
rom near char mes1c[] = " Enjoy the Project! ";
rom near char mes1d[] = "(c) 2006 by B. Brey ";
rom near char mes1[] = "   Set Hours:       ";
rom near char mes2[] = "   Set Minutes:     ";
rom near char mes3[] = " 0=24hr or 1=12hr   ";
rom near char mes4[] = " The PIC4550 rules! ";
rom near char mes5[] = "                    ";

/*    these are for a 2 x 16 display

rom near char mes1a[] = " PIC 4550 Demo   ";
```

```
rom near char mes1b[] = " Test Program   ";
rom near char mes1c[] = "   Enjoy!!       ";
rom near char mes1d[] = "(c)2006 by Brey ";
rom near char mes1[] = "  Set Hours:     ";
rom near char mes2[] = " Set Minutes:    ";
rom near char mes3[] = "0=24hr or 1=12hr ";
rom near char mes4[] = "The 4550 rules!  ";
rom near char mes5[] = "                 ";

*/

/*    global data memory variables */

                         // time of day
int milli;
char seconds;
char minutes;
char hours;

char buffer[21];         // LCD display line buffer
char displayTimeFlag;    // display time flag
char timeFormat;         // 0 = 24 hour, 1 = 12 hour

#pragma code

void ShowTime (void)
{
        char portDbuffer = PORTD;
        char temph = hours;
        if (displayTimeFlag != 0)
        {
                if (temph > 12 && timeFormat == 1)          // make AM/PM
                        temph -= 12;
                else if (temph == 0 && timeFormat == 1)
                        temph = 12;
                buffer[0] = temph/10 | 0x30;
                if (buffer[0] == 0x30 )
                        buffer[0] = 0x20;        // blank leading hour zero;
                buffer[1] = temph % 10 | 0x30;
                buffer[2] = ':';
                buffer[3] = minutes / 10 | 0x30;
                buffer[4] = minutes % 10 | 0x30;
                buffer[5] = ':';
                buffer[6] = seconds / 10 | 0x30;
                buffer[7] = seconds % 10 | 0x30;
                if (timeFormat == 1)
                {
                        buffer[8] = 0x20;
                        if (hours >= 12)
                                buffer[9] = 'P';
                        else
                                buffer[9] = 'A';
                        buffer[10] = 'M';
                        buffer[11] = 0;
                        AddrLCD(0xc4);              // display time
                }
                else
                {
                        buffer[8] = 0;
                        AddrLCD(0xc6);
                }
                putsLCD(buffer);
        }
        PORTD = portDbuffer;
}

void DoTime (void)
{
        milli++;
```

```
        if (delayms != 0)   // do ms time delay
        {
                delayms--;
                if (delayms == 0)
                        mflag = 0;
        }
        if (milli == 1000)
        {
                milli = 0;
                seconds++;
                if (delaySec != 0)         // do second time delay
                {
                        delaySec--;
                        if (delaySec == 0)
                                sflag = 0;
                }
                if (seconds == 60)
                {
                        seconds = 0;
                        minutes++;
                        if (minutes == 60)
                        {
                                minutes = 0;
                                hours++;
                                if (hours == 24)
                                        hours = 0;
                        }
                }
        }
        if (milli == 0)     // check for new second
        {
                INTCONbits.GIEH = 1;      // enable high-priority interrupts
                ShowTime();
        }
}

void HighInt(void)        // do interrupts
{
        if (PIR1bits.TMR1IF == 1)        // do 24-hour clock
        {
                PIR1bits.TMR1IF = 0;      // reenable Timer 1 interrupt
                WriteTimer1(-11810);      // reprogram biased count (190)
                DoTime();
        }
}

char ReadKey (void)
{
        char select = 0xef;
        char col = -4;
        char mask = 1;
        PORTD = 0;
        do
        {
                while ((PORTD & 0x0f) != 0x0f)
                        ClrWdt();
                DelayMs(15);    // 15 ms
        } while ((PORTD & 0x0f) != 0x0f);
        do
        {
                while ((PORTD & 0x0f) == 0x0f)
                        ClrWdt();
                DelayMs(15);
        } while ((PORTD & 0x0f) == 0x0f);
        do
        {
                PORTD = select;
```

```
                    select <<= 1;
                    col += 4;
            } while ((PORTD & 0x0f) == 0x0f);
            while ((PORTD & mask) == mask)
            {
                    col++;
                    mask <<= 1;
            }
            return keyLookUp[col];
}

void SetTime (char* clock, char maxTensDigit)
{
        char temp;
        do
        {
                *clock = 0;
                ShowTime();
                temp = ReadKey();
                if (temp <= maxTensDigit)
                        *clock += temp * 10;
                ShowTime();
                temp = ReadKey();
                if (temp < 10)
                        *clock += temp;
                ShowTime();
                temp = ReadKey();
        } while (temp != 13);
        AddrLCD(0x80);
        putrsLCD(mes2);
}

void InitializeSystem (void)
{
        ADCON1 = 0x3f;                          // all inputs are digital
        PORTE = 0x80;                           // turn on Port D pullups
        TRISD = 0x0f;                           // set up Port D inputs

        WDTCONbits.SWDTEN = 1;                  // enable watchdog timer

        milli = seconds = minutes = hours = 0;      // time = 00:00:00
        displayTimeFlag = 0;                    // don't display time
        delayms = delaySec = 0;                 // clear delays

        OpenTimer1 (TIMER_INT_ON &              // program Timer 1
                    T1_16BIT_RW &
                    T1_SOURCE_INT &
                    T1_PS_1_1 &
                    T1_OSC1EN_OFF &
                    T1_SYNC_EXT_OFF);

        WriteTimer1 (-12000);                   // set Timer 1 count (1.0 ms)

        RCONbits.IPEN = 1;                      // IPEN = 1
        IPR1bits.TMR1IP = 1;                    // make Timer 1 high priority
        INTCONbits.GIEH = 1;                    // enable high-priority interrupts

        OpenLCD();                              // initialize the LCD display
}

void ShowStrings (const rom char *line1, const rom char *line2, char delay)
{
        AddrLCD (0x80);
        putrsLCD (line1);
        AddrLCD (0xc0);
        putrsLCD (line2);
        DelaySec (delay);
}

void main (void)
```

```
{
        char a;

        InitializeSystem ();

        ShowStrings (mes1a, mes1b, 5 );
        ShowStrings (mes1c, mes1d, 5 );

        AddrLCD (0xc0);
        putrsLCD (mes5);

        AddrLCD (0x80);
        putrsLCD (mes1);
        timeFormat = 0;
        displayTimeFlag = 1;                    // show time

        SetTime (&hours, 2);                    // set hours
        AddrLCD (0x80);
        putrsLCD (mes2);

        SetTime (&minutes, 5);                  // set minutes
        AddrLCD (0x80);
        putrsLCD (mes3);

        AddrLCD (0x80);
        putrsLCD(mes3);
        a = ReadKey();
        if (a < 2)
                timeFormat = a;
        AddrLCD (0x80);
        putrsLCD (mes5);
        AddrLCD (0xc0);
        putrsLCD (mes5);
        ShowTime();
        for (a = 0; a < 20; a++)
        {
                AddrLCD(0x80 | a);
                WriteLCD(mes4[a]);
                DelayMs(150);
        }
        for (a = 0; a < 5; a++)
        {
                AddrLCD (0x80);
                putrsLCD (mes5);
                DelayMs (400);
                AddrLCD (0x80);
                putrsLCD (mes4);
                DelayMs (400);
        }
        while(1)
        {
                ClrWdt();                       // keep spot happy
        }
}
```

EXAMPLE D-2

```
/*
        Time delays and LCD control
        (c) 2006 by Barry B. Brey

        Contains:

        DelayMs (unsigned int count)            // causes 1-K to 64-K millisecond delay
        DelaySec (unsigned int count)           // causes 1-K to 64-K second delay

        OpenLCD()                               // open LCD display for use
        AddrLCD()                               // address a display position
                                                // 0x80 is line 1, 0xC0 is line 2
        putsLCD (char *buffer )                 // display a character string from RAM
```

```
                    putrsLCD (const rom char buffer)   // display a string from ROM
                    WriteLCD (char data )               // display a single character
    */

    #ifndef __BLCD_H
    #define __BLCD_H

    unsigned int delayms;                   // ms delay
    char mflag;
    unsigned int delaySec;                  // seconds delay
    char sflag;

    # pragma code

    void ShortDelay (void)
    {
            Nop();                          // wait a bit
            Nop();
            Nop();
            Nop();
            Nop();
            Nop();
            Nop();
            Nop();
            Nop();
            Nop();
    }

    // The mflag and sflag are needed below because the while (delaysm)
    //     instruction will at times be interrupted between halves of an integer
    //     number which causes it to malfunction so a byte flag is needed.

    void DelayMs (unsigned int count)       // up to 64-K milliseconds
    {
            delayms = count;                // delayms changed in DoTime()
            mflag = 1;
            while (mflag);
    }

    void DelaySec (unsigned int count)      // up to 64-K seconds
    {
            delaySec = count;               // dalaySec changed in DoTime()
            sflag = 1;
            while (sflag);
    }

    void SendLCD (char command, char data ) // internal use
    {
            char a;
            PORTEbits.RE0 = command;        // set or clear RS
            PORTD = data & 0xf0;            // d7 == d4 to LCD
            ShortDelay();                   // wait a bit
            PORTEbits.RE1 = 1;              // set E
            ShortDelay();                   // wait a bit
            PORTEbits.RE1 = 0;              // clear E
            ShortDelay();
            PORTD = data << 4;              // d3 - d0 to LCD
            ShortDelay();                   // wait a bit
            PORTEbits.RE1 = 1;              // set E
            ShortDelay();                   // wait a bit
            PORTEbits.RE1 = 0;              // clear E
            PORTE = 0x80;
            for (a = 0; a < 27; a++)
                    ShortDelay();           // wait 40 us
    }

    void OpenLCD (void)                     // Open LCD
    {
            char a;
            char reset[] = {0x28, 1, 12, 6}; // set up commands
            TRISE = TRISE & 0xfc;           // RE0 and RE1 are output
```

```
        TRISD = TRISD & 0x0f;            // RD4 -- RD7 are output
        PORTEbits.RE1 = 0;               // Clear E
        DelayMs(16);                     // wait 16 ms (power up time)
        for (a = 0; a < 3; a++)          // send 4 reset commands
        {
                SendLCD(0, 0x20);
                DelayMs(5);              // wait 5 ms
        }
        for (a = 0; a < 4; a++)          // send setup commands
        {
                SendLCD(0, reset[a]);
                DelayMs(2);
        }
}
void AddrLCD  (char addr)                // set display address
{                                        // line 1 is 0x80 through 0x87
        SendLCD (0, addr);               // line 2 is 0xc0 through 0xcf
}

void putsLCD  (char *buffer)             // display data memory string
{
        while (*buffer)
        {
                SendLCD(1, *buffer);
                buffer++;
        }
}

void putrsLCD (const rom char *buffer)   // display program memory string
{
        while (*buffer)
        {
                SendLCD(1, *buffer);
                buffer++;
        }
}

void WriteLCD (char data)                // display 1 character
{
        SendLCD (1, data);
}

#endif
```

Once the software from Example D-1 is placed onto the PIC, which requires a programmer such as the PICSTART PLUS, the system is powered by connecting it to the USB connector on a PC. The first thing you will see is the phrase "Set Hours:". Enter the time in "24 hour format" by typing on the keypad and striking the Enter key (or whatever key functions as an Enter key on your keyoad). The keypad may or may not have an Enter key. (Figure D-1 uses the D key for Enter with the keypad from the schematic).

The next step is to develop, or at least modify, the boot-block loader program available from Microchip. This is downloaded from the Microchip Website, and a compiled hex version is available. Before using the boot-loader program it must be modified and rewritten to function with your hardware. The system enters the boot loader whenever any key on the keypad is pressed and held before the system is reset or powered up. If the system is reset without holding down a key, the application loaded into the PIC board executes.

The only file needed for the system to function in this manner is the file listed in Example D-3. Open a new project and set it up as written in Example D-3. Make sure that the files indicated are added to the project from the Microchip download and that the project build opens are changed as indicated, or the resulting code will not fit in the boot block space of 2K bytes. Once everything is set up and the program is written, build the program (or download the hex file) and program the PIC for the last time with the PICSTART PLUS programmer.

EXAMPLE D-3

```
/*
To create this program
        1. Create a new program with the project wizard
        2. Add the 18f4550.1kr linker file
            the files/folders are in MCHPFSUSB\fw\Boot
        3. Copy the system and autofile folders to your directory
        4. Copy io_cfg.h to your directory
        5. Add the folowing source files to the project:
            boot.c  usb9.c usbctrltrf.c usbdrv.c usbdsc.c usbmmap.c
        6. Add the follwing header files to the project:
            boot.h usb.h  usbcfg.h  usbdsc.h usbmmap.h  io_cfg.h typedefs.h
        7. Under build options, select the project, choose the MPLAB C18 tab,
            click on alternate, and enter -scs
 Now build the project!
*/

#include <p18f4550.h>
#include "system\typedefs.h"
#include "system\usb\usb.h"
#include "io_cfg.h"

#pragma config WDT = OFF                // spot, the watchdog turned on
#pragma config WDTPS = 32768            // watchdog count (131 seconds)
#pragma config BOR = ON                 // brown-out voltage
#pragma config MCLRE = ON               // master clear on
#pragma config PWRT = OFF               // power up timer on
#pragma config PBADEN = OFF             // Port B digital
#pragma config PLLDIV = 4               // clock for full-speed USB
#pragma config FOSC = ECPLLIO_EC        // PIC USB clock is 48 MHz
#pragma config CPUDIV = OSC1_PLL2       // external oscillator with PLL
#pragma config USBDIV = 2               // high-speed USB
#pragma config VREGEN = ON              // turn on USB regulator
#pragma config STVREN = ON              // stack overflow reset
#pragma config LVP = OFF                // low-voltage program
#pragma config WRTB = ON                // boot block write protection

// interrupt vectors

#pragma code _HIGH_INTERRUPT_VECTOR = 0x000008
void _high_ISR (void)
{
        _asm goto 0x808 _endasm         // new vector address
}

#pragma code _LOW_INTERRUPT_VECTOR = 0x000018
void _low_ISR (void)
{
        _asm goto 0x818 _endasm         // new vector address
}

#pragma code

void main (void)
{
        char temp;                      // save reset state
        char temp1;
        char temp2;
        char temp3;
        temp2 = PORTE;
        PORTE = 0x80;                   // turn on Port D pull-ups
        temp = ADCON1;
        temp1 = TRISD;
        temp3 = PORTD;
        ADCON1 != 0x0F;                     // digital I/O
        TRISD = 0x0F;                   // set up keypad
        PORTD = 0;                      // select all keypad keys

    // check for bootload mode (any keypad key on a reset)
```

```
        if ((PORTD & 0x0f) == 0x0f)  // if no key is pressed, select user mode
        {
                PORTE = temp2;                          // restore reset values
                TRISD = temp1;
                ADCON1 = temp;
                PORTD = temp3;
                _asm goto 0x800 _endasm         // new reset address
                }

        // we have bootload mode

        mInitializeUSBDriver();                 // see usbdrv.h
        USBCheckBusStatus();                    // modified to always enable USB module
        while (1)
        {
                USBDriverService();             // see usbdrv.c
                BootService();                  // see boot.c
        }
}
```

After installing the boot loader, the system is now ready to be programmed from the USB connection to the PC. There is only one problem, however: the software must be written to start at location 0x800 instead of 0x000. This is accomplished by changing the linker file as illustrated in Example D-4, and it is a good idea to give it new name such as boot18f4550.lkr instead of 18f4550.lkr. The first two lines of CODEPAGE are new and the third is commented out. The codepage called boot is the protected boot loader and the page called vectors redefines the starting address as 0x800.

EXAMPLE D-4

```
// $Id: 18f4550.lkr,v 1.3 2004/08/23 18:08:22 curtiss Exp $
// File: 18f4550.lkr
// Sample linker script for the PIC18F4550 processor
LIBPATH .

FILES c018i.o
FILES clib.lib
FILES p18f4550.lib

CODEPAGE    NAME=boot       START=0x0       END=0x7FF       PROTECTED
CODEPAGE    NAME=vectors    START=0x800     END=0x0x829     PROTECTED
// CODEPAGE    NAME=vectors    START=0x0       END=0x29            PROTECTED
// CODEPAGE    NAME=page       START=0x2A      END=0x7FFF
CODEPAGE    NAME=page       START=0x82A     END=0x7FFF
CODEPAGE    NAME=idlocs     START=0x200000  END=0x200007    PROTECTED
CODEPAGE    NAME=config     START=0x300000  END=0x30000D    PROTECTED
CODEPAGE    NAME=devid      START=0x3FFFFE  END=0x3FFFFF    PROTECTED
CODEPAGE    NAME=eedata     START=0xF00000  END=0xF000FF    PROTECTED

ACCESSBANK NAME=accessram START=0x0       END=0x5F
DATABANK    NAME=gpr0       START=0x60      END=0xFF
DATABANK    NAME=gpr1       START=0x100     END=0x1FF
DATABANK    NAME=gpr2       START=0x200     END=0x2FF
DATABANK    NAME=gpr3       START=0x300     END=0x3FF
DATABANK    NAME=usb4       START=0x400     END=0x4FF       PROTECTED
DATABANK    NAME=usb5       START=0x500     END=0x5FF       PROTECTED
DATABANK    NAME=usb6       START=0x600     END=0x6FF       PROTECTED
DATABANK    NAME=usb7       START=0x700     END=0x7FF       PROTECTED
ACCESSBANK NAME=accesssfr START=0xF60      END=0xFFF       PROTECTED

SECTION     NAME=CONFIG     ROM=config

STACK SIZE=0x100 RAM=gpr3
```

Once these changes are made, any program written must use the boot18f4550.lkr file of Example D-4 in place of the 18f4550.lkr file. Example D-5 illustrates the changes that are made to D-1 so it functions in the boot environment. (The first section is removed from D-1 and the second section replaces it). This changes the location of the reset and interrupt high vector. This test software did not use the low interrupt vector.

EXAMPLE D-5

```
// original from Example D-1 (which is removed)
// must use modified linker file

void HighInt (void);       // prototypes

#pragma interrupt HighInt
#pragma code high_vector = 8

void high_vector (void)                    // high-priority vector
{
        _asm GOTO HighInt _endasm         // goto high software
}


// replaces original in Example D-1

void HighInt (void);

extern void _startup (void);               // see c018i.c in your C18 compiler dir

#pragma code _RESET_INTERRUPT_VECTOR = 0x800
void _reset (void)
{
        _asm goto _startup _endasm
}
#pragma interrupt HighInt
#pragma code _HIGH_INTERRUPT_VECTOR = 0x808
void _high_ISR (void)
{
        _asm goto HighInt _endasm          // goto high software
}
```

The hex files for the boot loader (which is programmed through a PICSTART PLUS) is called boot.hex on http://members.ee/net/brey, the test software from D-1 is called test.hex, and the test software modified for a boot is called boottest.hex. These can be downloaded to test the system.

INDEX

/